Lecture Notes in Mathematics

Edited by A. Dold and B. Eckmann

488

Representations of Algebras

Proceedings of the International Conference
Ottawa 1974

Edited by V. Dlab and P. Gabriel

Springer-Verlag
Berlin · Heidelberg · New York 1975

Editors
Prof. Vlastimil Dlab
Department of Mathematics
Carleton University
Ottawa K15 5B6 Canada

Prof. Peter Gabriel
Mathematisches Institut
der Universität Zürich
Freiestraße 36
8000 Zürich/Schweiz

AMS Subject Classifications (1970): 16A18, 16A58, 16A64, 20C

ISBN 3-540-07406-6 Springer-Verlag Berlin · Heidelberg · New York
ISBN 0-387-07406-6 Springer-Verlag New York · Heidelberg · Berlin

Printed in Germany

Offsetdruck: Julius Beltz, Hemsbach/Bergstr.

PREFACE

During the recent years, a number of significant advances have been made in the theory of representations of algebras. Therefore, a meeting reflecting these advances and exploring the relationship among the results in this area was desirable. Such a meeting, the International Conference on Representations of Algebras was held at Carleton University, Ottawa, on September 3 - 7, 1974. It is our pleasure to acknowledge with gratitude the financial assistance of the National Research Council of Canada to support the Conference.

The program of the conference included 18 invited addresses and 10 contributed papers. In accordance with the Springer Lecture Notes policy, the papers outside the scope of the Conference and abstracts do not appear in this volume; on the other hand, the papers of P. Gabriel, M. Loupias, L. A. Nazarova and A. V. Roiter - M. M. Kleiner, who were unable to attend the meeting, are included. The papers appear in the form submitted by the authors; only very few technical alterations have been made.

We wish to thank Carleton University for the support in organizing the Conference. In particular, we wish to express our sincere thanks to the Secretary of the Conference, Luis Ribes, for his efficiency and success in making the meeting run smoothly and to Donna Desaulniers and Susanne Greening for their most appreciated secretarial assistance.

Oberwolfach, May 1975 Vlastimil Dlab and Peter Gabriel

TABLE OF CONTENTS

LIST OF INVITED LECTURES

MAURICE AUSLANDER	Almost split sequences I
SHEILA BRENNER	Quivers with commutativity condition
M. C. R. BUTLER	On the classification of local integral representations of abelian p-groups
S. B. CONLON	Finite linear p-groups of degree p and the work of G. Szekeres
VLASTIMIL DLAB	Algebras, species and graphs
ANDREAS DRESS	Relative Grothendieck rings
KENT R. FULLER	On rings of finite representation type
LAURENT GRUSON	On rings with the decomposition property
H. JACOBINSKI	Unique decomposition of lattices over orders
GERALD J. JANUSZ	The local index of elements in the Schur group
HERBERT KUPISCH	Quasi-Frobenius algebras of finite representation type
GERHARD O. MICHLER	Green correspondence between blocks with cyclic defect group
IRVING REINER	Locally free class groups of orders
IDUN REITEN	Almost split sequences II
CLAUS MICHAEL RINGEL	The representation type of local algebras
K. W. ROGGENKAMP	The augmentation ideal of finite groups, an interesting module
WINFRIED SCHARLAU	Automorphisms and involutions of incidence algebras
HIROYUKI TACHIKAWA	Balancedness and left serial algebras of finite type

LIST OF CONTRIBUTED PAPERS

ANTHONY BAK (presented by W.Scharlau)	Integral representations of a finite group which preserve a nonsingular form
JON F. CARLSON	Free modules over group algebras of p-groups
RENATE CARLSSON	The Wedderburn principal theorem for associative triple systems
R. GOW	Simple components of the group algebras of some groups of Lie type
E. L. GREEN	Modules having waists
WOLFGANG HAMERNIK	Indecomposable modules with cyclic vertex
Y. IWANAGA	On rings whose proper homomorphic images are QF-3 rings
WOLFGANG MÜLLER	Indecomposable modules over a finite dimensional algebra with radical square zero
FRANK J. SERVEDIO	Principal irreducible Lie-algebra modules
EARL J. TAFT	Hopf algebras with non-semisimple antipode

LIST OF REGISTERED PARTICIPANTS

J. E. ADNEY, Michigan State University

MAURICE AUSLANDER, Brandeis University

J. A. BEACHY, Northern Illinois University

E. A. BEHRENS, McMaster University

W. D. BLAIR, Northern Illinois University

ROBERTA BOTTO MURA, University of Alberta

CHRIS BUHR, Queen's University

SHEILA BRENNER, University of Liverpool

M. C. R. BUTLER, University of Liverpool

VICTOR CAMILLO, University of Iowa

HUMBERTO CARDENAS, Universidad Nacional Autonoma de Mexico

JON F. CARLSON, University of Georgia

RENATE CARLSSON, Universität Hamburg

C. R. CHAPMAN, University of Guelph

S. B. CONLON, University of Sydney

W. H. DAVENPORT, College of Petroleum and Minerals, Saudi Arabia

J. D. DIXON, Carleton University

VLASTIMIL DLAB, Carleton University

ANDREAS DRESS, Institute for Advanced Study, Princeton

MARVIN B. ENS, Queen's University

CARL FAITH, Rutgers University

FRANK FIALA, Carleton University

H. K. FARAHAT, University of Calgary

RUDOLF FRITSCH, Universität Konstanz

KENT R. FULLER, University of Iowa

S. C. GOEL, Ohio University

ARACELI REYES DE GONZALEZ, Instituto de Matematicas, Mexico

E. G. GOODAIRE, Memorial University

R. GOW, Carleton University

E. L. GREEN, University of Pennsylvania

L. GRUSON, Université de Lille

VERENA GURUSWAMI

W. H. GUSTAFSON, Indiana University

W. HAMERNIK, Universität Giessen

W. HAUPTMANN, Universität Giessen

BARBARA HEIDECKER, Universität Tübingen

A. G. HEINICKE, University of Western Ontario

A. HORN, Universität Giessen

D. B. HUNTER, School of Mathematics, Bradford

GEORGE IVANOV, Institute for Advanced Study, Princeton

Y. IWANAGA, Tokyo Kyoiku Daigaku

H. JACOBINSKI, Chalmers University of Technology, Goteborg

S. K. JAIN, Ohio University

G. J. JANUSZ, University of Illinois, Urbana

ALFREDO JONES, University of Sao Paulo

S. KLASA, Carleton University

G. R. KRAUSE, University of Manitoba

H. KUPISCH, Universität Heidelberg

JOHN LAWRENCE, Carleton University

DAVID MADISON, Carleton University

ABDUL MAJEED, Carleton University

C. K. MARTIN, Georgia State University

G. MICHLER, Universität Giessen

R. A. MOLLIN, Queen's University

B. J. MUELLER, McMaster University

W. MÜLLER, Universität München

M. B. NATHANSON, Institute for Advanced Study, Princeton

B. OLTIKAR, Carleton University

J. OSTERBURG, University of Cincinnati

HERBERT PAHLINGS, Carleton University

M. H. PEEL, North East London Polytechnic

M. PLATZECK, Brandeis University

ANDY PLETCH, Carleton University

I. PRESSMAN, Carleton University

D. PRETZEL, Imperial College of Science & Technology, London

MICHEL RACINE, Université d'Ottawa

F. F. RAGGI, Universidad Nacional Autonoma de Mexico

DINESH RAJKUNDLIA

R. BAUTISTA RAMAS, Universidad Autonoma de Mexico

I. REINER, University of Illinois, Urbana

IDUN REITEN, University of Trondheim

R. RENTSCHLER, Université de Paris

LUIS RIBES, Carleton University

CLAUS M. RINGEL, Universität Bonn

G. ROBINSON, University of Toronto

K. W. ROGGENKAMP, Universität Stuttgart

W. SCHARLAU, Universität Münster

FRANK J. SERVEDIO, Dalhousie University

DORE SUBRAO

H. TACHIKAWA, Tokyo Kyoiku Daigaku

S. TAKAHASHI, Université de Montréal

E. J. TAFT, Institute for Advanced Study, Princeton

DON TAYLOR, La Trobe University, Melbourne

W. TETER

G. V. WOOD, University College of Swansea

H. YAHYA, University of Calgary

ALMOST SPLIT SEQUENCES I

Maurice Auslander

The main purpose of these talks is to introduce the notion of almost split sequences. The first talk is devoted to giving various consequences of their existence in order to indicate the diversity of their applicability. The second talk is devoted to a more detailed, but by no means definitive, examination of these sequences themselves. This is based on the expectation that almost split sequences will prove to be a useful invariant for studying indecomposable modules. Proofs for these results will appear elsewhere (see [1] and [2] for example).

Throughout this discussion all our rings are artin algebras. We recall that a ring Λ is said to be an artin algebra if it is finitely generated as a module over its center C and C is a commutative artin ring. A ring Λ is an artin algebra if and only if Λ has the structure of an R-algebra with R a commutative artin ring which has the additional property that Λ is a finitely generated R-module.

Clearly every artin algebra is a two-sided artin ring (the converse is not true). We now list some of the properties of artin algebras that we shall need which do not hold for arbitrary two-sided artin rings.

Suppose M is a finitely generated Λ-module. Then $\operatorname{End}_\Lambda(M)$, the endomorphism ring of M, is an artin algebra and hence a two-sided artin ring. Moreover an injective envelope $I(M)$ of M is a finitely generated Λ-module.

Let C be the center of Λ. If we denote by I the injective envelope over C of $C/\text{rad}\, C$, then $\operatorname{Hom}_C(M,I)$ is a finitely generated right Λ-module or, equivalently, a finitely generated Λ^{op}-module where Λ^{op} is the opposite ring of Λ. If we denote by $\text{mod}\,\Lambda$ and $\text{mod}\,\Lambda^{op}$, the category of finitely generated Λ-modules, then we obtain the well known duality $D\colon \text{mod}\,\Lambda \longrightarrow \text{mod}\,\Lambda^{op}$ given by $D(X) = \operatorname{Hom}_C(X,I)$ for all X in $\text{mod}\,\Lambda$ (all X in $\text{mod}\,\Lambda^{op}$).

Associated with category $\text{mod}\,\Lambda$ are the two important additive categories $\underline{\text{mod}}\,\Lambda$ and $\overline{\text{mod}}\,\Lambda$. The objects of $\underline{\text{mod}}\,\Lambda$ ($\overline{\text{mod}}\,\Lambda$) are the same as those of $\text{mod}\,\Lambda$ but the morphisms from A to B in $\underline{\mathrm{mod}}\ \Lambda$ ($\overline{\mathrm{mod}}\ \Lambda$) is the group $\mathrm{Hom}_\Lambda(A,B)$ modulo the subgroups consisting of those Λ-morphisms from A to B which factor through projective objects in mod Λ (which factor through injective objects in mod Λ). We denote this factor group by $\underline{\mathrm{Hom}}_\Lambda(A,B)$ (by $\overline{\mathrm{Hom}}_\Lambda(A,B)$) and denote by $\underline{f}$ ($\overline{f}$), the image in $\underline{\mathrm{Hom}}_\Lambda(A,B)$ ($\overline{\mathrm{Hom}}_\Lambda(A,B)$) of a morphism f in $\mathrm{Hom}_\Lambda(A,B)$. The composition in $\underline{\mathrm{mod}}\ \Lambda$ ($\overline{\mathrm{mod}}\ \Lambda$) is given by $\underline{g}\ \underline{f} = \underline{gf}$ ($\overline{g}\ \overline{f} = \overline{gf}$). If we let $\mathrm{mod}_P\Lambda$ ($\mathrm{mod}_I\Lambda$) be the full subcategory of mod Λ consisting of those Λ-modules with no non-zero projective summands (no non-zero injective summands), then the full subcategory $\underline{\mathrm{mod}}_P\Lambda$ ($\overline{\mathrm{mod}}_I\Lambda$) of $\underline{\mathrm{mod}}\ \Lambda$ ($\overline{\mathrm{mod}}\ \Lambda$) consisting of the objects in $\mathrm{mod}_P\Lambda$ ($\mathrm{mod}_I\Lambda$) is dense in $\underline{\mathrm{mod}}\ \Lambda$ ($\overline{\mathrm{mod}}\ \Lambda$), i.e. every object in $\underline{\mathrm{mod}}\ \Lambda$ ($\overline{\mathrm{mod}}\ \Lambda$) is isomorphic to something in $\underline{\mathrm{mod}}_P\Lambda$ ($\overline{\mathrm{mod}}_I(\Lambda)$). Finally, the duality $D\colon \mathrm{mod}\ \Lambda \longrightarrow \mathrm{mod}\ \Lambda^{op}$ induces a duality $\mathrm{mod}_P\Lambda \longrightarrow \mathrm{mod}_I\Lambda^{op}$ which in turn induces a duality $\underline{\mathrm{mod}}_P\Lambda \longrightarrow \overline{\mathrm{mod}}_I\Lambda^{op}$.

We now recall the duality $\mathrm{Tr}\colon \underline{\mathrm{mod}}_P\Lambda \longrightarrow \underline{\mathrm{mod}}_P\Lambda^{op}$ which plays an important role in our discussion. Let M be in $\mathrm{mod}_P\Lambda$ and let $P_1 \longrightarrow P_0 \longrightarrow M \longrightarrow 0$ be a minimal projective presentation for M. Then define $\mathrm{Tr}M$ to be the $\mathrm{Coker}(P_0^* \longrightarrow P_1^*)$ where $X^* = \mathrm{Hom}_\Lambda(X,\Lambda)$. Clearly $\mathrm{Tr}M$ is in $\mathrm{mod}_P\Lambda^{op}$. While this map on objects can not be extended to a functor from $\mathrm{mod}_P\Lambda$ to $\mathrm{mod}_P\Lambda^{op}$, it can be extended to a functor $\mathrm{Tr}\colon \underline{\mathrm{mod}}_P\Lambda \longrightarrow \underline{\mathrm{mod}}_P\Lambda^{op}$ which is called the transpose and is easily seen to be a duality. Finally the composite functors

$$\underline{\mathrm{mod}}_P\Lambda \xrightarrow{\mathrm{Tr}} \underline{\mathrm{mod}}_P\Lambda^{op} \xrightarrow{D} \overline{\mathrm{mod}}_I\Lambda$$

$$\overline{\mathrm{mod}}_I\Lambda \xrightarrow{D} \underline{\mathrm{mod}}_P\Lambda^{op} \xrightarrow{\mathrm{Tr}} \underline{\mathrm{mod}}_P\Lambda$$

are equivalences of categories which are inverses of each other. In particular we have the following

Proposition 0: For an M in $\mathrm{mod}_P \Lambda$ the following are equivalent:

i) M is indecomposable

ii) $\mathrm{Tr}M$ is indecomposable in $\mathrm{mod}_P \Lambda^{op}$

iii) $\mathrm{DTr}M$ is indecomposable in $\mathrm{mod}_I \Lambda$

iv) $\exists$ a unique indecomposable Y in $\mathrm{mod}_I \Lambda$ such that $M \cong \mathrm{TrD}(Y)$, namely $Y \cong \mathrm{TrD}M$

Clearly the operation of End M on M induces an $(\mathrm{End}\ M)^{op}$-module structure on $\mathrm{Ext}^1_\Lambda(M,\mathrm{DTr}M)$. Since the endomorphisms of M which factor through projectives operate as 0 on $\mathrm{Ext}^1_\Lambda(M,\mathrm{DTr}M)$, we can consider $\mathrm{Ext}^1_\Lambda(M,\mathrm{DTr}M)$ as a module over $\underline{\mathrm{End}}\ M^{op}$. This $\underline{\mathrm{End}}\ M^{op}$-module has the following properties:

Proposition 1.

a) $\mathrm{Ext}^1_\Lambda(M,\mathrm{DTr}M) \cong D(\underline{\mathrm{End}}\ M)$. Hence $\mathrm{Ext}^1_\Lambda(M,\mathrm{DTr}M) \cong I(\underline{\mathrm{End}}\ M^{op}/\mathrm{rad}\ \underline{\mathrm{End}}\ M^{op})$ and is thus an injective cogenerator for $\mathrm{Mod}(\underline{\mathrm{End}}\ M^{op})$.

b) If $f\colon \mathrm{DTr}M \longrightarrow X$ is a morphism in $\mathrm{Mod}\ \Lambda$, then $\mathrm{Ext}^1_\Lambda(M,f)\colon \mathrm{Ext}^1_\Lambda(M,\mathrm{DTr}M) \longrightarrow \mathrm{Ext}^1_\Lambda(M,X)$ is a monomorphism if and only if f is a splitable monomorphism, i.e. if and only if there is a $g\colon X \longrightarrow \mathrm{DTr}M$ such that $gf = \mathrm{id}_{\mathrm{DTr}M}$.

One particularly significant consequence of b) is

Theorem 2. Let X be in $\mathrm{mod}_I \Lambda$. Then the C-morphism

$$\mathrm{Hom}_\Lambda(X,\mathrm{DTr}M) \longrightarrow \mathrm{Hom}_{\underline{\mathrm{End}}(M)^{op}}(\mathrm{Ext}^1_\Lambda(M,X),\mathrm{Ext}^1_\Lambda(M,\mathrm{DTr}M))$$

given by $f \longmapsto \mathrm{Ext}^1_\Lambda(M,f)$ is an isomorphism while $\mathrm{Hom}_{\underline{\mathrm{End}}(M)^{op}}(\mathrm{Ext}^1_\Lambda(M,X),\mathrm{Ext}^1_\Lambda(M,\mathrm{DTr}M)) \cong \mathrm{Hom}_C(\mathrm{Ext}^1_\Lambda(M,X),I(C/\mathrm{rad}\ C))$ as C-modules.

Theorem 2 is particularly useful in studying the functor $F\colon \mathrm{Mod}\ \Lambda \longrightarrow \mathrm{Mod}\ \underline{\mathrm{End}}(M)^{op}$ given by $F(X) = \mathrm{Ext}^1_\Lambda(M,X)$ for all X in $\mathrm{Mod}\ \Lambda$. We now give some results concerning this functor.

Proposition 3. Let M be in $\mathrm{mod}_P \Lambda$. Then the functor $F\colon \mathrm{Mod}\ \Lambda \longrightarrow \mathrm{Mod}\ \underline{\mathrm{End}}\ M^{op}$ given by $F(X) = \mathrm{Ext}^1(M,X)$ has the following properties:

a) If X is a finitely generated Λ-module, then $F(X)$ is a finitely generated $\underline{\mathrm{End}}\ M^{op}$-module.

b) If $\mathrm{Ext}^1_\Lambda(M,\Lambda) = 0$, then there is a finitely generated Λ-module X such that $F(X)$ is a generator for $\mathrm{Mod}\ \underline{\mathrm{End}}\ M^{op}$.

c) If $\mathrm{Ext}^1_\Lambda(M,M) = 0 = \mathrm{Ext}^1_\Lambda(M,\Lambda)$, then given any $\underline{\mathrm{End}}\ M^{op}$-module Y, there is a Λ-module X such that $F(X) \cong Y$. If Y is finitely generated $\underline{\mathrm{End}}\ M^{op}$-module then X can be chosen to be finitely generated Λ-module. If Y is finitely generated and indecomposable, then X can be chosen to be finitely generated and indecomposable.

As an immediate consequence of b) we have

Proposition 4. Suppose $\mathrm{Ext}^1_\Lambda(M,M) = 0 = \mathrm{Ext}^1(M,\Lambda)$. Let $\mathcal{J}$ be the subcategory of $\mathrm{Mod}\ \Lambda$ consisting of those indecomposable X such that $F(X)$ is indecomposable and let $\mathcal{S}$ be the subcategory of indecomposable modules in $\mathrm{mod}\ \underline{\mathrm{End}}\ M^{op}$. Then the functor $F\colon \mathcal{J} \longrightarrow \mathcal{S}$ is dense. Thus the cardinality of the isomorphism classes of objects in $\mathcal{S}$ is at most the cardinality of the isomorphism classes of objects in $\mathcal{J}$. In particul if Λ is of finite representation type so is $\underline{\mathrm{End}}\ M^{op}$.

Note: There are examples of indecomposable Λ-modules M satisfying $\mathrm{Ext}^1_\Lambda(M,M) = 0 = \mathrm{Ext}^1_\Lambda(M,\Lambda)$ with Λ of finite representation type such that $\mathrm{End}\ M^{op}$ is not of finite representation type. It would be interesting to know if Proposition 4 can be used to give new examples of rings of finite representation type.

We now turn our attention to almost split sequences. Suppose M in $\mathrm{mod}_P \Lambda$ is indecomposable. Then $\mathrm{End}\ M^{op}$, and hence $\underline{\mathrm{End}}\ M^{op}$, is a local ring. Since by Proposit we know that the $\underline{\mathrm{End}}\ M^{op}$-modules $\mathrm{Ext}^1_\Lambda(M,\mathrm{DTr}M)$ and $I(\underline{\mathrm{End}}\ M^{op}/\mathrm{rad}\ \underline{\mathrm{End}}\ M^{op})$ are isomorphic it follows that $\mathrm{Ext}^1_\Lambda(M,\mathrm{DTr}M)$ has a simple socle. Then, on the basis of Proposition 1, we have

Proposition 4. For a nontrivial element $0 \longrightarrow DTrM \longrightarrow V \longrightarrow M \longrightarrow 0$ of $Ext^1_\Lambda(M,DTrM)$, the following properties are equivalent:

a) $0 \longrightarrow DTrM \longrightarrow V \longrightarrow M \longrightarrow 0$ generates the socle of $Ext^1_\Lambda(M,DTrM)$.

b) If Y is an arbitrary Λ-module and $g: DTrM \longrightarrow Y$ is not a splitable monomorphism then there is an $h: V \longrightarrow Y$ such that $g = hi$.

c) If X is an arbitrary Λ-module and $f: X \longrightarrow M$ is not a splitable epimorphism, then there is an $h: X \longrightarrow V$ such that $ph = f$.

More generally we have

Proposition 5. Let $0 \longrightarrow A \xrightarrow{i} B \xrightarrow{p} C \longrightarrow 0$ be a non-split exact sequence in mod Λ with A and C indecomposable. The following statements are equivalent:

a) There is a commutative exact diagram

$$\begin{array}{ccccccccc} & & 0 & & 0 & & & & \\ & & \downarrow & & \downarrow & & & & \\ 0 & \longrightarrow & A & \longrightarrow & B & \longrightarrow & C & \longrightarrow & 0 \\ & & \downarrow & & \downarrow & & \| & & \\ 0 & \longrightarrow & DTrC & \longrightarrow & V & \longrightarrow & C & \longrightarrow & 0 \\ & & \downarrow & & \downarrow & & & & \\ & & 0 & & 0 & & & & \end{array}$$

where $0 \longrightarrow DTrC \longrightarrow V \longrightarrow C \longrightarrow 0$ is a generator for the socle of $Ext^1_\Lambda(C,DTrC)$.

b) Given any generator $0 \longrightarrow DTrC \longrightarrow V \longrightarrow C \longrightarrow 0$ of the socle of $Ext^1_\Lambda(C,DTrC)$, there is a commutative exact diagram

$$\begin{array}{ccccccccc} & & 0 & & 0 & & & & \\ & & \downarrow & & \downarrow & & & & \\ 0 & \longrightarrow & A & \longrightarrow & B & \longrightarrow & C & \longrightarrow & 0 \\ & & \downarrow & & \downarrow & & \| & & \\ 0 & \longrightarrow & DTrC & \longrightarrow & V & \longrightarrow & C & \longrightarrow & \\ & & \downarrow & & \downarrow & & & & \\ & & 0 & & 0 & & & & \end{array}$$

c) Given any morphism $g: A \longrightarrow Y$ which is not a splitable monomorphism, then there is an $h: B \longrightarrow Y$ such that $g = hi$.

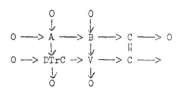

c') Same as c) except that Y is assumed to be finitely generated.

d) Given any morphism $f: X \longrightarrow C$ which is not a splitable epimorphism, then there is an $h: X \longrightarrow B$ such that $f = ph$.

d') Same as d) except X is assumed to be finitely generated.

An exact sequence $0 \longrightarrow A \longrightarrow B \longrightarrow C \longrightarrow 0$ in $\text{mod } \Lambda$ is said to be an almost split sequence if it has the following properties: a) it is not a splitable exact sequence; b) A and C are indecomposable and c) it satisfies any of the equivalent properties stated in Proposition 5. On the basis of our previous remarks, it is not difficult to establish

Proposition 6. a) An exact sequence $0 \longrightarrow A \longrightarrow B \longrightarrow C \longrightarrow 0$ in $\text{mod } \Lambda$ is an almost split sequence if and only if the exact sequence $0 \longrightarrow D(C) \longrightarrow D(B) \longrightarrow D(A) \longrightarrow 0$ in $\text{mod } \Lambda^{op}$ is an almost split sequence.

b) Given any indecomposable A in $\text{mod}_I \Lambda$, there is an almost split sequence $0 \longrightarrow A \longrightarrow B \longrightarrow C \longrightarrow 0$.

c) Given any indecomposable C in $\text{mod}_P \Lambda$, there is an almost split sequence $0 \longrightarrow A \longrightarrow B \longrightarrow C \longrightarrow 0$.

d) For two almost split sequences $0 \longrightarrow A \longrightarrow B \longrightarrow C \longrightarrow 0$ and $0 \longrightarrow A' \longrightarrow B' \longrightarrow C' \longrightarrow 0$, the following are equivalent:

i) The sequences are isomorphic

ii) $A \approx A'$

iii) $C \approx C'$.

The rest of this talk is devoted to giving several applications of the existence of almost split sequences in diverse settings. The first result connects finitely generated Λ-modules with large, i.e. not finitely generated, Λ-modules.

Proposition 7. For a finitely generated indecomposable Λ-module M, the following statements are equivalent:

a) $\mathrm{Hom}_\Lambda(M,N) \neq 0$ for an infinite number of non-isomorphic finitely generated indecomposable Λ-modules N.

b) There is a denumerably generated large indecomposable Λ-module N such that $\mathrm{Hom}_\Lambda(M,N) \neq 0$.

c) There is a Λ-module N not having any finitely generated summands such that $\mathrm{Hom}_\Lambda(M,N) \neq 0$.

As a consequence of this result we have the following

Proposition 8. For an artin algebra Λ the following statements are equivalent.

a) Λ is not of finite representation type.

b) There are large indecomposable Λ-modules.

We now give an application to finite group theory of the existence of almost split sequences.

Let G be a finite group of order n, k a field whose characteristic divides the order of G and $k[G]$, the group ring of G over k. Further assume that $0 \longrightarrow L \longrightarrow P_2 \longrightarrow P_1 \longrightarrow P_0 \longrightarrow k \longrightarrow 0$ is an exact sequence of $k[G]$-modules with $P_2 \longrightarrow P_1 \longrightarrow P_0 \longrightarrow k \longrightarrow 0$ the beginning of a minimal projective resolution of k.

Propostion 9. For a finitely generated $k[G]$-module A we have the following:

a) $A \approx L$

b) A is indecomposable and there exists a non-trivial group extension $\{1\} \longrightarrow A \longrightarrow E \longrightarrow G \longrightarrow \{1\}$ with the property that given any non-trivial group extension $\{1\} \longrightarrow B \longrightarrow E' \longrightarrow G \longrightarrow \{1\}$ with B a $k[G]$-module, there is a commutative diagram

$$\begin{array}{ccccccccc} \{1\} & \longrightarrow & B & \longrightarrow & E' & \longrightarrow & G & \longrightarrow & \{1\} \\ & & \downarrow & & \downarrow & & \| & & \\ \{1\} & \longrightarrow & A & \longrightarrow & E & \longrightarrow & G & \longrightarrow & \{1\} \end{array}$$

c) There is a non-trivial group extension $\{1\} \longrightarrow A \longrightarrow F \longrightarrow G \longrightarrow \{1\}$ with the property that given any $k[G]$-morphism $A \longrightarrow B$ which is not a splitable monomorphism, then in the pushout diagram

$$\begin{array}{ccccccccc} \{1\} & \longrightarrow & A & \longrightarrow & F & \longrightarrow & G & \longrightarrow & \{1\} \\ & & \downarrow & & \downarrow & & \| & & \\ \{1\} & \longrightarrow & B & \longrightarrow & E' & \longrightarrow & G & \longrightarrow & \{1\} \end{array},$$

the bottom row is a trivial group extension.

Note: If a $k[G]$-module A satisfies any of the above conditions, then the group extensions described in b) and c) are isomorphic. Further it is not difficult to see that $H^2(G,L) \cong k$ and so all non-trivial extensions in $H^2(G,L)$ are isomorphic and thus satisfy conditions a) and b).

We end our examples of applications with a criterion for relative projectivity of Λ-modules. Suppose $\Gamma \subset \Lambda$ is a subalgebra of Λ. We recall that a Λ-module M is relatively projective over Γ if and only if the natural morphism of Λ-modules $f\colon \Lambda \otimes_\Gamma M \longrightarrow M$ given by $f(\lambda \otimes m) = \lambda m$ is a splitable epimorphism of Λ-modules.

Proposition 10. Suppose $\Gamma \subset \Lambda$ is a subalgebra of Λ. Let C be a non-projective Λ-module and $0 \longrightarrow A \longrightarrow B \longrightarrow C \longrightarrow 0$ an almost split sequence. C is relatively Γ-projective if and only if the sequence $0 \longrightarrow A \longrightarrow B \longrightarrow C \longrightarrow 0$ does not split as a sequence of Γ-modules.

Bibliography

[1] M. Auslander and I. Reiten, Representation Theory of Artin Algebras III, Communications in Algebra, 3, (1975) p.pgs 239-294.

[2] M. Auslander, Large modules over artin algebras, A collection of Papers in Honour of Samuel Eilenberg, Academie Press, 1975.

Brandeis University
Waltham, Massachusetts 02154

ALMOST SPLIT SEQUENCES II

Maurice Auslander and Idun Reiten

In [1] various applications of the existence and uniqueness of almost split sequences for artin algebras were given. These results motivate trying to get some more information on what the almost split sequences look like for artin algebras. This is far from being known in general. Below we discuss different ways of obtaining such information. To illustrate how our information can be applied, we give a result about periodic modules for self-injective algebras.

We shall assume throughout this paper that Λ is an artin algebra and all our modules will be finitely generated. mod Λ will denote the category of finitely generated left Λ-modules. We recall that a non-split exact sequence $0 \to A \overset{f}{\to} B \overset{g}{\to} C \to 0$ of Λ-modules is <u>almost split</u> if A and C are indecomposable Λ-modules, and if $h: X \to C$ is not a splitable epimorphism, then there is a map $j: X \to B$ such that $gj = h$. We now explain the content of the paper, section by section. Proofs are for the most part omitted, and a more complete and detailed version will be published elsewhere.

In section 1 we study the almost split sequence $0 \to A \overset{f}{\to} B \overset{g}{\to} C \to 0$ by studying the map $g: B \to C$ or the map $f: A \to B$. This approach leads maturally to the notion of <u>irreducible maps</u>. We discuss these maps and their connection with almost split sequences. These maps also give rise to interesting invariants for indecomposable modules.

In section 2 we discuss a method for constructing new almost split sequences from given ones, based upon equivalences between module categories modulo projectives or modulo injectives. At the end of this section we apply these results together with some results from section 1 to prove the following result about periodic modules: Let Λ be a self-injective algebra and C an indecomposable non-projective periodic Λ-module. (i.e. $\Omega^n C \cong C$ for some $n \geq 1$, where $\Omega^n C$ denotes the nth syzygy module for C) such that an infinite number of non-isomorphic indecomposable Λ-modules have a non-zero map to C. Then there is an infinite number of non-isomorphic indecomposable periodic Λ-modules. This result is related to the following result of Alperin: Let k be a field of characteristic p algebraic over its prime field and G a finite group such that p divides the order of G. If kG is of infinite representation type, then an infinite number of indecomposable kG-modules have a periodic resolution.

In section 3 we give some information on the nature of the sequence of a different kind. We show that in most cases the almost split sequence can be described as the pushout of a certain diagram. Further we discuss what it means for the Λ-module A in the almost split sequence $0 \to A \xrightarrow{f} B \xrightarrow{g} C \to 0$ to be simple.

§ 1

It is useful to study the almost split sequence $0 \to A \xrightarrow{f} B \xrightarrow{g} C \to 0$ by studying the map $g: B \to C$ or the map $f: A \to B$. This will also enable us to work inside non-abelian categories, in particular inside $\underline{\mathrm{mod}\Lambda}$, the category $\mathrm{mod}\Lambda$ modulo projectives (see [1]), which will be useful later. We start by making the following

Definition: i) Let C be an indecomposable Λ-module. A map $g: B \to C$ which is not a splitable epimorphism is *right almost split* if whenever $h: X \to C$ is not a splitable epimorphism, there is a map $j: X \to B$ such that $gj = h$, and <u>minimal right almost split</u> if in addition for any proper summand B' of B, the restriction map $g': B' \to C$ is not right almost split.

ii) Let A be an indecomposable Λ-module. A map $f: A \to B$ which is not a splitable monomorphism is <u>left almost split</u> if whenever $h: A \to X$ is not a splitable monomorphism, there is a map $j: B \to X$ such that $jf = h$, and <u>minimal left almost split</u> if in addition for any proper summand B' of B the projection map $f': A \to B'$ is not left almost split.

We have the following connection with the almost split sequences.

<u>Proposition 1.1</u>: Let C be an indecomposable Λ-module.

i) If C is not projective, then $g: B \to C$ is minimal right almost split if and only if $0 \to \mathrm{Ker}g \to B \overset{g}{\to} C \to 0$ is an almost split sequence.

ii) If C is projective, then $g: B \to C$ is minimal right almost split if and only if g is a monomorphism and $B \cong \underline{r}\, C$, where $\underline{r}$ denotes the radical of Λ .

<u>Proposition 1.1'</u>: Let A be an indecomposable Λ-module.

i) If A is not injective, then $f: A \to B$ is minimal left almost split if and only if $0 \to A \overset{f}{\to} B \to \mathrm{Coker}\, f \to 0$ is an almost split sequence.

ii) If A is injective, then $f: A \to B$ is minimal left almost split if and only if f is an epimorphism and $B \cong {}^{A}/\mathrm{soc}A$, where $\mathrm{soc}A$ denotes the socle of A.

Closely connected with the above maps are the irreducible maps which we now define.

Definition: A map $g: B \to C$ which is neither a splitable monomorphism nor a splitable epimorphism is irreducible if whenever we have a commutative diagram $\begin{array}{ccc} & X & \\ {}^{f}\nearrow & & \searrow^{h} \\ B & \xrightarrow[g]{} & C \end{array}$, then either f is a splitable monomorphism or h is a splitable epimorphism.
The useful connection between irreducible maps and almost split maps and sequences is given by our next result.

Proposition 1.2: Let C be an indecomposable Λ-module.
i) A map $g: B \to C$ is irreducible if and only if there is a map $g': B' \to C$ such that $(g,g'): B \amalg B' \to C$ is minimal right almost split.
ii) If C is not projective, then $g: B \to C$ is irreducible if and only if there is a map $g': B' \to C$ such that $0 \to \operatorname{Ker}(g,g') \to B \amalg B' \to C \to 0$ is almost split.

Proposition 1.2': Let A be an indecomposable Λ-module.
i) A map $f: A \to B$ is irreducible if and only if there is a map $f': A \to B'$ such that $(f,f'): A \to B \amalg B'$ is minimal left almost split.
ii) If A is not injective, then $f: A \to B$ is irreducible if and only if there is a map $f': A \to B'$ such that $0 \to A \to B \amalg B' \to \operatorname{Coker}(f,f') \to 0$ is almost split.
In both cases, ii) is a direct consequence of i) and Proposition 1.1 (1.1').
From the above it is clear that information about irreducible maps gives informaiion about the structure of almost split sequences and conversely. For example, we conclude that for an indecomposable Λ-module B, there is only a finite number of

indecomposable Λ-modules A such that there is an irreducible map f: A → B, and only a finite number of indecomposable Λ-modules C such that there is an irreducible map g: B → C.

On the other hand, it is not hard to prove that if g: B → C is irreducible, then g is either an epimorphism or a monomorphism. Hence if $0 \to A \xrightarrow{f} B \xrightarrow{g} C \to 0$ is an almost split sequence, then for any summand B' of B the restriction map g': B'→ C is either an epimorphism or a monomorphism.

Another interesting feature of the irreducible maps and the almost split sequences is that they give rise to invariants for indecomposable modules C: If $0 \to A \xrightarrow{f} B \xrightarrow{g} C \to 0$ is an almost split sequence, what significance has the number of non-isomorphic summands of B, or the number of copies of each summand? What does it mean that for all irreducible maps g: B → C with B indecomposable g is always an epimorphism, or always a monomorphism? To illustrate, we explain what the situation is in an easy special case, namely if Λ is Nakayama (generalized uniserial): The irreducible maps between indecomposable modules are all of the type $0 \to \underline{r}C \to C$ or $B \to {}^{B}/\mathrm{soc}B \to 0$. If C is indecomposable and not projective or simple, the middle term B in the almost split sequence $0 \to A \xrightarrow{f} B \xrightarrow{g} C \to 0$ is the direct sum $B_1 \amalg B_2$ of two indecomposable modules, where one of the restriction maps $g_1: B_1 \to C$ is a monomorphism and the other one $g_2: B_2 \to C$ is an epimorphism. If C is simple and not projective, the middle term B is indecomposable of length two.

Another interesting class of modules suggested by irreducible maps are the ones of the form C=Cokerf, where f: A → B is an irreducible monomorphism and A and B indecomposable. It would be interesting to classify these modules. For a Nakayama algebra Λ we see by the above that we get exactly the simple Λ-modules. It is interesting that one can prove in general that such modules C=Cokerf have the property that for all irreducible maps $g: B \to C$, g is an epimorphism. But it is not conversely true in general as for Nakayama algebras, that if C is an indecomposable Λ-module such that all irreducible maps $g: B \to C$ are epimorphisms, then $C \cong \text{Coker} f$, where $f: A \to B$ is an irreducible monomorphism between indecomposable modules A and B.

§ 2

In this section we discuss a method for computing new almost split maps and sequences and irreducible maps from given ones. We shall use the categories $\underline{\text{mod}}\Lambda$ (modΛ modulo projectiees) and $\overline{\text{mod}}\Lambda$ (modΛ modulo injectives) (see[1]). These categories are not abelian, so we can not talk about almost split sequences here. But our definition of (minimal) right and left almost split maps and irreducible maps makes good sense in $\underline{\text{mod}}\Lambda$ and $\overline{\text{mod}}\Lambda$. For $g: B \to C$ a map of Λ-modules we consider the corresponding map $g: B \to C$ in $\underline{\text{mod}}\Lambda$, and compare with respect to the above types of maps. One of our useful results is

Proposition 2.1: Let C be an indecomposable non-projective Λ-module, B a Λ-module with no projective summands and $g: B \to C$ a map. Then $g: B \to C$ is a minimal right almost split map if and only if $B \amalg P \to C$ is a minimal right almost split map, where P is a projective cover for Coker g.

Similar results hold for minimal left almost split maps and for irreducible maps, and we can also consider $\overline{\text{mod}\Lambda}$ instead of $\underline{\text{mod}\Lambda}$.

Assume that F: $\underline{\text{mod}\Lambda} \to \underline{\text{mod}\Lambda'}$ is an equivalence of categories (i.e. Λ and Λ' are by definition stably equivalent), where Λ' may be isomorphic to Λ . The idea is then to use the functor F and Proposition 2.1 to construct an almost split sequence of Λ'-modules from an almost split sequence of Λ-modules. Dlab and Ringel have described a method for describing the almost split sequences for hereditary algebras of finite representation type (deducable from [6]). Hence we can use our method to construct almost split sequences for algebras stably equivalent to hereditary algebras, in particular for algebras of finite representation type with radical square zero, since they are stably equvalent to hereditary algebras of finite representation type [3] , [4] .

Also when $\Lambda = \Lambda'$, there are interesting examples of such functors F. For example, if Λ is a self-injective algebra, it is well known and not hard to prove that Ω^1: $\underline{\text{mod}\Lambda} \to \underline{\text{mod}\Lambda}$ is an equivalence of categories (see [7],[2]). For a Λ-module M with no projective summands, $\Omega^1 M$ denotes the first syzygy module for M, i.e. is determined by the exact sequence $0 \to \Omega^1 M \to P \to M \to 0$, where P is a projective cover for M. In this case there is a simple way of constructing the new almost split exact sequence directly, from an almost split sequence $0 \to A \to B \to C \to 0$, as indicated by the following diagram with exact rows and columns.

$$\begin{array}{ccccccccc}
 & & 0 & & 0 & & 0 & & \\
 & & \downarrow & & \downarrow & & \downarrow & & \\
0 & \to & \Omega^1 A & \to & \Omega^1 B \amalg P & \to & \Omega^1 C & \to & 0 \\
 & & \downarrow & & \downarrow & & \downarrow & & \\
0 & \to & P_A & \to & P_A \amalg P_B & \to & P_C & \to & 0 \\
 & & \downarrow & & \downarrow & & \downarrow & & \\
0 & \to & A & \to & B & \to & C & \to & 0 \\
 & & \downarrow & & \downarrow & & \downarrow & & \\
 & & 0 & & 0 & & 0 & &
\end{array}$$

Here P_A denotes a projective cover for A, P_C a projective cover for C, and P is some projective Λ-module, which is zero if and only if $P_A \amalg P_C \to B$ is a projective cover.

Apart from using the above to compute new almost split sequences for its own sake, we can also give an application to periodic modules over self-injective algebras. We recall that a Λ-module M with no projective summands is periodic if $\Omega^n M \cong M$ for some $n \geq 1$, where $\Omega^n M$ denotes the nth syzygy module for M.

Proposition 2.2: Let Λ be a self-injective algebra and $g: B \to C$ an irreducible map between indecomposable non-projective Λ-modules. If C has a periodic resolution, then so does B.

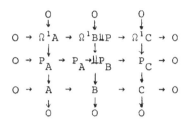

Proof: By Proposition 1.2 we have an almost split sequence $0 \to K \to B \amalg X \amalg P \to C \to 0$ where P is projective and X has no projective summands. By assumption, $\Omega^n C \cong C$ for some $n \geq 1$. By repeated application of the above we get an almost split sequence $0 \to \Omega^n K \to \Omega^n B \amalg \Omega^n X \amalg Q \to \Omega^n C \to 0$, where Q is projective. Since the almost split sequence is uniquely determined by the right hand module [1][5], we conclude that $\Omega^n B \amalg \Omega^n X \cong B \amalg X$. This means that $B \amalg X$ is periodic, and consequently B is periodic, since B is a summand of $B \amalg X$.

One can use this last result to prove the following

Theorem 2.3: Let Λ be a self-injective algebra and C an indecomposable periodic non-projective Λ-module such that an infinite number of non-isomorphic Λ-modules have a non-zero map to C. Then there is an infinite number of indecomposable non-isomorphic periodic Λ-modules.

We end the section by mentioning that the functor DTr: $\underline{\text{mod}\Lambda} \to \overline{\text{mod}\Lambda}$ (see[1]) is a useful example of a functor which helps us construct new almost split sequences.

§ 3

In this section we give some more information on what the almost split sequences look like. The approach of studying almost split sequences via irreducible maps had to do with describing the middle term in terms of its indecomposable summands. Here we give a different type of description, as a pushout of a certain diagram [5] .

Let C be an indecomposable Λ-module and $0 \to K \to P \to C \to 0$ an exact sequence, where P is a projective cover for C. Consider the exact sequence $0 \to K/\underline{r}K \to P/\underline{r}K \to C \to 0$. Then in general one has that $K/\underline{r}K \cong \text{soc DTr } C$ (see [5]). So we have a diagram of the following type

$$\begin{array}{ccccccccc} & & 0 & & & & & & \\ & & \downarrow & & & & & & \\ 0 & \to & K/\underline{r}K & \to & P/\underline{r}K & \to & C & \to & 0 \ . \\ & & \downarrow & & & & & & \\ & & \text{DTr}C & & & & & & \end{array}$$

Proposition 3.1: With the above notation, assume that for at least one simple Λ-module S with $Ext^1(C,S) \neq 0$, End (S) has more than two elements. Then the almost split sequence is a pushout for some diagram

$$\begin{array}{ccccccccc} & & 0 & & & & & & \\ & & \downarrow & & & & & & \\ 0 & \to & K/\underline{r}K & \to & P/\underline{r}K & \to & C & \to & 0 \\ & & \downarrow & & & & & & \\ & & DTrC & & & & & & \end{array} .$$

In [5] we give an example, where Λ is an algebra over a field with two elements, where it is not possible to obtain the almost split sequence as a pushout of a diagram of the above type.

Finally, we mention another type of result, which on one hand helps compute the almost split sequences when A (or C) is simple (see [5]). This result is interesting also because it shows how one can characterize certain types of modules in terms of the almost split sequence.

Propostion 3.2: A non-injective indecomposable module A is simple if and only if in the almost split sequence $0 \to A \xrightarrow{f} B \xrightarrow{g} C \to 0$, g is an essential epimorphism.

We can specialize this result to get several equivalent characterizations of when B is a projective cover (or equivalently, B is projective). For example, this is the case if and only if for any indecomposable Λ-module X, any map $h: X \to A$ factors through an injective module. Here a certain class of simple modules is characterized in terms of the almost split sequence, and also this information can be used to do some computations. For example, if the square of the radical of Λ is zero, it is not hard to see that all simple non-injective modules S have the property that any map h: X → S factors through an injective module. Given A, it is possible to describe $^{C}/\underline{r}C$ (as we earlier could describe socA in terms of C). Hence we can deseribe B when g: B → C is a projective cover. To actually construct the sequence, it is useful to use irreducible maps.

References

1. M. Auslander, Almost split sequences I, these Proceedings.

2. M. Auslander, M. Bridger, Stable module theory, Mem. Amer. Math. Soc. 94, 1969.

3. M. Auslander, I. Reiten, Stable equivalence of artin algebras, Proc. of the Conf. on orders, group rings and related topics, Springer Lecture Notes 353 (1973), 8-70.

4. M. Auslander, I. Reiten, Stable equivalence of dualizing R-varieties III: Dualizing R-varieties stably equivalent to hereditary dualizing R-varieties, Adv. in Math.,17 (1975)

5. M. Auslander, I. Reiten, Representation theory of artin algebras III: Almost split sequences, Comm. in Algebra,3, (1975), 239-293.

6. V.Dlab, C.M. Ringel, Representations of graphs and algebras, Carleton Math. Notes No. 8, August 1974.

7. A. Heller, The loop space functor in homological algebra, Trans. Amer. Math. Soc. 96 (1960), 382-394.

M.Auslander
Brandeis University
Waltham, Massashusetts 02154

I.Reiten
University of
Trondheim , Norway

MODULES HAVING WAISTS

M. Auslander*, E.L. Green**, I. Reiten

We introduce a new class of indecomposable modules which we call modules having waists. A more complete version of the following results and their proofs will be published elsewhere. We would like to thank all those involved with the organization of the ICRA, especially Professor V. Dlab.

§1.

Let R be a ring. Unless otherwise stated, all modules will be left modules. An R-module M <u>has a waist</u> if there is a non-trivial proper submodule M' of M such that every submodule of M contains M' or is contained in M'. In this case, we say that M' <u>is a waist in</u> M. One immediately has that

a) if M has a waist then M is indecomposable.

b) if M' is a waist in M and $X \subsetneq M' \subsetneq Y \subseteq M$ then M'/X is a waist in Y/X. In particular, Y/X is indecomposable.

c) if M is noetherian and has a waist, it has a unique maximal waist.

d) if M" is a waist in M' and if M' is a waist in M then M" is a waist in M.

e) if M" is a waist in M', and $M' \subseteq M$ with M'/M" a waist in M/M" then M' is a waist in M.

* Partially supported by NSF Grant GP 33406X1
** Partially supported by NSF Grant GP 29429A3

Although the study of modules having waists may be of interest for more general rings, we will henceforth assume R is a left Artin ring with radical $\underline{r}$.

Let M be a left R-module. Recall that the Loewy length of M , denoted $\ell\ell(M)$ is the smallest integer m such that $\underline{r}^m M = 0$. The lower Loewy series for M is the sequence $0 \neq S_0(M) \subseteq S_1(M) \subseteq \ldots \subseteq S_n(M) = M$ where $S_0(M) = \mathrm{soc}(M)$ and $S_i(M) = \pi_i^{-1}(\mathrm{soc}(M/S_{i-1}(M)))$ where $\pi_i : M \to M/S_{i-1}(M)$, for $i \geq 1$.

<u>Proposition 1</u>. Suppose M' is a waist in M . Then

1) $M' = \underline{r}^i M$ for some $i \geq 1$.

2) $M' = S_j(M)$ for some j .

For i and j in (1) and (2) we have $i + j + 1 = \ell\ell(M)$.

It is easy to see that the class of modules having waists includes non-simple modules having either a unique maximal submodule or unique minimal submodule. Hence non-simple indecomposable projective, injective and uniserial modules all have waists.

We have the following useful classifications of when a module has a waist.

<u>Theorem 2</u>. Let R be an Artin ring. Suppose M' is a submodule of M . Then the following statements are equivalent:

1) M' is a waist in M

2) if $X \subsetneqq M' \subsetneqq Y \subseteq M$ then Y/X is indecomposable.

3) $M'/\underline{r}M'$ is a waist in $M/\underline{r}M'$.

4) let $\pi : M \longrightarrow M/M'$ be the canonical surjection. Then M' is a waist in $\pi^{-1}(\mathrm{soc}(M/M'))$.

Proposition 3. If $M' \subseteq M$ such that $M'/\underline{r}M'$ is simple and $M'/\underline{r}M' \cong \mathrm{soc}(M/\underline{r}M')$ then

a) if $\underline{r}M' \neq 0$, $\underline{r}M'$ is a waist in M .

b) if $M' \neq M$, M' is a waist in M .

Given a module M , we call $M/\underline{r}M$ the top of M and denote it by top(M). The special kinds of modules having waists described in Proposition 3, namely those where top(M') is simple, play an important role in the study of modules having waist in left Artin rings R such that $R/\underline{r}^2$ is of finite representation type. In a sense, they are the only modules having waists that occur. This will be made more precise in §2.

If R is an Artin algebra, i.e., R is a finitely generated module over its center C which is an Artin ring, then there is a duality between left R-modules and right R-modules. Namely, if E is the C-injective envelope of $C/\mathrm{rad}(C)$, then $D(X) = \mathrm{Hom}_C(X,E)$ for X a left or right R-module.

Proposition 4. If R is an Artin algebra, then M' is a waist in $M \Leftrightarrow D(M/M')$ is a waist in $D(M)$.

We now consider the relation between the representation type of R and modules having waists.

Proposition 5. If R is an Artin algebra then the length of modules having waists is bounded. In particular, if M has a waist then the length of M , denoted $\ell(M)$, $\leq \max\{\ell(P)+\ell(Q):$ P is an indecomposable left projective R-module and Q is an indecomposable right projective R-module$\}$.

Corollary 6. If R is an Artin algebra and every indecomposable R-module has a waist or is simple then R is of finite representation type.

§2.

We now study modules having waists in rings with $\underline{r}^2 = 0$. This especially of interest because by Theorem 2 if M' is a waist in an R-module, where R is an arbitrary Artin ring, with $M' = \underline{r}^i M = S_j(M)$ then

1) $M'/\underline{r}M'$ is a waist in $\underline{r}^{i-1}M/\underline{r}M'$ and

2) $M'/S_{j-1}(M)$ is a waist in $S_{j+1}(M)/S_{j-1}(M)$.

Thus, given a module having a waist in an Artin R , it induces in general two different modules having waists over $R/\underline{r}^2$. Finally if we consider the method of constructing new waists discussed in §3, we can at times, knowing the waists for $R/\underline{r}^2$, create new ones of larger Loewy lengths.

For the rest of this section we assume R is an Artin

algebra with $\underline{r}^2 = 0$. We have

Proposition 6. Suppose R is of finite representation type. Then every waist has a simple top or simple socle.

Thus, as mentioned after Proposition 3, if Λ is an arbitrary Artin algebra such that $\Lambda/\underline{r}^2$ is of finite representation type and M' is a waist in M , then at least one of the following modules must be simple: top(M), soc(M), top(M'), soc(M/M').

Proposition 7. Let R be an Artin algebra with $\underline{r}^2 = 0$. If there exists an indecomposable projective of length ≥ 4 then there exists an indecomposable R-module not having a simple top or simple socle.

From Propositions 5, 6 and 7, we easily get the following theorem.

Theorem 8. Let R be an Artin algebra with $\underline{r}^2 = 0$. Then the following statements are equivalent:

1) Every indecomposable left R-module has simple top or simple socle.

2) Every indecomposable left R-module has a waist or is simple.

3) Every indecomposable left R-module is either projective,

injective or uniserial.

1') Every indecomposable right R-module has simple top or simple socle.

2') Every indecomposable right R-module has a waist or is simple.

3') Every indecomposable right R-module is either projective, injective or uniserial.

If one considers the separated diagram for an Artin algebra R with $\underline{r}^2 = 0$ (see [1]), one sees that R satisfies (1)-(6) of Theorem 8 if and only if the separated diagram for $R/\underline{r} + \underline{r}$ is composed of disjoint copies of the following types of diagrams:

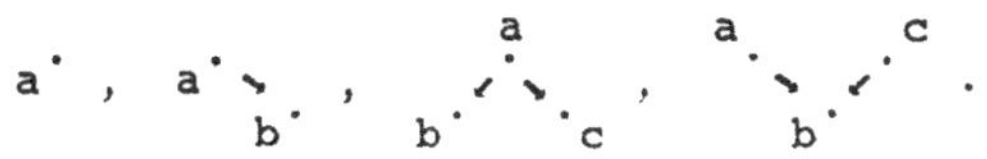

More generally, if Λ is a factor ring of a tensor algebra associated to a k-species $\mathcal{S} = (K_i, {}_iM_j)_{i,j \in \mathcal{C}\ell}$, with each $K_i = k$, then each non-simple Λ-module has a waist if the diagram associated to $\mathcal{S}$ (see [2] for definitions) is composed of disjoint diagrams of the following types:

$$\dot{a}_1 \to \dot{a}_2 \to \cdots \to \dot{a}_m, \ m \geq 1 \ ; \quad \dot{a}_n \to \dot{a}_{n-1} \to \cdots \to \dot{a}_0 \leftarrow \dot{b}_1 \leftarrow \cdots \leftarrow \dot{b}_m ,$$

$$n \geq 1 , \ m \geq 1$$

and

$$\dot a_n \leftarrow \dot a_{n-1} \leftarrow \cdots \leftarrow \dot a_0 \rightarrow \dot b_1 \rightarrow \cdots \rightarrow b_m \ , \ n \geq 1 \ , \ m \geq 1 \ .$$

§3.

We begin by describing a general technique of creating new modules from old.

<u>Theorem 9</u>. Let S be an arbitary ring. Let $A \subseteq B$ and $C \subseteq D$ be S-modules and suppose there is an isomorphism $\alpha : B/A \xrightarrow{\approx} C$. Then the following statements are equivalent:

1) There is a module X with $B \subseteq X$ and an isomorphism $\beta : X/A \xrightarrow{\approx} D$ such that the following diagram commutes:

$$\begin{array}{ccc} X/A & \xrightarrow[\beta]{\approx} & D \\ | \cup & & | \cup \\ B/A & \xrightarrow[\alpha]{\approx} & C \end{array}$$

2) If $\varphi : \mathrm{Ext}^1(D,A) \longrightarrow \mathrm{Ext}^1(C,A)$ is induced from $C \subseteq D$ then the exact sequence $0 \longrightarrow A \longrightarrow B \xrightarrow{\alpha\circ\pi} C \longrightarrow 0$, where $\pi : B \longrightarrow B/A$, is in the image of φ.

3) If $\psi : \mathrm{Ext}^1(D/C,B) \longrightarrow \mathrm{Ext}^1(D/C,C)$ is induced from $B \xrightarrow{\alpha\circ\pi} C$ then the exact sequence $0 \longrightarrow C \longrightarrow D \longrightarrow D/C \longrightarrow 0$ is in the image of ψ.

Given $A \subseteq B$, $C \subseteq D$ and $\alpha : B/A \xrightarrow{\approx} C$ as in Theorem 9, if conditions (1)-(3) hold, we say that we can paste B and D by α . X is called the pasted module.

Corollary 10. Given $A \subseteq B$, $C \subseteq D$ and $\alpha : B/A \xrightarrow{\approx} C$. Then if $\mathrm{pd}_S(D/C) \leq 1$ or injective $\mathrm{dimension}_S(A) \leq 1$ then we can paste B and D by α . In particular, if S is hereditary we can always paste modules.

We hasten to remark that even if both B and D have waists, the pasted module X need not even be indecomposable. Nevertheless the following theorem gives a way to create new waists from old.

Theorem 11. Let R be a left Artin ring. Let M and N be R-modules with $\ell\ell(M) = m$ and $\ell\ell(N) = n$.

1) Suppose that there is an isomorphisms $\alpha : M/\underline{r}M \xrightarrow{\approx} \underline{r}^{n-1}N$ and we can paste M and N by α . If N has a waist then so does the pasted module X .

2) Suppose that there is an isomorphism $\beta : M/S_{m-2}(M) \xrightarrow{\approx} \mathrm{soc}(N)$ and we can paste M and N by β . If M has a waist then so does the pasted module X .

As an important application we state

Proposition 12. Let R be a left Artin ring. Let M and N be R-modules such that

i) $M/\underline{r}M$ is simple and

ii) there is an isomorphisms $\alpha : M/\underline{r}M \xrightarrow{\approx} \mathrm{soc}(N)$.

If we can paste M and N by α then the pasted module has a waist.

We give an example of how these results may be applied. Let Λ be a hereditary Artin algebra such that $\Lambda/\underline{r}^2$ is of finite representation type. Let S be a simple Λ-module. Let P be the projective cover of S and E be the injective envelope of S. Applying Corollary 10 and Proposition 12 there is a module X_S having a waist such that $P \subseteq X_S$ and $X_S/\underline{r}P \simeq E$. From §2 we see that if Y is a Λ-module with non-simple top and non-simple socle having a waist, then Y is a submodule of a factor of X_S for some simple module S.

Finally, we mention that we have results concerning the construction of modules having waists in radical square zero algebras with non-simple tops and socles.

References

1. Dlab, V., and Ringel, C., On algebras of finite representation type, Carleton University Lecture Notes No. 2, 1973.
2. Green, E.L., The representation theory of tensor algebras, J. of Algebra, to appear.

M. Auslander
Brandeis University
U.S.A.

E.L. Green
University of Pennsylvania
U.S.A.

I. Reiten
Trondheim University
Norway

QUIVERS WITH COMMUTATIVITY CONDITIONS AND SOME PHENOMENOLOGY OF FORMS

Sheila Brenner

1. Introduction

This paper is written in the spirit of experimental science. It reports some observed regularities and suggests that there should be a theory to explain them.

The observations concern the relationship between the representation type of certain algebras finitely generated over a field, and the definiteness of a suitably defined (quadratic) form. This relationship is well established in the case of representations of quivers and the relevant results are outlined in Section 2.

In Section 3 the concept of quiver with commutativity conditions is introduced to generalise the problem of linear representations of partially ordered sets. It is indicated that the relationship between representation type and (in an appropriate sense) definiteness of form holds also in these cases.

Section 4 gives some further instances, and Section 5 outlines a naive explanation which may afford some insight into the phenomena.

2. Preliminaries

A quiver is a finite connected directed graph. We denote by P_Q and A_Q (respectively) the set of points and set of arrows of the quiver Q. If $\gamma = (r, s) \in A_Q$, we write $i(\gamma) = r$, $f(\gamma) = s$.

A representation V of a quiver Q over a field k is an assignment to each $r \in P_Q$ of a finite dimensional vector space V(r), and to each $\gamma = (r, s) \in A_Q$, a homomorphism $V(\gamma) : V(r) \to V(s)$. The representation V is said to be indecomposable if $V(r) = V_1(r) \oplus V_2(r)$ for all $r \in P_Q$ and $V(r, s)V_i(r) \subset V_i(s)$ $(i = 1, 2)$ for all $(r, s) \in A_Q$ implies either $V_1(r) = 0$ for all $r \in P_Q$ or $V_2(r) = 0$ for all $r \in P_Q$.

The quiver Q is said to be

(i) of finite type if there exists only a finite number of isomorphism classes of indecomposable representations,

(ii) of tame type if there are infinitely many isomorphism classes of indecomposable representations but these classes can be parametrised by a finite set of integers together with a polynomial irreducible over k,

(iii) of wild type if, given a finite dimensional k-algebra E, there exist infinitely many pairwise non-isomorphic representations of Q with endomorphism algebra isomorphic to E.

[These types are clearly exclusive. They are also exhaustive. It would be nice to have definitions of wild and tame which made this immediately apparent, and which covered the slight generalisations required in some other cases. See e.g. [3].]

It turns out that this classification of quivers is independent of arrow direction. The quivers of finite type are the Dynkin diagrams

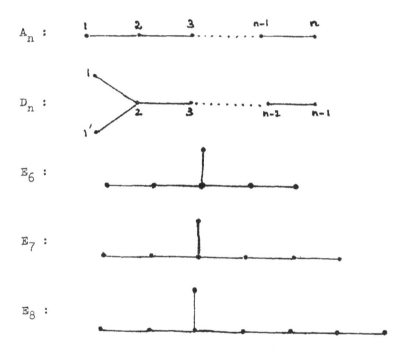

A_n : 

D_n : 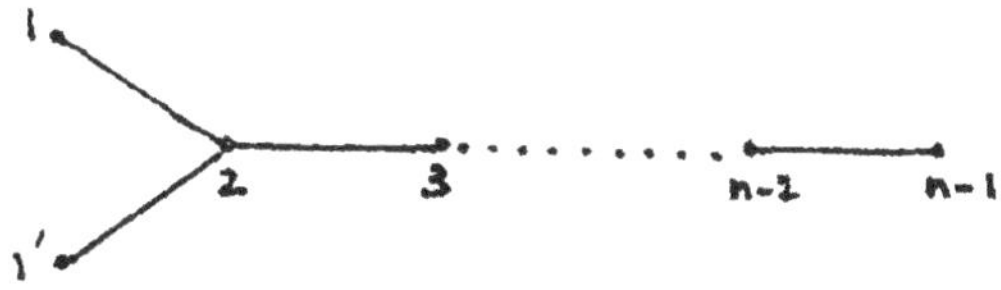

E_6 : 

E_7 : 

E_8 : 

Those of tame type are the extended Dynkin diagrams

$\hat{A}_n$: 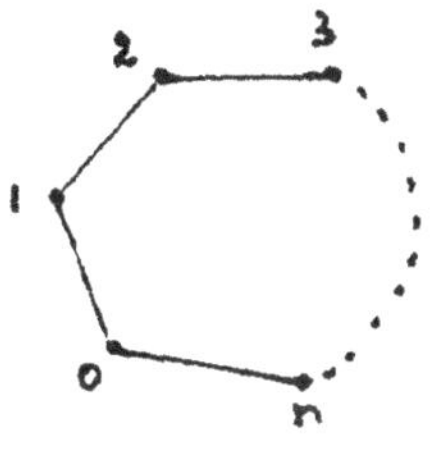

(including $\hat{A}_0$ 

)

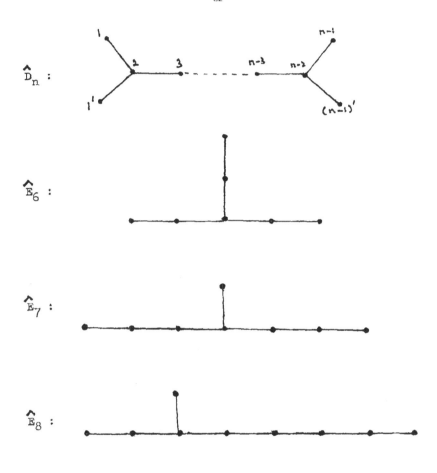

$\hat{D}_n$:

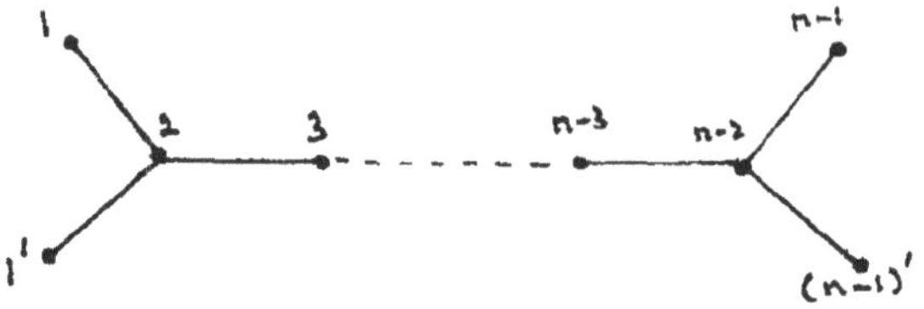

$\hat{E}_6$:

$\hat{E}_7$:

$\hat{E}_8$:

Corresponding to the quiver Q we may define a quadratic form B(Q) on $\mathbb{Z}^{|P_Q|}$ (with components indexed by P_Q) by

$$B(Q) = \sum_{r \in P_Q} d^2_r - \sum_{\gamma \in A_Q} d_{i(\gamma)} d_{f(\gamma)}.$$

The results above may now be stated in the form

[6, 7, 12]:

Q is of finite, tame or wild type according to whether B(Q) is positive definite, semi-definite or indefinite.

The following argument due to Tits, and quoted in [7], shows that, for an infinite field k, if Q is of finite type, then B(Q) must be positive definite.

Let V be a representation of Q. Define its dimension $d(V) = (d_r)_{r\epsilon P_Q} \in \mathbb{Z}^{|P_Q|}$ by $d_r = \dim V(r)$. The representations of Q with dimension d form a manifold of dimension $\sum_{\gamma\epsilon A_Q} d_{i(\gamma)} d_{f(\gamma)}$. Acting on this manifold is the automorphism group G_0, which is the product of the automorphism groups of the V(r) and therefore has dimension $\sum_{r\epsilon P_Q} d_r^2$. However, within this is the stability subgroup $G_1 \simeq k$ which leaves invariant each representation V with $d(V) = d$. Thus we have $G = G_0/G_1$ acting faithfully on M, and M is covered by orbits of G, each orbit corresponding to an isomorphism class of representations of Q with dimension d. Thus we must have, if Q is of finite type,

$$\dim M \leqslant \dim G$$

i.e. $$\sum_{r\epsilon P_Q} d_r^2 - \sum_{\gamma\epsilon A_Q} d_{i(\gamma)} d_{f(\gamma)} \geqslant 1.$$

Bernstein, Gelfand and Ponomarev [1] have shown how to extend this argument to the case of finite fields k.

3. Quivers with Commutativity Condition

Several authors [8, 11, 13] have studied the problem of representing a partially ordered set by sets of subspaces of a vector space ordered by inclusion. Such representations are representations of the appropriate quiver in which all the arrows are represented by inclusions. The inclusion condition imposes the requirement that certain sub-quivers are commutative diagrams and this suggests the following definitions.

A commutativity condition on a quiver Q is a pair of points s, t $\in P_Q$ and two subsets $\{\lambda_j : 1 \leqslant j \leqslant \ell\}$, $\{\rho_j : 1 \leqslant j \leqslant r\}$ of A_Q satisfying

$$i(\lambda_1) = i(\rho_1) = s;\ f(\lambda_\ell) = f(\rho_r) = t;$$

$$i(\lambda_{j+1}) = f(\lambda_j),\ 1 \leqslant j \leqslant \ell - 1;\ i(\rho_{j+1}) = f(\rho_j),\ 1 \leqslant j \leqslant r - 1.$$

A representation of the quiver Q with this commutativity condition is a representation V of Q which satisfies

$$V(\rho_r)V(\rho_{r-1})\ldots V(\rho_2)V(\rho_1) = V(\lambda_\ell)V(\lambda_{\ell-1})\ldots V(\lambda_1). \quad (3.1)$$

Thus the problem of classifying representations of quivers with commutativity conditions lies between the problems of classifying representations of quivers and of classifying representations of partially ordered sets.

Representations of quivers with commutativity conditions occur in some approaches to the study of integral representations of finite p-groups [4]. The 'figure of eight'

diagram of vector spaces

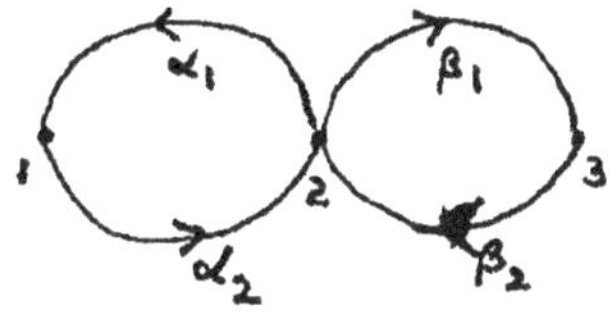

with $\beta_2\beta_1 = \alpha_2\alpha_1$, raised by Gelfand [10], is also a representation of a quiver with commutativity condition. Its indecomposable representations have been found by Nazarova and Roiter [14]. It is of tame type.

Consideration of the significance attributed to the quadratic form by the argument of Tits given above leads to the construction of a quadratic form $B(Q_C)$ for a quiver Q with set C of commutativity conditions. We observe that the condition (3.1) on representations V of Q of dimension $d = (d_r : r \in P_Q)$ may be considered as defining a submanifold[1] M′, of codimension $d_s d_t$, of the manifold M of all representations of Q with dimension d. Thus $B(Q_C)$ is obtained from B(Q) by adding the appropriate term of form $d_s d_t$ for each commutativity condition.

Thus for the 'figure of eight' diagram above

$$B(Q) = d_1^2 + d_2^2 + d_3^2 - 2d_1d_2 - 2d_2d_3 = (d_1 - d_2)^2 + (d_3 - d_2)^2 - d_2^2$$

---

[1]This is an oversimplification. Generally we should consider a union of submanifolds. However (see §4) the procedure given here leads to the form which, in simple cases at least, should correspond to representation type. See also §5.

and $\qquad B(Q_C) = B(Q) + d_2{}^2 = (d_1 - d_2)^2 + (d_3 - d_2)^2.$

The fact that $B(Q_C)$ is semidefinite while $B(Q)$ is indefinite corresponds to the fact that, with the commutativity condition, the 'figure of eight' is tame while, without the condition, it is wild.

It seems that the nature of the form gives representation type of a diagram in quite general circumstances. For example, denote by $\underset{n}{\overset{\ell}{m\Diamond p}}$ the diagram of Figure 1 with commutativity condition $\beta_1\alpha_1 = \beta_2\alpha_2$. (No directions have been indicated for the arrows of the 'arms' of this diagram since the representation type is clearly independent of these directions.) Then

$$B\left(\underset{n}{\overset{\ell}{o\Diamond o}}\right) = \sum_{r=0}^{\ell-1} \frac{r+2}{2(r+1)} \left(a_{\ell-r} - \frac{r+1}{r+2}\, a_{\ell-r-1}\right)^2$$

$$+ \sum_{r=0}^{n-1} \frac{r+2}{2(r+1)} \left(c_{n-r} - \frac{r+1}{r+2}\, c_{n-r-1}\right)^2$$

$$+ (b_0 - \tfrac{1}{2}a_0 - \tfrac{1}{2}c_0)^2 + (d_0 - \tfrac{1}{2}a_0 - \tfrac{1}{2}c_0)^2$$

$$+ \frac{1}{2(\ell+1)}\, a_0{}^2 + \frac{1}{2(n+1)}\, c_0{}^2,$$

which is positive definite for all ℓ and n. Correspondingly $\underset{n}{\overset{\ell}{o\Diamond o}}$ is of finite type for all ℓ and n (see [7]).

However

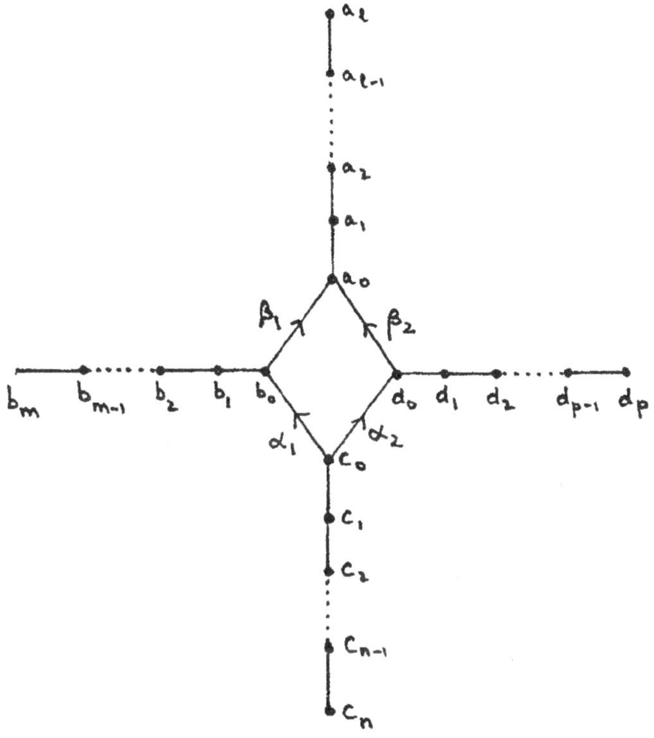
a_ℓ
$a_{\ell-1}$
a_2
a_1
a_0
β_1
β_2
b_m
b_{m-1}
b_2
b_1
b_0
d_0
d_1
d_2
d_{p-1}
d_p
α_1
α_2
c_0
c_1
c_2
c_{n-1}
c_n

Figure 1

$$B\left(m \diamond p\right) = \sum_{r=0}^{m-1} \frac{r+2}{2(r+1)}\left(b_{m-r} - \frac{r+1}{r+2}\, b_{m-r-1}\right)^2$$

$$+ \sum_{r=0}^{p-1} \frac{r+2}{2(r+1)}\left(d_{p-r} - \frac{r+1}{r+2}\, d_{p-r-1}\right)^2$$

$$+ \left(a_0 + \tfrac{1}{2}c_0 - \tfrac{1}{2}b_0 - \tfrac{1}{2}d_0\right)^2 + \frac{3}{4}\left(c_0 - \frac{1}{3}b_0 - \frac{1}{3}d_0\right)^2$$

$$+ \frac{m+4}{6(m+1)}\left(b_0 - \frac{2(m+1)}{m+4}d_0\right)^2$$

$$+ \frac{4 - mp}{2(p+1)(m+4)}d_0^2,$$

which is positive definite, semidefinite or indefinite according as $mp <, =$ or > 4.

We now show that this does indeed give the representation type of $m \diamond p$. Suppose that V is an indecomposable representation of this diagram. Then it is easy to see that V satisfies at least one of the following conditions:

(i) $\dim V(a_0) = 1$, $V(r) = 0$ if $r \neq a_0$;

(ii) $V(a_0) = 0$, $V(\beta_1) = V(\beta_2) = 0$, in which case V is a representation of a quiver corresponding to the Dynkin diagram A_{m+p+3};

(iii) $V(\alpha_1)$ and $V(\alpha_2)$ are injective and $V(a_0) = \operatorname{im} V(\beta_1) + \operatorname{im} V(\beta_2)$.

There are only a finite number of isomorphism classes of

indecomposables satisfying (i) or (ii). We shall show that the problem of classifying indecomposable representations satisfying (iii) is equivalent to the problem of classifying indecomposable representations of the partially ordered set $I_{m+1,p+1,1} = \{b_0 \geq b_1 \geq \ldots \geq b_m, d_0 \geq d_1 \geq \ldots \geq d_p, e\}$. It is known that $I_{m+1,p+1,1}$ has the required property [6, 7, 8, 12, 3].

Suppose that

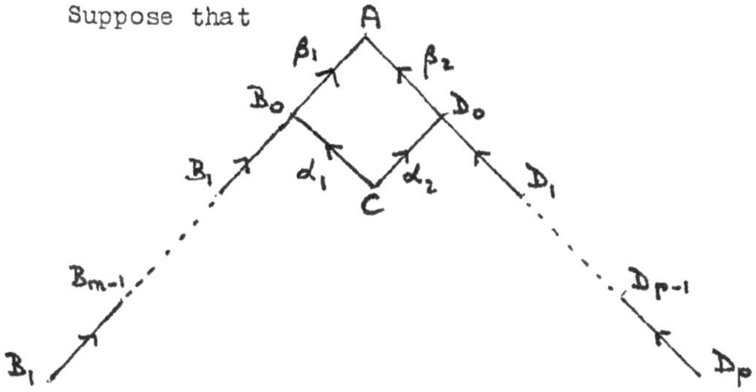

with $\beta_1\alpha_1 = \beta_2\alpha_2$ is a representation of 

with $\ker \alpha_1 = \ker \alpha_2 = 0$ and $A = \operatorname{im} \beta_1 + \operatorname{im} \beta_2$. Let U be the pushout of α_1 and α_2 and let $h : U \to A$ be the unique map defined by β_1 and β_2 in the usual way. Then the representation

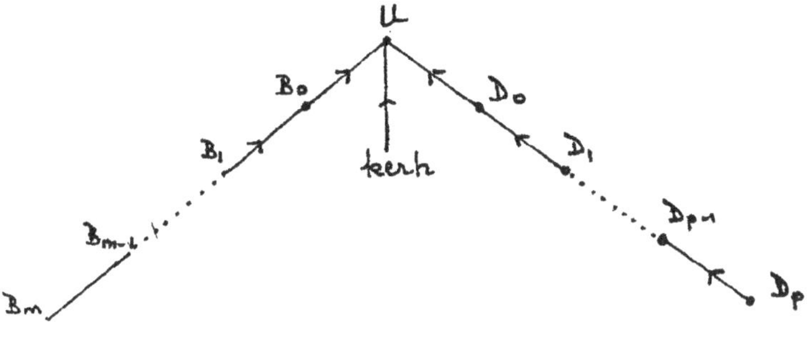

of $I_{m+1,p+1,1}$ has the same endomorphism algebra as the original representation of $m\overset{o}{\underset{o}{\Diamond}}p$.

Suppose now that

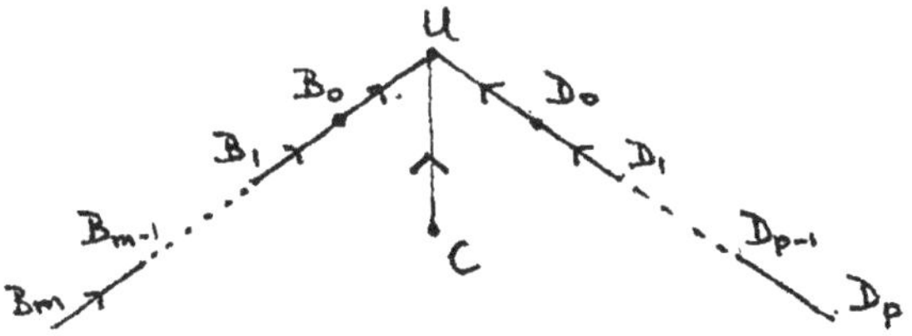

is a representation of $I_{m+1,p+1,1}$ with $B_0 + D_0 = U$. (This condition must be satisfied by any indecomposable representation of $I_{m+1,p+1,1}$ [2].) Then

with β_1, β_2 the natural maps induced by inclusion followed by projection onto $^U/C$, and α_1, α_2 inclusions, is a representation of $m\overset{o}{\underset{o}{\Diamond}}p$.

The constructions above are functorial and the composite functions are naturally equivalent to the identities on the appropriate categories.

The diagrams $m\overset{\ell}{\underset{n}{\Diamond}}o$, which are of finite, tame or wild type according as $m(\ell + n)$ <, = or > 4, and

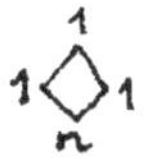

and

, which are tame or wild according as n = or > 0, can be treated by broadly similar methods. In all cases the representation type is given by the definiteness of the corresponding form.

A fuller account of representations of quivers with commutativity conditions will be given elsewhere. However it is interesting to look at two further examples:

I:

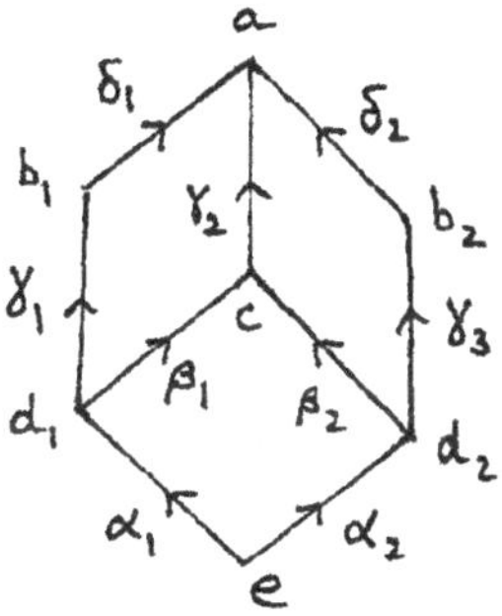

with $\beta_1\alpha_1 = \beta_2\alpha_2$, $\delta_1\gamma_1 = \gamma_2\beta_1$ and $\delta_2\gamma_2 = \gamma_2\beta_2$.

II:

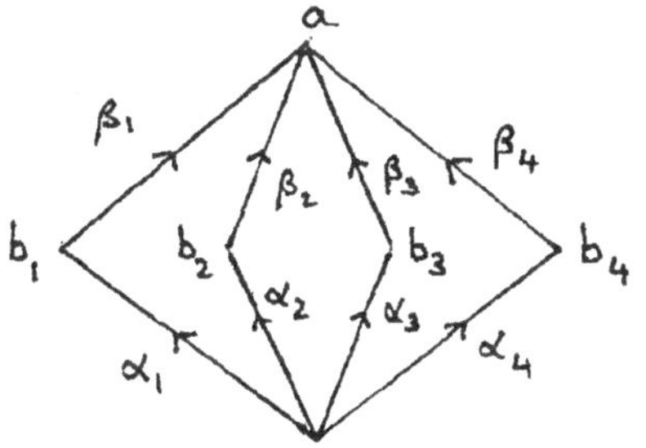

with $\beta_i\alpha_i = \beta_j\alpha_j$, $1 \leqslant i, j \leqslant 4$.

The first is of finite type and the second of tame type. The corresponding forms are

$$B(I) = (b_1 - \tfrac{1}{2}a - \tfrac{1}{2}d_1)^2 + (b_2 - \tfrac{1}{2}a - \tfrac{1}{2}d_2)^2 + (c + \tfrac{1}{2}e - \tfrac{1}{2}a - \tfrac{1}{2}d_1)^2$$
$$+ \frac{1}{4}(e + a)^2 + \frac{1}{2}(e - \tfrac{1}{2}d_1 - \tfrac{1}{2}d_2)^2 + \frac{3}{8}(d_1 - d_2)^2$$

and

$$B(II) = \sum_{i=1}^{4} (b_i - \tfrac{1}{2}a - \tfrac{1}{2}c)^2 + \frac{1}{4}(a + f)^2 - \frac{1}{4}(a - f)^2.$$

At first sight it looks as if these examples contradict the assertion that the representation type is given by the definiteness of the corresponding form. However the Tits argument shows that we should really be looking at forms on $\mathbb{Z}^{+|P_Q|}$ (where $\mathbb{Z}^+$ denotes the non-negative integers). As a form on $\mathbb{Z}^{+7}$, $B(I)$ is positive definite; and as a form on $\mathbb{Z}^{+6}$, $B(II)$ is semidefinite.

Similarly the diagram

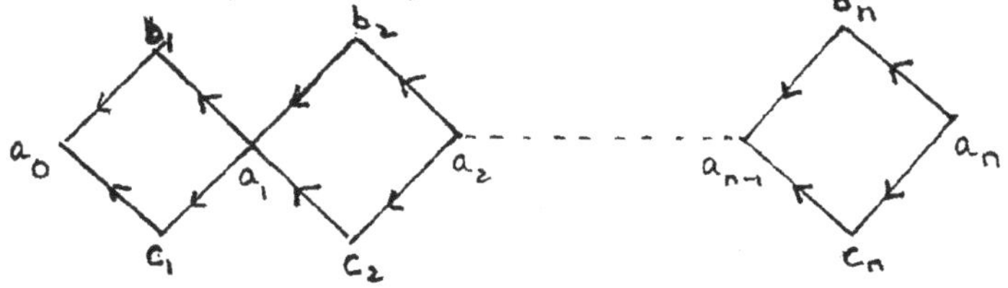

with each diamond commutative, is certainly of infinite type, and almost certainly tame, for $n \geq 2$. The corresponding partially ordered set is, however, of finite type. The appropriate form is

$$\sum_{i=1}^{n} (b_i - \tfrac{1}{2}a_{i-1} - \tfrac{1}{2}a_i)^2 + \sum_{i=1}^{n} (c_i - \tfrac{1}{2}a_{i-1} - \tfrac{1}{2}a_i)^2 + \tfrac{1}{2}a_0^2 + \tfrac{1}{2}a_n^2.$$

This is semidefinite on $\mathbb{Z}^{+(3n+1)}$, but becomes positive definite if we require the components of $d \in \mathbb{Z}^{+(3n+1)}$ to have the partial ordering imposed by the condition that all the maps are injective.

It is because of these phenomena that we have used the term 'definiteness' rather than signature in describing the required properties of forms.

Finally it is amusing to note that there is a surprisingly large intersection between the finite type quivers with commutativity conditions and the diagrams used by Carter[1] [5] in describing conjugacy classes of elements of the classical Weyl groups, especially if one of each pair of 'duplicates' in Carter's list is dropped.

---

[1] I am grateful to Alun Morris for drawing this paper to my attention.

4. Further Phenomenology of Forms

Gelfand and Ponomarev [9] have shown that the algebra generated over a field k by nilpotent elements x and y satisfying $xy = yx = 0$ is of tame type. Consider a representation of this algebra on a vector space V, with x and y representated by the endomorphisms x and y of V. Write $V_1 = \ker x \cap \ker y$ and let V_2 and V_3 be complements of V_1 in ker x and ker y, respectively, V_4 be a complement of ker x + ker y in V. Corresponding to this decomposition of V we may write x and y in the block matrix forms

$$x = \begin{pmatrix} 0 & 0 & x_1 & x_2 \\ 0 & 0 & 0 & 0 \\ 0 & 0 & x_3 & x_4 \\ 0 & 0 & 0 & 0 \end{pmatrix}, \quad y = \begin{pmatrix} 0 & y_1 & 0 & y_2 \\ 0 & y_3 & 0 & y_4 \\ 0 & 0 & 0 & 0 \\ 0 & 0 & 0 & 0 \end{pmatrix}$$

with x_3 and y_3 nilpotent. These matrices satisfy the conditions on x and y and we have lost no generality in writing x and y in this form. We write $d_i = \dim V_i$ $(1 \leqslant i \leqslant 4)$, $d = (d_1, d_2, d_3, d_4)$ and show how to construct a form which corresponds to the Tits interpretation.

First we note that the condition that an $n \times n$ matrix is nilpotent does not impose n^2 conditions on its elements. However we can find automorphisms of V_2 and V_3 which make x_3 and y_3 strictly upper triangular and so automatically nilpotent. With this further condition we calculate the dimension of the manifold M of representations of the required type and dimension d to be

$$\dim M = (d_1 + d_4)(d_2 + d_3) + 2d_1d_4 + \tfrac{1}{2}\{d_2(d_2 - 1) + d_3(d_3 - 1)\}.$$

We now calculate the automorphism group G acting on M. This is the quotient G_0/G_1, where G_0 is the group of automorphisms of V which preserves the conditions on our representations (i.e. preserves the zero's in the block forms for x and y and leaves x_3 and y_3 in upper triangular form), and the stability subgroup G_1 is the group of automorphisms of V which leaves invariant each representation satisfying the conditions. Thus G_0 is the group of automorphisms of form

$$\begin{pmatrix} \alpha & \xi & \eta & \lambda \\ 0 & \beta & 0 & \mu \\ 0 & 0 & \gamma & \nu \\ 0 & 0 & 0 & \delta \end{pmatrix}$$

with β and γ upper triangular, and G_1 is the group of automorphisms of form

$$\begin{pmatrix} k'1 & 0 & 0 & \lambda \\ 0 & k'1 & 0 & 0 \\ 0 & 0 & k'1 & 0 \\ 0 & 0 & 0 & k'1 \end{pmatrix}$$

where $k' \in k$, 1 is the identity on the appropriate space and $\lambda \in \hom(V_4, V_1)$. Hence

$$\dim G = d_1^2 + d_4^2 + (d_1 + d_4)(d_2 + d_3) + \tfrac{1}{2}\{d_2(d_2 + 1) + d_3(d_3 + 1)\} - 1$$

and so the Tits form is

$$\begin{aligned} B &= \dim G - \dim M + 1 \\ &= (d_1 - d_4)^2 + d_2 + d_3. \end{aligned}$$

This is semi-definite on $\mathbb{Z}^{+4}$.

It is known [3] that the algebra generated over k by commuting variables x and y satisfying $x^n = y^m = 0$ is wild unless $n = m = 2$. It is not easy to see how to write down a 'general' representation in this case. However these algebras always have as homomorphic image the algebra generated over k by x' and y' satisfying $x'y' = y'x'$, $x'^3 = y'^2 = x'^2y' = 0$. We can then write x' and y' in the block matrix forms

$$x' = \begin{pmatrix} 0 & 0 & \alpha_1 & 0 & 0 & 0 \\ 0 & 0 & 0 & \alpha_2 & 0 & 0 \\ 0 & 0 & 0 & 0 & \alpha_3 & 0 \\ 0 & 0 & 0 & 0 & 0 & \alpha_4 \\ 0 & 0 & 0 & 0 & 0 & 0 \\ 0 & 0 & 0 & 0 & 0 & 0 \end{pmatrix},$$

$$y' = \begin{pmatrix} 0 & 0 & 0 & \beta_1 & 0 & 0 \\ 0 & 0 & 0 & 0 & \beta_2 & 0 \\ 0 & 0 & 0 & 0 & 0 & \beta_3 \\ 0 & 0 & 0 & 0 & 0 & 0 \\ 0 & 0 & 0 & 0 & 0 & 0 \\ 0 & 0 & 0 & 0 & 0 & 0 \end{pmatrix}$$

with $\alpha_1\beta_3 = \beta_1\alpha_4$. It is not hard to see that such a representation is equivalent to the quiver

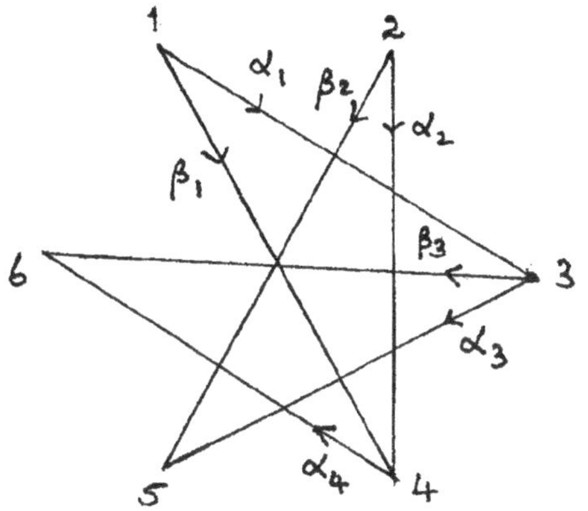

with commutativity condition $\alpha_1\beta_3 = \beta_1\alpha_4$. The corresponding form is

$$(d_2 - \tfrac{1}{2}d_4 - \tfrac{1}{2}d_5)^2 + \tfrac{3}{4}(d_5 - \tfrac{1}{3}d_4 - \tfrac{2}{3}d_3)^2 + \tfrac{2}{3}(d_3 - \tfrac{1}{4}d_4 - \tfrac{3}{4}d_1 - \tfrac{3}{4}d_6)^2$$

$$+ \tfrac{5}{8}(d_1 - d_4 + \tfrac{1}{5}d_6)^2 + \tfrac{2}{5}(d_6 - \tfrac{5}{6}d_4)^2 - \tfrac{5}{12}d_4^2,$$

and is indefinite on $\mathbb{Z}^{+6}$.

As a further example we consider the case of a quiver in which two maps are required to have composite equal to zero.[1] Suppose that the quiver Q contains the subquiver

Q':

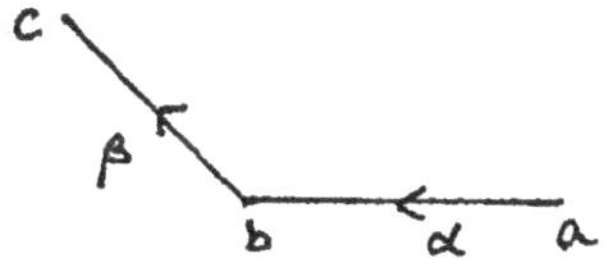

with the condition $\beta\alpha = 0$ and such that there are no other arrows in Q entering or leaving b. Let V be a representation of Q. Choose complements A of $\ker \alpha = K$ in V(a), B of $\operatorname{im} \alpha$ in $\ker \beta$ and L of $\ker \beta$ in V(b). Then we may write

[1] I am indebted to C. M. Ringel for suggesting this example.

α and β in the forms

$$\alpha = \begin{pmatrix} 0 & 1 \\ 0 & 0 \\ 0 & 0 \end{pmatrix} \qquad \beta = (0 \quad 0 \quad \beta').$$

The components $\theta \in \mathrm{Aut}\, V(a)$ and $\phi \in \mathrm{Aut}\, V(b)$ of an element of the automorphism group of V must then have the block forms

$$\theta = \begin{pmatrix} \theta_1 & \theta_{12} \\ 0 & \theta_{22} \end{pmatrix}, \qquad \phi = \begin{pmatrix} \theta_{22} & \phi_{12} & \phi_{13} \\ 0 & \phi_{22} & \phi_{23} \\ 0 & 0 & \phi_{33} \end{pmatrix}.$$

Since there are no arrows other than α and β entering or leaving V(b), the stability subgroup contains all elements which, for some $k' \in k$, are multiplication by k' on V(r) if $r \neq b$ and of form

$$\begin{pmatrix} k'1 & \phi_{12} & \phi_{13} \\ 0 & \phi_{22} & \phi_{23} \\ 0 & 0 & k'1 \end{pmatrix}$$

on $\underline{V}(\underline{b})$.

Thus the contribution of Q′ to the form B(Q) is

$$k^2 + a(a - k) + \ell^2 + c^2 - c\ell$$

where $k = \dim K$, $a = \dim V(a)$, $c = \dim V(c)$, $\ell = \dim L$. This, as we expect, is exactly the contribution we would get if we replaced Q′ by

Q″ :

in Q.

A similar analysis for

Q''' :

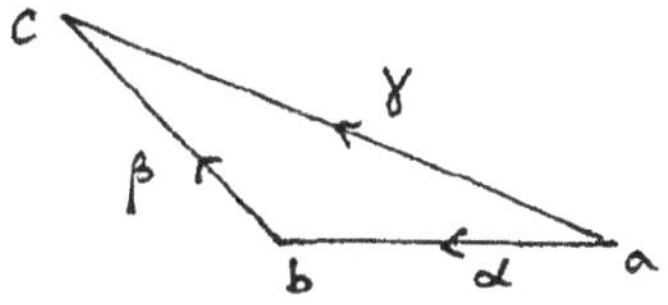

with the commutativity condition $\beta\alpha = \gamma$ leads to the contribution

$$a^2 + b^2 + c^2 - ab - bc + k_1{}^2 + k^2 - k_1(a - b) - k_2(b - c)$$

where $k_1 = \dim \ker \alpha$ and $k_2 = \dim \ker \beta$. Since we require, $k_1 \geqslant a - b$, $k_2 \geqslant b - c$ as well as $k_1 \geqslant 0$, $k_2 \geqslant 0$, the least contribution of Q''' to $B(Q)$ (for fixed a, b, c) is

$$a^2 + b^2 + c^2 - ab - bc$$

as is suggested in Section 3. It is this 'least' contribution which we should expect to give the representation type.

5. A Naive View of the Form B

The relationship between representation type and the definiteness of a suitable form may be considered in terms of a 'general position' argument. No proofs are offered here, though it seems possible that the methods of algebraic geometry may prove fruitful.

For simplicity we consider the case of a quiver Q. (Only minor changes are required in other cases similar to those considered in this paper.) Suppose V is a representation of Q. An endomorphism of V is an element

$\theta = (\theta_r)_{r \epsilon P_Q} \epsilon \prod_{r \epsilon P_Q} \text{End } V(r)$ which satisfies for each

$\gamma = (r, s) \epsilon A_Q$,

$$V(\gamma)\theta_r = \theta_s V(\gamma). \tag{5.1}$$

If V has dimension $d = (d_r)$ and we consider a matrix presentation of V, then (5.1) affords $\sum_{\gamma \epsilon A_Q} d_{i(\gamma)} d_{f(\gamma)}$ equations for the $\sum_{r \epsilon P_Q} d_r^2$ components of θ. Clearly these components are only determined up to scalar multiples so, if V is in general position (so that the equations (5.1) are of maximal rank), we have just enough equations to determine θ (modulo scalars) if

$$B_V(Q) = \sum_{r \epsilon P_Q} d_r^2 - \sum_{\gamma \epsilon A_Q} d_{i(\gamma)} d_{f(\gamma)} = 1.$$

It seems (see also §4 of [1]) that if $B_V(Q) > 1$, then the 'flabbiness' is sufficient to allow an idempotent in End V.

If B(Q) is positive definite, then there are at most just enough equations to determine θ (modulo scalars). However, if B(Q) is semidefinite, there are situations in which we have an equation to spare. If, instead of thinking of (5.1) as equations for scalar components of the θ_i, we think of them as equations for $n \times n$ blocks, then we can use the 'extra' equation to make End V isomorphic to the commutant

$$C_W(\alpha) = \{\phi \in \text{End } W : \alpha\phi = \phi\alpha\}$$

of a prescribed endomorphism α of an n-dimensional space W.

If B(Q) is indefinite, we can have two (or more) equations to spare and so construct a representation V of Q with

$$\text{End } V \simeq C_W(\alpha_1, \alpha_2) = \{\phi \in \text{End } W : \phi\alpha_i = \alpha_i\phi, \ i = 1, 2\}.$$

This is sufficient to ensure that Q is of wild type [3].

6. Conclusion

There are many circumstances in which the representation type of a finitely generated algebra is given by the definiteness of a suitably defined form. The evidence suggests that there should be a formal argument relating the two properties.

If this result is established, it may be hard to apply it except in simple cases where it is easy to write down the 'right' form. It would be useful to have some systematic procedure for refining the presentation of a representation so as to make the new form 'less positive' than the old, and a way of deciding when the process may be terminated.

References

1. I.N. BERNSTEIN, I.M. GELFAND, V.A. PONOMAREV Coxeter functors and Gabriel's theorem. Uspechi Mat.Nauk 28 (1973), 19-33.
2. Sheila BRENNER. Endomorphism Algebras of Vector Spaces with Distinguished Sets of Subspaces. J.Alg. 6 (1967) 100-114
3. Sheila BRENNER. Decomposition Properties of some Small Diagrams of Modules. Symposia Matematica. In press.
4. M.C.R. BUTLER. On the Classification of Local Integral Representations of Abelian p-Groups. These Proceedings, 4,01-4,18.
5. R.W. CARTER. Conjugacy Classes in the Weyl Group. Compositio Mathematica 25 (1972) 1-59.
6. Peter DONOVAN and M.R. FREISLICH. The Representation Theory of Finite Graphs and Associated Algebras. Carleton Lecture Notes No.5, Ottawa (1973).
7. Peter GABRIEL. Unzerlegbare Darstellungen I. Manuscripta math. 6 (1972) 71-103.
8. Pierre GABRIEL. Réprésentations Indécomposables des Ensembles Ordonnés. Seminaire Dubreil (Algebre) 26e année 1972/3 13 1301-1304.

9. I.M. GELFAND and V.A. PONOMAREV. Indecomposable representations of the Lorentz group. Uspechi Mat.Nauk 23 (1968) 3-60.

10. I.M. GELFAND. Cohomologies of infinite dimensional Lie algebras; some questions of integral geometry. Proc. ICM Nice (1970).

11. M.M. KLEINER. Partially ordered sets of finite type. Zap. Naučn. Sem. Leningrad. Otdel. Mat. Inst. Steklov. (LOMI) 28 (1972) 32-41.

12. L.A. NAZAROVA. Representation of quivers of infinite type. Izv. Akad. Nauk SSSR, ser.mat. 37 (1973) 752-791.

13. L.A. NAZAROVA and A.V. ROITER. Representations of partially ordered sets. Zap. Naučn. Sem. Leningrad. Otdel. Mat. Inst. Steklov. (LOMI) 28 (1972) 5-31.

14. L.A. NAZAROVA and A.V. ROITER. On a problem of I.M. Gelfand. Funkc. Anal. i Priloz. 7 (1973) 54-69.

University of Liverpool,

Liverpool, England

ON THE CLASSIFICATION OF LOCAL INTEGRAL REPRESENTATIONS OF FINITE ABELIAN p - GROUPS

M. C. R. Butler

§1 Introduction

1.1 Lattices over orders and integer group rings are notoriously complicated objects. A theorem of Dade's [6] shows that 'most' orders have infinite representation type (i.e. infinitely many non-isomorphic indecomposable lattices); and, as the many papers cited in §§14, 15 of Reiner's survey paper [15] indicate, the classification of the lattices over orders of finite representation type can be a lengthy, 'ad-hoc', process. This paper develops further a strategy which was shown in [4] to work nicely for the 2-adic integral representations of the Four Group $C_2 \times C_2$. The leading idea is to relate lattice categories to other, better understood categories, primarily, to the categories of vector space representations of quivers or of partially ordered sets [1, 2, 3, 7, 8, 9, 10, 13, 14]. It will appear, however, that the specific relationship studied here has an inbuilt limitation - for some orders, it can be used to classify only a subclass of the class of all lattices, so it is probably too coarse a relation on which to base, for example, an alternative derivation of Jacobinski's list of commutative orders of finite representation type [11].

1.2 We consider only lattices over group rings $\Lambda = RG$, with G an abelian p-group of finite order $|G| > 1$, and R a discrete valuation ring with quotient field K of characteristic 0 and residue field k of characteristic p. Let $e_0, e_1, \ldots, e_{r-1}$ denote the primitive idempotents in the commutative semi-simple K-algebra $L = KG$. To each Λ-lattice M and each idempotent e_i, we may associate a lattice $e_i \otimes_\Lambda M$ over the order $\Lambda_i = \Lambda e_i$, and consider the positions inside $L \otimes_\Lambda M$ of the objects $1 \otimes_\Lambda M$, $e_0 \otimes_\Lambda M$, ..., $e_{r-1} \otimes_\Lambda M$. This leads to our main theoretical construct, a functor Δ from the category $\mathcal{L}_0$ of Λ-lattices which have no direct summands isomorphic to $\Lambda, \Lambda_0, \Lambda_1, \ldots$, or Λ_{r-1}, into a certain category $\underline{V}$ of R-torsion, Λ-module representations of the quiver

1.2(a)

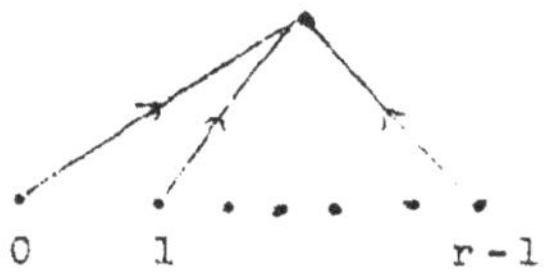

Our main theorem 3.2 identifies a large full subcategory $\mathcal{L}_f$ of $\mathcal{L}_0$ such that Δ is a 'representation equivalence' of $\mathcal{L}_f$ with $\underline{V}$; that is, $\Delta : \mathcal{L}_f \to \underline{V}$ induces a bijection of isomorphism classes of objects, preserves and reflects idempotent morphisms, and for each lattice M in $\mathcal{L}_f$ induces a ring surjection of $\mathrm{End}_\Lambda(M)$ onto $\mathrm{End}_{\underline{V}}(\Delta(M))$ with a quasi-regular ideal as its kernel.

1.3 There are two characterisations of $\mathcal{L}_f$. The more interesting is that its objects are the lattices isomorphic to those obtained by applying a certain construction $\Phi : \underline{V} \to \mathcal{L}_0$ (but Φ is not a functor), and it is the fact that $\Delta \circ \Phi \cong Id_{\underline{V}}$ which enables the morphism lifting assertions of 3.2 to be proved. The other characterisation in §3.1, is intrinsic to $\mathcal{L}_0$, and makes it evident that $\mathcal{L}_f$ can be different from $\mathcal{L}_0$. With 3.2, this circumstance is the limitation on the method mentioned in §1.1. Nevertheless, there are interesting orders for which $\mathcal{L}_f = \mathcal{L}_0$. This is so if $\Lambda_0, \Lambda_1, \ldots, \Lambda_{r-1}$ are all discrete valuation rings, for example, if K is a splitting field for G, or if p is unramified in R (i.e. pR is prime in R).

1.4 The rest of the paper is concerned essentially with the categories $\underline{V}$. We show that $\underline{V}$ contains a category $\mathfrak{D}_r(k)$ of k-vector space representations of the quiver 1.2(a) which is of known representation type. It is of finite type for $r = 2, 3$, of 'tamely infinite' type for $r = 4$, and of 'wildly infinite' type for $r \geqslant 5$ (§3.4). These statements can be used to prove and refine (for our orders) the above mentioned theorem of Dade. Thereafter we concentrate on some orders with $r \leqslant 4$, so that $G = C_p, C_{p^2}, C_{p^3}$, or $C_2 \times C_2$, and obtain new proofs of a number of known results by analysing $\underline{V}$. The most interesting result is for $G = C_{p^2}$ with p unramified, for which $\underline{V}$ is shown in §4 to be representation equivalent to a category of quiver representations of the Dynkin diagram D_{2p}. Finally we consider the case $G = C_8$ and 2R prime, for which Jakovlev [12] has recently determined all the indecomposable lattices (at least, when R is complete) by different methods. Our discussion is incomplete, and concludes with a conjecture that $\mathcal{L}_0$ is representation equivalent to a category of k-vector space representations of the partially ordered set depicted in §5.1.

§2 The functor Δ

2.1 Throughout this paper, π denotes a prime element in R, ℓ the ramification degree of p in R, so that $pR = \pi^{\ell}R$, and

$$J = \Lambda\pi + \sum_{g \in G} R(1 - g)$$

the Jacobson radical of $\Lambda = RG$. We assume that e_0 is the trivial idempotent,

$$e_0 = \frac{1}{|G|} \sum_{g \in G} g ,$$

so that $\Lambda_0 = \Lambda e_0 = Re_0$ is the rank-1 trivial Λ-lattice. The assumptions on R and G imply the following easily verified facts.

2.1 (a) $\exists\, q \geq 1$ such that $J^q \subset |G|\Lambda$.

2.1 (b) for each e_i, the Jacobson radical of Λ_i is $J\Lambda_i$ and $\Lambda_i / J\Lambda_i \cong k$.

2.1 (c) for each e_i, $\exists\, n_i \in Z$ such that $|G|(e_i - n_i e_0) \in \Lambda\pi$.

Λ-lattices may be defined abstractly as finitely generated Λ-modules which are torsion free, hence free, as R-modules. However, it is more convenient for us to view them as finitely generated Λ-submodules of $L = KG$-modules, and to note that lattice morphisms are simply the restrictions of L-module morphisms. In particular, we shall write $e_i M$ and KM for the objects $e_i \otimes_\Lambda M$ and $L \otimes_\Lambda M$ associated in §1.2 to a given Λ-lattice M.

Next, we recall that, with our assumptions on R and G, the lattice M has a non-zero free summand if and only if $|G| e_0 M \not\subset \pi M$. In view of 2.1 (c) and the definition of $\mathcal{E}_0$, it follows that

2.1 (d) if M is a lattice in $\mathcal{L}_0$, then

$$\pi^{-1}|G|e_iM \subset M \quad \text{for} \quad i = 0,1,\ldots, \text{ and } r-1 .$$

2.2 We now define the quiver category $\underline{V}$. An object of $\underline{V}$ is a sequence $V_* = (V|V_0,\ldots,V_{r-1}) = (V|V_i)$ (for brevity) such that

2.2 (a) V is a finitely generated Λ-module and $\pi^{-1}|G|V = 0$;

2.2 (b) $V_0, V_1,\ldots, V_{r-1}$ are Λ-submodules of V such that, for each $i \in \{0,1,\ldots, r-1\}$, $\sum_{j\neq i} V_j = V$;

2.2 (c) for each $i \in \{0,1,\ldots, r-1\}$, the action of Λ on V_i factorises through the canonical ring surjection $\lambda \to \lambda e_i$ of Λ onto Λ_i.

Finally, a morphism $V_* \to V'_*$ in $\underline{V}$ is defined to be an element ϕ of $\mathrm{Hom}_\Lambda(V,V')$ such that $V_i\phi \subset V'_i$ for each i.

2.3 Let M be a Λ-lattice. Then KM has a sublattice

$$e_*M = e_0M + e_1M + \ldots + e_{r-1}M$$

which is a direct sum and, since $\Sigma e_i = 1$, can be expressed in the form

$$e_*M = M + \sum_{j \neq i} e_jM$$

for each i. Consider the Λ-modules

$$V = e_*M/M \quad \text{and} \quad V_i = (e_iM + M)/M \cong e_iM/M \cap e_iM .$$

When M is in $\mathcal{L}_0$, so that 2.1 (d) holds, we see at once that $V_* = (V|V_i)$ is an object in $\underline{V}$, and we define it to be the value $\Delta(M)$ of the proposed functor $\Delta : \mathcal{L}_0 \to \underline{V}$. Let $\theta : M \to N$ be a lattice morphism.

It is restriction of an L-morphism $KM \to KN$, and since this commutes with the idempotents, it induces a morphism $\Delta(\theta) : \Delta(M) \to \Delta(N)$ in $\underline{V}$. Obviously Δ is an additive functor. For any 2 lattices M, N in $\mathcal{L}_0$, it induces a morphism

$$\Delta_{M,N} : \mathrm{Hom}_\Lambda(M,N) \to \mathrm{Hom}_{\underline{V}}(\Delta(M), \Delta(N)); \quad \theta \to \Delta(\theta) .$$

2.4 Proposition. The following assertions hold for any 2 lattices M and N in $\mathcal{L}_0$.

2.4 (a) If $e_i M$ is Λ_i-free for $0 \leq i < r$, then $\Delta_{M,N}$ is surjective.

2.4 (b) If $M \cap e_i M \subset J e_i M$ for $0 \leq i < r$, then $\mathrm{Ker}(\Delta_{M,M})$ is a quasi-regular ideal in $\mathrm{End}_\Lambda(M)$.

2.4 (c) If the hypotheses of 2.4 (a) and (b) hold for both M and N, then $M \cong N$ if and only if $\Delta(M) \cong \Delta(N)$.

Proof of 2.4 (a). Suppose that each $e_i M$ is Λ_i-free, and let $\phi : \Delta(M) \to \Delta(N)$ be a $\underline{V}$-morphism. For each i, ϕ restricts to a Λ_i-morphism ϕ_i from $e_i M/_{M \cap e_i M}$ into $e_i N/_{N \cap e_i N}$, so it lifts to a Λ_i-morphism $\theta_i : e_i M \to e_i N$. It is trivial to verify that the map $\theta : e_* M \to e_* N$ coinciding with θ_i on $e_i M$ maps M into N, and that $\Delta(\theta) = \phi$.

Proof of 2.4 (b). For $\theta \in \mathrm{End}_\Lambda(M)$, it follows from the definitions that

$$\Delta_{M,M}(\theta) = 0 \iff (e_* M)\theta \subset M$$

$$\iff (e_i M)\theta \subset M \cap e_i M \quad \text{for } 0 \leq i < r .$$

Let $\theta \in \mathrm{Ker}(\Delta_{M,M})$ and assume the hypothesis of 2.4 (b). Then, for each i, 2.1 (a) shows that $(e_iM)\theta^q \subset J^q e_i M \subset |G| e_i M$, so that by 2.1 (d),

$$M\theta^q \subset (e_*M)\theta^q \subset |G| e_* M \subset \pi M .$$

Now M is free over R and πR is the prime ideal of the discrete valuation ring R, so $1 - \theta^q \in \mathrm{Aut}_\Lambda(M)$. Therefore θ is quasi-regular in $\mathrm{End}_\Lambda(M)$.

Proof of 2.4 (c). Assume its hypotheses and let $\phi : \Delta(M) \to \Delta(N)$ be an isomorphism. By 2.4 (a), there exist $\theta : M \to N$ and $\theta' : N \to M$ such that $\Delta(\theta) = \phi$ and $\Delta(\theta') = \phi^{-1}$. By 2.4 (b), $1_M - \theta\theta'$ and $1_N - \theta'\theta$ are quasi-regular elements of $\mathrm{End}_\Lambda(M)$ and $\mathrm{End}_\Lambda(N)$, respectively, from which it follows at once that θ and θ' are both isomorphisms.

§3 The main theorem

3.1 Definition. Let $\mathcal{L}_f$ denote the full subcategory of $\mathcal{L}_0$ containing the lattices M with the following two properties:

3.1(a) e_iM is a free Λ_i-module $(0 \leq i < r)$;

3.1(b) $M \cap e_iM \subset Je_iM$ $(0 \leq i < r)$.

The conclusions of §2.4 therefore hold for any two lattices in $\mathcal{L}_f$. The following main theorem supplements these conclusions.

3.2 Theorem. The functor $\Delta : \mathcal{L}_f \to \underline{V}$ induces a bijection of isomorphism classes of objects and preserves and reflects idempotents. For each lattice M in $\mathcal{L}_f$, $\mathrm{End}_\Lambda(M)$ is an extension by $\mathrm{End}_{\underline{V}}(\Delta(M))$ of its quasi-regular ideal $\mathrm{Ker}(\Delta_{M,M})$.

Since 2.4 holds, we already know that Δ induces an injective mapping of isomorphism classes of objects, that it maps non-zero idempotents to non-zero idempotents, and that the last sentence of the theorem is true. It remains to be shown that Δ is surjective on isomorphism classes of objects and idempotents. The proofs follow in 3.2(a) and (b) below, and depend on the construction $\Phi : \underline{V} \to \mathcal{L}_f$ described in 3.2(a).

3.2(a) Let V_* be an object in $\underline{V}$. There is a lattice $\Phi(V_*)$ in $\mathcal{L}_f$ and an isomorphism $\Delta(\Phi(V_*)) \tilde{\to} V_*$.

Proof. By 2.1(b), V_i/JV_i is a vector space over k. Let d_i be its dimension, $F_i = \Lambda_i^{d_i}$ the free Λ_i-module on d_i generators, and $f_i : F_i \to V_i$ a Λ_i-epimorphism inducing an isomorphism of F_i/JF_i onto V_i/JV_i. Therefore

$$\mathrm{Ker}(f_i) \subset JF_i.$$

Now let $F = \bigoplus_{i=0}^{r-1} F_i$ and observe that by 2.2(b) and (c), the map

$f = \sum_i f_i : F \to V$ is a surjective Λ-homomorphism. Let $M = \mathrm{Ker}(f)$.

Clearly, M is a Λ-lattice, and we can show that M has the properties required of $\Phi(V_*)$. We prove first that

$$e_i M = F_i .$$

Since $M \subset F$, the inclusion $e_i M \subset e_i F = F_i$ is obvious. Conversely, let $x = e_i x \in F_i$. By 2.2(b), $\exists x_j \in F_j$ $(j \neq i)$ such that $f(x) = \sum_{j \neq i} f(x_j)$, so by the definition of M, $x \in \sum_{j \neq i} x_j + M$.

Hence $x = e_i x \in e_i M$, and this proves the opposite inclusion, $F_i \subset e_i M$.

It follows that 3.1(a) holds and, since $\mathrm{Ker}(f_i) = M \cap F_i = M \cap e_i M$, so also does 3.1(b). To show that M is in $\mathcal{L}_f$, we have only to verify that it has no summands isomorphic to Λ, Λ_0, ..., or Λ_{r-1}. Since $\pi^{-1}|G|V = 0$ (2.2(a)), it follows that $\pi^{-1}|G|e_0 M = \pi^{-1}|G|F_0 \subset M$, so M has no free summands. Suppose, for some i and some $x \in M$, that $\Lambda_i x$ is a non-zero summand of M, then it is a summand of $M \cap e_i M = \mathrm{Ker}(f_i)$ and of $e_i M = F_i = \Lambda_i^{d_i}$, which facts contradict the assumption that f_i induces an isomorphism of F_i/JF_i onto V_i/JV_i. These arguments complete the proof that M is in $\mathcal{L}_f$, and a simple calculation shows that $f : F = e_* M \to V$ induces a $\underline{V}$-isomorphism $f_* : \Delta(M) \xrightarrow{\sim} V_*$.

3.2(b) Let $\epsilon \in \mathrm{End}_{\underline{V}}(\Delta(M))$ and $\epsilon = \epsilon^2$. For $M \in \mathcal{L}_f$, $\exists\, \tau = \tau^2 \in \mathrm{End}_\Lambda(M)$ such that $\Delta(\tau) = \epsilon$.

Proof. By 3.2(a), there exist lattices $P = \Phi(\mathrm{Ker}(\epsilon))$; $Q = \Phi(\mathrm{Im}(\epsilon))$ in $\mathcal{L}_f$ and isomorphisms $\phi : \Delta(P) \xrightarrow{\sim} \mathrm{Ker}(\epsilon)$ and $\psi : \Delta(Q) \xrightarrow{\sim} \mathrm{Im}(\epsilon)$. Let

$N = P \oplus Q$. By 2.4(c), there is a Λ-isomorphism $\theta : N \to M$ such that $\Delta(\theta) = \phi \oplus \psi$, and then it is clear that $\epsilon = \Delta(\tau)$, where $\tau = \theta^{-1}(0_P \oplus 1_Q)\theta$ is an idempotent on M.

3.2(a) and (b) complete the proof of the theorem. We have also obtained a new characterisation of $\mathcal{L}_f$:

3.2(c) A lattice is in $\mathcal{L}_f$ if and only if it is isomorphic to $\Phi(V_*)$, for some V_* in $\underline{V}$.

3.3(a) Proposition. $\mathcal{L}_f = \mathcal{L}_0$ if each of $\Lambda_0, \Lambda_1, \ldots, \Lambda_{r-1}$ is a discrete valuation ring.

Proof. Let Λ_i be a discrete valuation ring with prime ideal $J\Lambda_i = \pi_i\Lambda_i$ $(0 \leqslant i < r)$. Let M be a lattice in $\mathcal{L}_0$. Then 3.1(a) certainly holds. Suppose that 3.1(b) fails, and let $x = e_i x \in (M \cap e_i M) \setminus (\pi_i e_i M)$. Then $\Lambda x = \Lambda_i x \subset M \subset e_* M$ and is a summand of $e_i M$ and, hence, of $e_* M$. Thus M has a summand isomorphic to Λ_i, which contradicts the choice of M from $\mathcal{L}_0$. So 3.1(b) also holds and M belongs to $\mathcal{L}_f$.

3.3(b) Corollary. $\mathcal{L}_f = \mathcal{L}_0$ if either K is a splitting field for G or p is unramified in R.

3.4 Let $\mathfrak{D}_r(k)$ denote the full subcategory of $\underline{V}$ whose objects V_* satisfy the condition $JV = 0$. Thus V_* is just a sequence $(V|V_i)$ consisting of a vector space V of finite dimension over k (with trivial G-action) and r subspaces $V_0, \ldots, V_{r-1}$ such that any $r-1$ of them span V. These categories are of known representation type :

3.4(a) $\mathfrak{D}_2(k)$ and $\mathfrak{D}_3(k)$ are of finite representation type.

3.4(b) $\mathfrak{D}_4(k)$ is of tamely infinite representation type ; that is, the infinitely many distinct indecomposables are fully classified [1, 3, 9, 10].

3.4(c) For $r \geq 5$, $\mathfrak{D}_5(k)$ is of wildly infinite representation type; that is, for each finite-dimensional k-algebra E, there exist infinitely many dimensions d for which there is an object V_* in $\mathfrak{D}_r(k)$ such that $\dim V = d$ and $\mathrm{End}_{\underline{V}}(V_*) \cong E$ [3].

The lattices M in $\mathcal{L}_f$ such that $\Delta(M)$ is in $\mathfrak{D}_r(k)$ define a full subcategory $\mathcal{L}_t$, and are characterised by the condition $M \cap e_i M = J e_i M$ for $i = 0, 1, \ldots, r-1$. Obviously, $\Delta : \mathcal{L}_t \to \mathfrak{D}_r(k)$ is a representation equivalence, in the sense that 3.2 conveys, so that 3.4(b) and (c) imply the following refinements of Dade's theorem [6].

3.4(d) For $r = 4$, $\mathcal{L}_t$ is of tamely infinite representation type.

3.4(e) For $r \geq 5$, $\mathcal{L}_t$ is of wildly infinite representation type (in the wider sense that for each finite-dimensional k-algebra E, there exist infinitely many ranks d for which there is a lattice M in $\mathcal{L}_t$ with $\dim_K(KM) \geq d$ and $\mathrm{End}_{\underline{V}}(\Delta(M)) \cong E$).

3.5 Since r is the number of irreducible representations of the abelian group G over the field F of characteristic 0, Artin's theorem [5, §39] shows that it is not less than the number of different cyclic subgroups of G. 3.4(e) excludes the possibility of classifying Λ-lattices when $r \geq 5$. It is trivial to show that $r \leq 4$ can only occur for $G = C_p, C_{p^2}, C_{p^3}$, or $C_2 \times C_2$.

We list some easy results on $\underline{V}$, $\mathcal{L}_f$ and $\mathcal{L}$.

3.5(a) $G = C_p$ and p unramified. Then $\pi^{-1}|G|$ is a unit in R, so $\underline{V} = \{0\}$. The only indecomposable lattices are $\Lambda, \Lambda_0, \Lambda_1$.

3.5(b) $G = C_2$ and $2R = \pi^{\ell}R$ $(\ell \geqslant 2)$. By 3.3(a), $\mathcal{L}_f = \mathcal{L}_0$. Since $r = 2$, an object V_* in $\underline{V}$ satisfies $V = V_0 = V_1$, and since G acts trivially on V_0, it acts trivially on V. Therefore $\underline{V}$ is (isomorphic to) the category of finitely generated $R/\pi^{\ell-1}R$-modules, and its indecomposables are the $\ell - 2$ cyclic modules $R/\pi R, \ldots, R/\pi^{\ell-2}R$. So there are exactly $\ell + 1$ indecomposable lattices.

3.5(c) $G = C_2 \times C_2$ and 2 unramified. By 3.3(a), $\mathcal{L}_f = \mathcal{L}_0$. Now $r = 4$ and (as shown in [4]), $\underline{V} = \mathfrak{D}_4(k)$. Therefore $\mathcal{L}_0$ is of tamely infinite representation type.

3.5(d) $G = C_p$ and $pR = \pi^{\ell}R$ $(\ell \geqslant 2)$ and $(X^p - 1)/(X - 1)$ is irreducible over K. Then $r = 2$ and (as in 3.5(b)) $\underline{V}$ has just $\ell - 2$ indecomposables. So also, then, has $\mathcal{L}_f$, but the relationship of $\mathcal{L}_f$ to $\mathcal{L}_0$ can depend on K.

§4 The cyclic group of order p^2.

The lattices for $G = C_{p^2}$ and p unramified in R are known and Λ has finite representation type (see [15], §14, for a bibliography of the problem). We show now how to obtain these results from an analysis of $\underline{V}$, noting that by 3.3(a), $\ell_0 = \ell_f$ so that $\Delta : \ell_0 \to \underline{V}$ is a representation equivalence.

Let g generate G and $\phi(X) = (X^p - 1)/(X - 1)$. Since $\phi(X)$ is irreducible over R when p is unramified, we have $r = 3$ and

$$e_0 = \frac{1}{p^2}\phi(g)\phi(g^p), \quad e_1 = \frac{1}{p^2}(p - \phi(g))\phi(g^p), \quad e_3 = \frac{1}{p}(p - \phi(g^p)).$$

Let $t = 1 - g$. Then

$$te_0 = 0 \quad \text{and} \quad t^{p-1}e_1 \in p\Lambda_1 .$$

Now let V_* be an object in $\underline{V}$. Since $p^{-1}|G| = p$, V is a vector space over k and V_0, V_1, V_2 subspaces such that $V_i + V_j = V$ for $i < j$. Also t is an endomorphism of V, leaving each of V_0, V_1, V_2 invariant, and the action of t on each subspace must satisfy 2.2 (c). Hence, $t = 0$ on V_0 and $t^{p-1} = 0$ on V_1, so it follows that $t^{p-1} = 0$ on V and V_2 also. Notice that

$$t^nV = t^nV_1 = t^nV_2 \subset V_1 \cap V_2 \quad \text{for} \quad n \geq 1.$$

We show now that $\underline{V}$ is representation equivalent to the category Σ_{2p} of those finite dimensional k-space representations

$\underset{\sim}{X}$

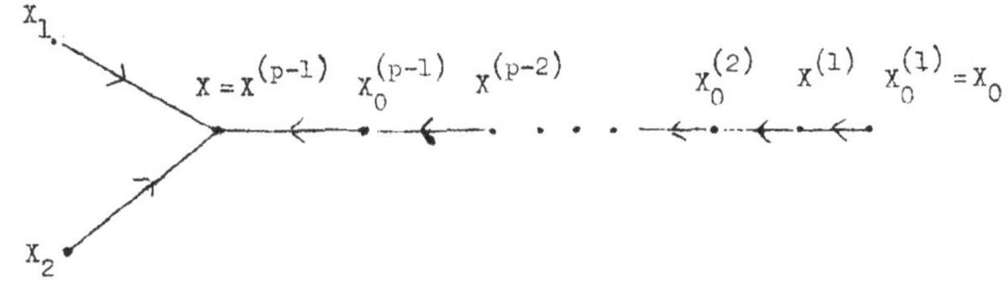

of the (directed) Dynkin diagram D_{2p} in which all the arrows represent inclusion maps and $X_i + X_j = X$ for $i < j$. To an object V_* in $\underline{V}$, we associate the representation $\underset{\sim}{X} = \underset{\sim}{X}(V_*)$ given by

$$X_0^{(n)} = \{t^{-(n-1)}(V_0) + tV\}/tV \quad \text{for} \quad 1 \leqslant n \leqslant p-1,$$

$$X^{(n)} = \{t^{-n}(0) \qquad + tV\}/tV \quad \text{for} \quad 1 \leqslant n \leqslant p-1,$$

$$\text{and} \quad X_i \quad = V_i/tV \qquad\qquad \text{for} \quad i = 1,2 .$$

According to this definition, each cyclic $k[t]$-module summand of V of nilpotence index precisely n determines a 1-dimensional subspace of $X^{(n)}/X^{(n-1)}$; bearing this in mind, it is quite straightforward to obtain a construction from Σ_{2p} to $\underline{V}$ and to deduce by arguments like those of §3 that $\underline{V} \to \Sigma_{2p}$ is a representation equivalence.

The representations of D_{2p} have been determined, [1,8], and the crucial point about the indecomposables in Σ_{2p} is that $\dim X = 1$ or 2 only. There are $4p - 3$ different indecomposables in Σ_{2p}, hence also in $\underline{V}$ and in $\mathcal{L}_0$. Including $\Lambda, \Lambda_0, \Lambda_1, \Lambda_2$, we obtain a total of exactly $4p + 1$ indecomposable Λ-lattices.

§5 The cyclic group of order 8

Jakovlev has classified the infinitely many indecomposable lattices for $G = C_8$ and R the ring of 2-adic integers [11]. We summarise briefly here some evidence supporting the conjecture that, for this group and for a discrete valuation ring R in which 2 is prime, the categories $\mathcal{L}_0$ and $\underline{V}$ are representation equivalent to a category of k-vector space representations of the partially ordered set

5.1 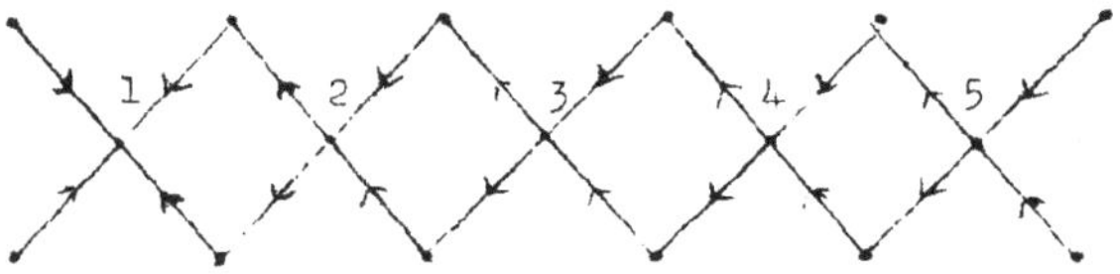

Since 2 is unramified, $r = 4$ and $\mathcal{L}_0 = \mathcal{L}_f$ is representation equivalent to $\underline{V}$. Let g generate G and $t = 1 - g$. An analysis of the action of t on the primitive idempotents shows that, in an object $V_* = (V|V_0, V_1, V_2, V_3)$ of $\underline{V}$, V is a finitely generated $R[t]$-module on which $4 = 2t = t^3 = 0$. Also

5.2 $$V_0 \subset \mathrm{Ker}(t), \quad V_1 \subset \mathrm{Ker}(t-2), \quad V_2 \subset \mathrm{Ker}(t^2-2), \quad V_3 \subset \mathrm{Ker}(2),$$

and, as usual, each V_i is t-invariant and any 3 of them span V. Hence V has a filtration,

5.3 $$V \supset tV \supset 2V \supset t^2V \supset (t^2V) \cap 2(t^{-1}(0)) \supset 0 .$$

There are 2 related pieces of evidence for the conjecture above. Firstly, there is a special type of V_* in which the inclusions of 5.2 are equalities, and it is not difficult to show that these are direct sums of exactly 5 types of indecomposables. We find that these 5 indecomposables can be distinguished from one another by specifying which term of the filtration 5.3 is the last non-zero term for each. Secondly, the conjecture is consistent with a description we can prove of objects V_* such that $t^2V = 0$. Such an object is determined up to isomorphism by the following partially ordered set of subspaces of $\bar{V} = V/tV$; we have written $S = 2^{-1}(0)$ and $\bar{U} = (U + tV)/tV$ for any $U \subset V$:

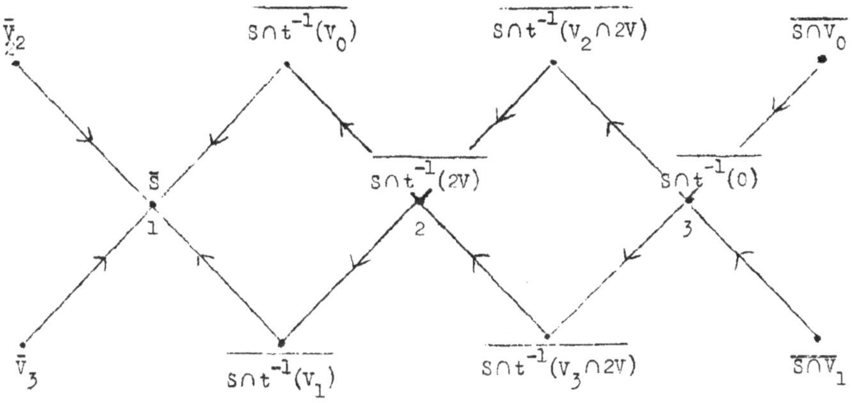

If $2V = 0$, then $S = V$ and vertices 2 and 3 coalesce; if $tV = 0$, then vertices 1, 2 and 3 coalesce (in fact V_* is in $\mathcal{D}_4(k)$). It seems likely that for arbitrary V_*, the two extra factors from the filtration 5.3 simply give rise to two more 'diamonds' as shown in 5.1.

References

[1] Bernstein, I.N., Gelfand, I.M., Ponamarev, B.A. Coxeter functors and Gabriel's theorem. Uspechi Mat. Nauk 28, 19-38 (1973), and Russian Math. Surveys 28, 17-32 (1973).

[2] Brenner, Sheila. Decomposition properties of some small diagrams of modules. To appear in Symposia Mathematica.

[3] Brenner, Sheila. On four subspaces of a vector space. J. Algebra 29 (1974) 587-599.

[4] Butler, M.C.R. The 2 - adic representations of Klein's Four Group, to appear in the Proc. Second International Conference on Group Theory, Canberra, 1973.

[5] Curtis C.W., Reiner, I. Representation theory of finite groups and associative algebras. Interscience, 1962.

[6] Dade, E. C. Some indecomposable group representations, Ann. of Math. (2) 77 (1963), 406-412.

[7] Donovan, P., Freislich, M.R. The representation theory of finite graphs and associated algebras. Carleton Lecture Notes No.5, 1973.

[8] Gabriel, P. Unzerlegbare Darstellungen I. Manuscripta Mathematica, 6(1972) 71 - 103.

[9] Gabriel, P. Indecomposable representations, II. Symposia Mathematica, XI (1973), 81-104.

[10] Gelfand, I.M., Ponamarev, B.A. Problems of linear algebra and the classification of quadruples of subspaces in a finite-dimensional vector space. Coll. Math. Soc. Bolyai 5, Tihany (1970) 163-237.

[11] Jacobinski, H. Sur les ordres commutatifs avec un nombres fini de réseaux indécomposables. Acta Math. 118 (1967) 1-31.

[12] Jakovlev, A.V. Classification of the 2-adic representations of the cyclic group of order 8. Zap. Nauc. Sem. Leningrad. Otdel. Mat. Inst. Steklov. (LOMI) 28(1972) 93-129.

[13] Nazarova, L.A. Representations of quivers of infinite type. Izv. Akad. Nauk. SSSR. Ser. Mat. 37 (1973) 752-791.

[14] Nazarova, L.A., Roiter, A.V. Representations of partially ordered sets. Zap. Nauc. Sem. Leningrad. Otdel. Mat. Inst. Steklov. (LOMI) 28 (1972) 5-31.

[15] Reiner, I. A survey of integral representation theory. Bull. Amer. Math. Soc., 76 (1970) 159-227.

M. C. R. Butler,
Department of Pure Mathematics,
University of Liverpool,
England.

FINITE LINEAR p-GROUPS OF DEGREE p AND THE WORK OF G. SZEKERES

S. B. Conlon

Let C be a field of characteristic not equal to the prime p and which contains all p power roots of 1. Abelian p-subgroups of $GL_p(C)$ are diagonalisable and we confine our attention to finite nonabelian p-subgroups of $GL_p(C)$.

Abstractly, these p-groups can be described as those finite nonabelian p-groups P with cyclic centre $Z(P)$ and which have an abelian maximal subgroup A. For it is well known that a p-group P has a faithful irreducible representation iff $Z(P)$ is cyclic. If P has an abelian maximal subgroup A, any representation π of P is a component of $(\pi_A)^P$ and so has degree 1 or p. Conversely, a nonabelian p-subgroup P of $GL_p(C)$ can be presented monomially with the corresponding permutation matrices being powers of a cycle of order p; it readily follows that P is isomorphic to a subgroup of $Z_{p^r} \text{ wr } Z_p$, where p^r is the highest order of a p power root of unity appearing in the monomial matrices, and so P has an abelian maximal subgroup, as $Z_{p^r} \text{ wr } Z_p$ has one.

G. Szekeres [3] enumerated all the groups G which are non-abelian and have a normal abelian subgroup A with G/A cyclic and such that no prime divisor of $|A|$ occurs to the 2nd power in $|G/A|$. This includes the case of nonabelian p-groups P with an abelian maximal subgroup A. We illustrate his techniques by looking closely at this last case. However attention should be drawn to this paper, as it seems to be generally unknown and the work has recently been rediscovered by other authors.

We take any element $x_\infty \in P - A$ and so $x_\infty^p = H \in A$ and x_∞ gives an automorphism of A of order p by conjugation. If $n \in Z$, then $n: A \longrightarrow A,\ a \longmapsto a^n$ is an endomorphism of the abelian group A. If X is the automorphism of A given by x_∞, then $Z[X]$ acts as a ring of operators on A. If the exponent of A is p^ℓ, then actually A becomes an R-module, where

$$R = Z[X]/(p^\ell,\ X^p - 1).$$

The original group P is then described by the R-module structure of A and the value of $x_\infty^p = H \in A$. There are more details to be tidied up - for instance, the possibility of several abelian maximal subgroups A and making sure that each isomorphic type occurs only once. This is done by giving a canonical description to indecomposable R-modules and normalising the possible value for $x_\infty^p = H \in A$.

By means of writing

$$\phi = X - 1 \text{ and } \pi = 1 + X + \dots + X^{p-1},$$

we see that

$$R \approx Z[\phi,\pi]/(\phi^{\ell(p-1)+1},\ \pi^{\ell+1},\ \pi\phi,\ p^\ell,\ p + \tbinom{p}{2}\phi + \dots + \phi^{p-1} - \pi).$$

Thus A is an abelian group of exponent p^ℓ with two nilpotent mutually annihilating operators ϕ,π on it. The last relation means that $\pi \equiv p \pmod{\phi}$, and so π is a modified form of raising to the pth power. More accurately

$$a^\phi = [a, x_\infty],$$

and

$$a^{\pi} = a^{\binom{p}{1}} . [a,x_{\infty}]^{\binom{p}{2}} \dots [a,x_{\infty}, \dots^{(p-1)} \dots, x_{\infty}]^{\binom{p}{p}} .$$

The endomorphism π has been used by A. Wiman [4] in classifying all p-groups of maximal class which have abelian maximal subgroups.

It is natural to describe the R-module structure of the indecomposable modules A by diagrams which give the paths generators follow under the action of powers of ϕ and π. The top vertices represent the module generators. There are no images within the apex as $\phi\pi = 0$.

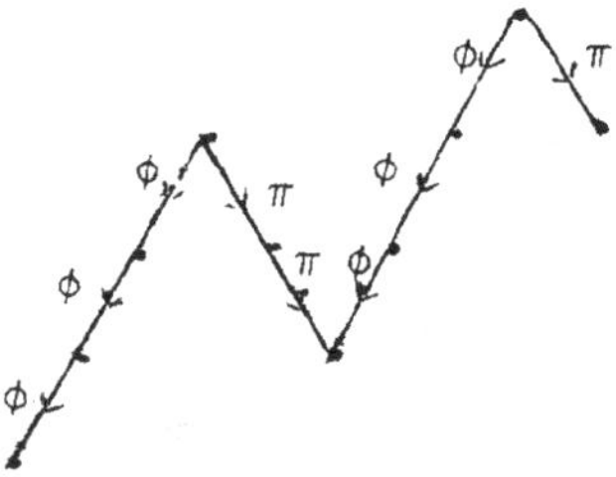

The lengths of the different arms are invariants. There are two types of strings, open, as illustrated above and closed:

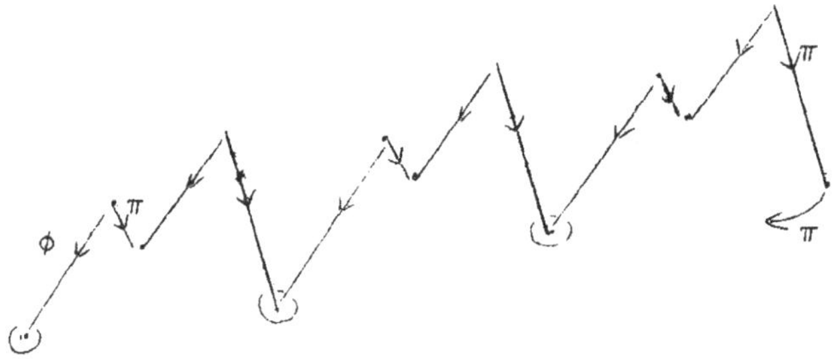

This latter string has a periodicity and the multiplication by π of the right most element carries it to a linear combination of the periodic elements (circled), the coefficients being those of a power of an irreducible polynomial over GF(p).

This is very similar to the classification of mutually annihilating operators on a vector space, recently obtained by I. M. Gel'fand, and V. A. Ponomarev, [1]. There is a second relevant paper by L. A. Nazarova, A. V. Roiter, V. V. Sergeitchuk and V. M. Bondarenko [2]. Szekeres's paper could be said to have anticipated both of these papers.

Returning to linear p-groups of degree p, we must seek out those P above which have cyclic centre. For $a \in Z(P)$, $a^{\pi} = a^{p}$. Also $a \in Z(P)$ iff $a^{\phi} = [a,x_{\infty}] = 1$. $\Omega_1 Z(P)$ must have order p and so there must be at most one "low" point in the corresponding string diagram. Thus we obtain the following diagrams:

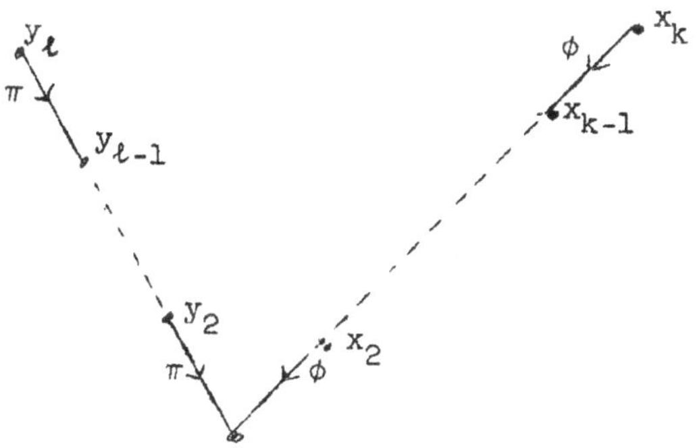

$$U_{k\ell 0}(k\geq 2,\ell\geq 1) = \langle x_1,\ldots,x_k,y_1,\ldots,y_\ell,x_\infty \mid y_i^p=y_{i-1}, [x_j,x_\infty] = x_{j-1},$$

$$x_1=y_1, y_\ell \text{ central}, \langle x_1,\ldots,x_k\rangle \text{ abelian}, H = x_\infty^p = 1,$$

$$x_k^{\pi} = x_k^{\binom{p}{1}} . x_{k-1}^{\binom{p}{2}} \cdots x_{k-p+1}^{\binom{p}{p}} = 1\rangle$$

(We assume $x_i = y_j = 1$ if $i,j \leq 0$).

$$U_{k\ell\infty}(k\geq 2,\ell\geq 1) = \langle x_1,\ldots,x_k,y_1,\ldots,y_\ell,x_\infty \mid \text{ relations as for } U_{k\ell 0}$$

$$\text{except that } H = x_\infty{}^p = y_\ell\rangle.$$

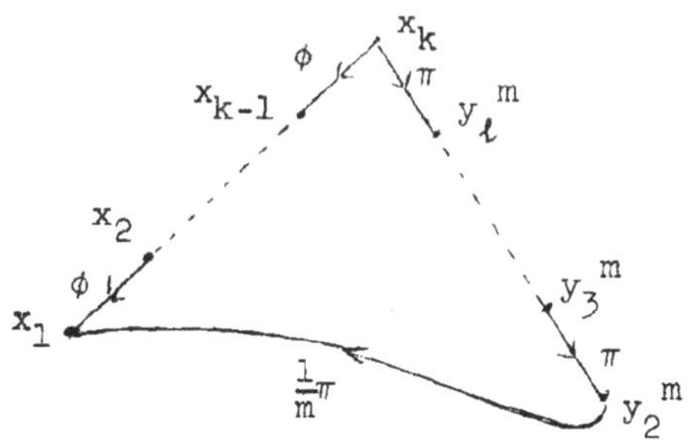

$(0 < m < p)$.

$U_{k\ell m}(k\geq 3, \ell\geq 1,\ 0<m<p) = \langle x_1,\ldots,x_k,y_1,\ldots y_\ell,x_\infty \mid$ relations as for $U_{k\ell 0}$, except that $x_k^\pi = x_k^{\binom{p}{1}} \ldots x_{k-p+1}^{\binom{p}{p}} = y_\ell^m\rangle$.

The only isomorphisms among the $U_{k\ell m}$ come from a certain equivalence relation on the irreducible polynomial associated to the closed strings and we have for $0 < m,\ m' < p$,

$U_{k\ell m} \approx U_{k\ell m'}$ iff $m \equiv m' r^{k-1} \mod p$ for some integer r.

Thus for fixed $k \geq 3$ and $\ell \geq 1$ there are $2 + (k-1,p-1)$ nonisomorphic groups and for $k = 2$ and fixed $\ell \geq 1$ there are 2 nonisomorphic groups.

$|U_{k\ell m}| = p^{k+\ell}$, $U_{k\ell m}$ has class k and $|Z(U_{k\ell m})| = p^\ell$.

Thus $U_{k\ell m}$ has maximal class iff $\ell = 1$. These last p-groups of maximal class and with an abelian maximal subgroup are those discussed by A. Wiman [4]. It turns out that each $U_{k\ell m}$ has $(p-1)p^{k+\ell-3}$ nonequivalent faithful irreducible representations. All irreducible representations and automorphism groups $\mathrm{Aut} = \mathrm{Aut}(U_{k\ell m})$ can be written down. It is also found that the different faithful irreducible representations π, π' of a given $U_{k\ell m}$ are conjugate under an automorphism $\theta \in \mathrm{Aut}(U_{k\ell m})$, i.e. π^θ is equivalent to π'. For a given faithful irreducible representation π of $U_{k\ell m}$ one can consider the similarity automorphism group ($SAut = SAut_\pi$) consisting of those elements of Aut realizable by similarity (conjugation) in $GL_p(C)$. Then in Aut we have $SAut_{(\pi^\theta)} = (SAut_\pi)^\theta$ and so SAut is well defined up to conjugacy in Aut. Moreover $|Aut/SAut| = (p-1)p^{k+\ell-3} =$ the number of distinct faithful irreducible representations. Note also that $SAut \approx N_{GL}(U_{k\ell m})/C_{GL}(U_{k\ell m})$. SAut includes the group IAut of inner automorphisms as a normal subgroup and so we can speak of the outer similarity automorphism group $OSAut = SAut/IAut$. Now $|IAut| = p^k$ and so the orders $|Aut|$ and $|SAut|$ can be readily obtained from the following values of orders of OSAut:

$U_{k\ell 0}$ $(k>2)$ and $U_{210}(p=2)(\approx D_8)$	$p(p-1)$(metacyclic)
$U_{k\ell\infty}$ $(\neq U_{21\infty}(p=2)(\approx Q_8))$	p (cyclic)
$U_{k\ell m}$ $(k>2,\ 0<m<p)$	$(k-1,\ p-1)$(cyclic)
$U_{2\ell 0}$ $(\neq U_{210}(p=2)(\approx D_8))$ and $U_{21\infty}(p=2)(\approx Q_8)$	$p(p^2-1)(SL_2(\mathbb{F}_p))$

Note that $|OSAut|_p = 1$ or p, which is why the constructive method of finding the $U_{k\ell m}$ is manageable.

Another consequence of the conjugacy of faithful irreducible representations of a given $U_{k\ell m}$ is that two nonabelian p-subgroups P,P' of $GL_p(C)$ are isomorphic iff they are conjugate as subgroups of $GL_p(C)$, i.e. there exists a $T \in GL_p(C)$ such that $T^{-1}PT = P'$.

We obtain realisations of the $U_{k\ell m}$ as follows. Let ω be a primitive p^{r+1} root of 1. Write x_∞ for the p-cycle permutation matrix with 1 in the top right hand corner and 1 in each subdiagonal place. Write $x_{r(p-1)+1} = \text{diag}\ (\omega,\omega,\ldots,\omega,\omega^{1-p})$. Then $U_{r(p-1)+1,1,0} \approx \langle x_{r(p-1)+1}, x_\infty\rangle$. For $k \leq r(p-1)+1$, set $x_k = [x_{r(p-1)+1}, x_\infty, \ldots^{(r(p-1)+1-k)} \ldots, x_\infty]$. Then $U_{k10} \approx \langle x_k, x_\infty\rangle$. x_1 is central and is equal to the scalar matrix $\omega^{(-p)^r}I$. Let ζ be a p^ℓth root of $\omega^{(-p)^r}$ and set $y_{\ell+1} = \zeta I$ and $y_\ell = \zeta^p I$. Then we have

$$U_{k\ell 0} \approx \langle x_k, x_\infty, y_\ell\rangle\ (\approx U_{k10} Y\, Z_{p^\ell}), U_{k\ell\infty} \approx \langle x_k,\ x_\infty' = x_\infty y_{\ell+1}\rangle \text{ and}$$

$$U_{k\ell m} = \langle x_k^{(m)} = x_k y_{\ell+1}^m, x_\infty\rangle\ (0<m<p).$$

References

[1] I. M. Gel'fand and V. A. Ponomarev, 'Indecomposable representations of the Lorentz group', Uspehi Mat. Nauk. 23 (1968), no. 2 (140), 3-60.

[2] L. A. Nazarova, A. V. Roiter, V. V. Sergeitchuk and V. M. Bondarenko, 'Application of the theory of modules over dyads to the classification of finite p-groups with an abelian subgroup of index p, and to the classification of pairs of annihilating operators', Sem. Math. V. A. Steklov Math. Inst. Leningrad 28 (1972), 69-92.

[3] G. Szekeres, 'Determination of a certain family of finite metabelian groups', Trans. Amer. Math. Soc. 66 (1949), 4-43.

[4] A. Wiman, 'Über mit Diedergruppen verwandte p-Gruppen', Arkiv För Mathematik, Astronomi och Fysik, 33A, (1946), no. 6.

ON RELATIVE GROTHENDIECK RINGS

Andreas W. M. Dress[1]
University of Bielefeld, Germany
and
The Institute for Advanced Study, Princeton, N. J.

Abstract: Green's rather abstract theory of G-functors is used to derive quite formally many results concerning the structure and the functorial behaviour of relative Grothendieck rings starting with a basic induction theorem. These results are developed and discussed with particular regard to the closely related results proved by I. Reiner and T. Y. Lam with module theoretic methods. The proof of the induction theorem is based on the theory of Burnside rings and multiplicative induction techniques.

Introduction: Between 1968 and 1971 T. Y. Lam and I. Reiner developed and studied the concept of relative Grothendieck rings. Their aim was to provide a flexible K-theoretic tool for the study and classification of modular representations of finite groups. By using quite sophisticated and elaborate ad hoc methods based on very specific module theoretic considerations and constructions, they proved a considerable number of basic results and studied various conjectures concerning the structure and the functorial behaviour of those rings. Their work has been summarized very conveniently in Bill Gustafson's talk, [8], at the Columbus, Ohio, conference on group rings and orders in May, 1972.

The present paper tries to pursue another approach to the theory. This approach originated from a close analysis of R. G. Swan's use of induced

---

[1] Supported in part by National Science Foundation grant GP-36418X2.

representations in modular and integral representation theory (cf. [17], [18]) and was announced already in [2]. Meanwhile several versions of abstract axiomatic induction theory have been worked out, the first one being T. Y. Lam's thesis [10], so it seems reasonable to rely on one of those versions.

Whereas the notion of Mackey functors, developed in [4], appears to be very smooth in dealing with the abstract part of the theory, it still seems to look a bit strange to representation theorists when applied to the usual setting of representation theory. Thus I have decided rather to use the more or less equivalent notion of G-functors as defined by Green in [7], only indicating once and again how matters could be rephrased using the notion of Mackey functors. However, since the particular facts concerning G-functors we will have to use in this paper (i.e., Prop. 3.2 and 3.3) do not seem to be mentioned explicitly in Green's work, a self-contained proof of these facts is given below.

Once the abstract machinery is developed one just has to fill in a basic induction theorem concerning relative Grothendieck rings to end up with a considerable number of general results which are closely related to those of Reiner and Lam mentioned above. So it remains to prove the induction theorem. Since only slight modifications of the procedure presented in [4] would lead to such a proof, it may have been sufficient just to indicate these modifications. However, to make this paper more readable I have tried to give a more or less self-contained proof. Thus instead of quoting the methods and results developed in [4] I have tried to translate the arguments given there into the more common language adopted in this paper.

The sections are as follows. The basic definitions are given in §1. The results of Lam and Reiner are recalled in §2. The theory of G-functors is discussed in connection with relative Grothendieck rings in §3 and the basic induction theorem is stated there as well. In §4 various applications of the induction theorem are given--the most intricate one probably concerning Lam's and Reiner's excision theorem (cf. [15]). In §5 the induction theorem is proved and in an appendix some further results concerning the abstract theory of G-functors are discussed.

§1. Basic Definitions.

Let Ω be a fixed field of characteristic p $(p \neq 0)$, G a finite group, and consider the category $\underline{\underline{G\text{-mod}}}$ of finitely generated left ΩG-modules ("G-modules" for short). For any finite left G-set S ("G-set" for short) we want to define the Grothendieck group $k_S(G)$ of "S-projective" G-modules and the Grothendieck ring $a_S(G)$ of G-modules "relative to S". For that purpose let us consider the functor

$$F_S : \underline{\underline{G\text{-mod}}} \to \underline{\underline{G\text{-mod}}} : M \mapsto M[S]$$

which associates to any G-module M the "generalized permutation module"

$$M[S] = \{ \sum_{s \in S} m_s \cdot s \mid m_s \in M \}$$

consisting of the set of all formal sums $\sum_{s \in S} m_s \cdot s$ $(m_s \in M)$ made into a G-module in the obvious way by defining $x(\Sigma m_s s) + x'(\Sigma m'_s s) = \Sigma(x m_s + x' m'_s)s$ for $x, x' \in \Omega$, $g(\Sigma m_s s) = \Sigma(g m_s)(gs) = \Sigma(g m_{g^{-1}s})s$ for $g \in G$, and to any (G-module-) homomorphism $\varphi : M \to M'$ the homomorphism

$$F_S(\varphi) = \varphi[S] : M[S] \to M'[S] : \Sigma m_s s \mapsto \Sigma \varphi(m_s)s.$$

Clearly F_S is an exact covariant functor, naturally equivalent to

$$-\otimes \Omega[S] : \underline{\underline{G\text{-mod}}} \to \underline{\underline{G\text{-mod}}} : M \mapsto M \underset{\Omega}{\otimes} \Omega[S]$$

(with Ω denoting the trivial ΩG-module in this formula).

Now we can introduce the

Definition: An (exact) sequence

$$\mathcal{E} : 0 \to M' \to M \to M'' \to 0$$

is called S-split if the sequence

$$F_S(\mathscr{C}) : 0 \to M'[S] \to M[S] \to M''[S] \to 0$$

is split exact. A G-module P is called S-projective, if it is projective with respect to S-split sequences, i.e., if any S-split sequence $0 \to M' \to M \to P \to 0$ splits.

To give an example of an S-split sequence let $I_M(S)$ denote the kernel of the homomorphism

$$\mathscr{S}_M : M[S] \to M : \Sigma m_s s \mapsto \Sigma m_s .$$

Then

$$0 \to I_M(S) \to M[S] \xrightarrow{\mathscr{S}_M} M \to 0$$

is S-split, since

$$\mathscr{S}_M[S] : (M[S])[S] \to M[S] : \sum_t (\Sigma m_{s,t} s)t \mapsto \sum_t (\sum_s m_{s,t})t$$

has a right inverse, given by the "diagonal"

$$\Delta_M : M[S] \to (M[S])[S] : \sum_s m_s s \mapsto \sum_t (\Sigma m_{s,t} s)t$$

with $m_{s,t} = \begin{cases} m_s & \text{for } s = t \\ 0 & \text{for } s \neq t . \end{cases}$

Particularly M must be isomorphic to a direct summand of M[S], if it is S-projective. On the other hand, one can show that M[S] is S-projective for any G-module M: if

$$0 \to M' \to M'' \xrightarrow{\lambda} M[S] \to 0$$

is S-split and if

$$\psi : (M[S])[S] \to M''[S]$$

is a right inverse of $\lambda[S]$, then the fact that Δ_M is as well a right inverse of

$$\mathscr{S}_{M[S]} : (M[S])[S] \to M[S] : \sum_t (\sum_s m_{s,t} s)t \mapsto \sum_s (\sum_t m_{s,t})s$$

and the commutativity of

$$\begin{array}{ccc} M''[S] & \xrightarrow{\lambda[S]} & (M[S])[S] \\ \downarrow \mathcal{G}_{M''} & & \downarrow \mathcal{G}_{M[S]} \\ M'' & \xrightarrow{\lambda} & M[S] \end{array}$$

together imply that

$$\mathcal{G}_{M''} \circ \psi \circ \Delta_M : M[S] \to (M[S])[S] \to M''[S] \to M''$$

is a right inverse of λ:

$$\lambda \circ \mathcal{G}_{M''} \circ \psi \circ \Delta_M = \mathcal{G}_{M[S]} \circ \lambda[S] \circ \psi \circ \Delta_M = \mathcal{G}_{M[S]} \circ \Delta_M = Id_{M[S]}.$$

Since, moreover, obviously $M' \oplus M''$ is S-projective if and only if M' and M'' are S-projective, the above observations imply easily

Proposition 1.1: P is an S-projective G-module, if and only if P is isomorphic to a direct summand of $P[S]$.

Let us mention two consequences:

Corollary 1: If P is an S-projective and M an arbitrary G-module, then $M \underset{\Omega}{\otimes} P$ is S-projective.

Proof: If P is isomorphic to a direct summand of $P[S]$, then $M \otimes P$ is isomorphic to a direct summand of $M \otimes P[S] \cong (M \otimes P)[S]$.

Corollary 2: If $S^G = \{s \in S \mid gs = s \text{ for all } g \in G\} \neq \phi$, then any G-module M is S-projective.

Proof: If $s_0 \in S^G$, then $S = \{s_0\} \cup (S \setminus \{s_0\})$ is a decomposition of S into disjoint G-subsets. Thus, Corollary 2 follows from

$$M[S_1 \cup S_2] \cong M[S_1] \oplus M[S_2],$$

$M[\{s_0\}] \cong M$ and Proposition 1.1.

Remark: Of course the above definitions are closely related to D. G. Higman's notion of relatively projective G-modules (cf. [1] or [9]) and include it as a special case: if $H \leq G$ and $S = G/H = \{gH | g \in G\}$, then $\mathscr{E}: 0 \to M' \to M \xrightarrow{\lambda} M'' \to 0$ is S-split, if and only if it is H-split, i.e., the restricted sequence

$$\mathscr{E}|_H : 0 \to M'|_H \to M|_H \to M''|_H \to 0$$

is split exact. For if $s_0 = H \in G/H = S$ and if $\psi : M''[S] \to M[S]$ is a right inverse of $\lambda[S] : M[S] \to M''[S]$, then one gets an H-homomorphism from M'' into M which is right inverse to λ by mapping $m'' \in M''$ onto the coefficient $(\psi(m''s_0))_{s_0}$ (in M) of s_0 in $\psi(m''s_0) \in M[S]$, whereas the functorial isomorphism

$$\Omega G \underset{\Omega H}{\otimes} M \xrightarrow{\sim} M[S] : g \otimes m \mapsto (gm)gH$$

shows that the splitting of $\mathscr{E}|_H$ implies the splitting of $F_S(\mathscr{E})$.

Thus a G-module P is G/H-projective if and only if it is relatively H-projective.

The same way one gets the following more general result:

Proposition 1.2: Let $\mathfrak{U}(S) = \{U \leq G | S^U \neq \phi\}$ with $S^U = \{s \in S | us = s$ for all $u \in U\}$. Then $\mathscr{E}: 0 \to M' \to M \xrightarrow{\lambda} M'' \to 0$ is S-split if and only if $\mathscr{E}|_U$ is split for any $U \in \mathfrak{U}(S)$. Thus P is S-projective if and only if P is $\mathfrak{U}(S)$-projective in the sense of [3].

Proof: Again if $\psi : M''[S] \to M[S]$ is a right inverse of $\lambda[S]$ and if $s \in S^U$ $(U \in \mathfrak{U}(S))$, then $M'' \to M : m'' \mapsto (\psi(m'' \cdot s))_s$ (with $\psi(m''s) = \sum_{t \in S} (\psi(m'' \cdot s))_t \cdot t$, $(\psi(m''s))_t \in M)$ is a U-homomorphism which is right inverse to λ.

On the other hand $F_S(\mathfrak{E})$ definitely splits if $F_{S'}(\mathfrak{E})$ splits for any transitive subset $S' \subseteq S$. But $S' \cong G/U$ for some $U \in \mathfrak{U}(S)$, which reduces everything to the case already considered above (cf. also §5, Lemma 4).

It is well known that in case $H \leq U \leq G$ the restriction $P|_U$ of a relatively H-projective G-module P is not necessarily a relatively H-projective U-module and that inducing an H-split sequence $\mathfrak{F}: 0 \to N' \to N \to N'' \to 0$ of U-modules to G does not necessarily lead to an H-split sequence $\Omega G \underset{\Omega U}{\otimes} \mathfrak{F}$ of G-modules. Thus the following result concerning the effect of restriction and induction on S-split sequences and S-projective modules may be viewed as a justification for using G-sets rather than subgroups in the definition of relatively split sequences and relatively projective modules.

So let $U \leq G$ and--by abuse of notation--write just S again rather than $S|_U$ for the U-set one gets by restricting the action of G on the G-set S to U. Then we can state

Proposition 1.3: (a) If $\mathfrak{E}: 0 \to M' \to M \to M'' \to 0$ is an S-split sequence of G-modules, then $\mathfrak{E}|_U$ is an S-split sequence of U-modules. If $\mathfrak{F}: 0 \to N' \to N \to N'' \to 0$ is an S-split sequence of U-modules, then the induced sequence $\Omega G \underset{\Omega U}{\otimes} \mathfrak{F}$ is an S-split sequence of G-modules.

(b) If P is an S-projective G-module, then $P|_U$ is an S-projective U-module. If Q is an S-projective U-module, then $\Omega G \underset{\Omega U}{\otimes} Q$ is an S-projective G-module.

Proof: The first statement follows from the trivial functorial isomorphism

$$M[S]|_U \cong M|_U[S|_U] = M|_U[S].$$

The second statement follows from the functorial isomorphism

$$(\Omega G \underset{\Omega U}{\otimes} N)[S] \cong \Omega G \underset{\Omega U}{\otimes} (N[S|_U])$$

which is a special case of the Frobenius reciprocity law.

The rest can be derived from the first two statements by using the fact that restriction is left and right adjoint to induction.

Now let $a(G)$ denote the representation ring of G-modules as defined by Green [6], i.e., $a(G)$ is the universal (Grothendieck) ring associated with the "half ring" $a^+(G)$ consisting of the set of isomorphism classes $[M]$ of G-modules M, made into a "half ring" by putting $[M]+[M'] = [M \oplus M']$, $[M] \cdot [M'] = [M \underset{\Omega}{\otimes} M']$.

Let

$$k_S(G) = \langle [P] \in a(G) \mid P \text{ S-projective} \rangle$$

denote the additive subgroup of $a(G)$ generated by the isomorphism classes of S-projective modules and let

$$i_S(G) = \langle \chi_{\mathscr{E}} = [M']-[M]+[M''] \in a(G) \mid \mathscr{E} : 0 \to M' \to M \to M'' \text{ S-split} \rangle$$

denote the additive subgroup of $a(G)$ generated by the Euler characteristics $\chi_{\mathscr{E}}$ of S-split sequences $\mathscr{E}$.

The following result is more or less obvious:

Proposition 1.4: $k_S(G)$ and $i_S(G)$ are ideals in $a(G)$, which annihilate each other:

$$k_S(G) \cdot i_S(G) = 0.$$

If $S^G \neq \phi$, then $k_S(G) = a(G)$ and $i_S(G) = 0$.

Proof: $i_S(G)$ is an ideal, because for any G-module M_1 and any

S-split sequence $\mathcal{E} : 0 \to M' \to M \to M'' \to 0$ the sequence $M_1 \otimes \mathcal{E} : 0 \to M_1 \otimes M' \to M_1 \otimes M \to M_1 \otimes M'' \to 0$ is S-split (since $F_S(M_1 \otimes \mathcal{E}) \cong M_1 \otimes F_S(\mathcal{E})$!) and thus we have

$$[M_1] \circ \chi_{\mathcal{E}} = \chi_{M_1 \otimes \mathcal{E}} \in i_S(G).$$

$k_S(G)$ is an ideal by Corollary 1 to Proposition 1.1. The same corollary implies together with the above observations that for any S-projective G-module P and any S-split sequence $\mathcal{E} : 0 \to M' \to M \to M'' \to 0$ the sequence $P \otimes \mathcal{E}$ is split (since it is S-split with all terms, thus particularly the final term being S-split) and therefore has a vanishing Euler characteristic. Thus $k_S(G) \cdot i_S(G) = 0$.

If finally $S^G \neq \phi$, then $k_S(G) = a(G)$ by Corollary 2 of Proposition 1.1 and thus $i_S(G) = a(G) \cdot i_S(G) = k_S(G) \cdot i_S(G) = 0$.

Now we define the relative Grothendieck ring of G-modules with respect to S to be the factor ring

$$a_S(G) = a(G)/i_S(G).$$

By Proposition 1.4 $a_S(G)$ acts naturally on $k_S(G)$. Moreover, we have a natural map

$$c_S = c_S(G) : k_S(G) \hookrightarrow a(G) \twoheadrightarrow a_S(G),$$

the Cartan map, which is an $a_S(G)$-module-homomorphism, thus its image is an ideal.

For $U \leq G$ of course we put $k_S(U) = k_{S|_U}(U)$ and $a_S(U) = a_{S|_U}(U)$.

Then Proposition 1.3 implies immediately that the restriction map $a(G) \to a(U) : [M] \mapsto [M|_U]$ maps $k_S(G)$ into $k_S(U)$ and $i_S(G)$ into $i_S(U)$ and thus induces a ring homomorphism $a_S(G) \to a_S(U)$, whereas the induction map

$a(U) \to a(G) : [N] \mapsto [\Omega G \otimes_{\Omega U} N]$ induces additive maps $k_S(U) \to k_S(G)$, $i_S(U) \to i_S(G)$, $a_S(u) \to a_S(G)$. More generally, let $U, V \leq G$, $g \in G$ and $gUg^{-1} \subseteq V$. It is well known that the homomorphism $\varphi_g : U \to V : u \mapsto gug^{-1}$ induces a restriction functor

$$\underline{\underline{res}}_g = \underline{\underline{res}}(U, g, V) : \underline{\underline{V\text{-mod}}} \to \underline{\underline{U\text{-mod}}}$$

defined by considering a V-module M as a U-module via $u \circ m = (gug^{-1})m$ ($u \in U$, $m \in M$) as well as an induction functor

$$\underline{\underline{ind}}_g = \underline{\underline{ind}}(V, g, U) : \underline{\underline{U\text{-mod}}} \to \underline{\underline{V\text{-mod}}}$$

which is left and right adjoint to $res(U, g, V)$ and maps a U-module N onto the V-module $\Omega V \otimes_{\Omega U} N$, where ΩV is considered as a right ΩU-module via $x \circ u = xgug^{-1}$ ($x \in \Omega V$, $u \in U$). Moreover, $\underline{\underline{res}}_g$ induces a ring homomorphism denoted by $res_g = res(U, g, V) : a(V) \to a(U)$, and $\underline{\underline{ind}}_g$ induces an additive map $ind_g = ind(V, g, U) : a(U) \to a(V)$.

We claim

<u>Proposition</u> 1.5: The above maps $res_g : a(V) \to a(U)$ and $ind_g : a(U) \to a(V)$ induce maps $res_g : k_S(V) \to k_S(U)$, $ind_g : k_S(U) \to k_S(V)$,

$res_g : i_S(V) \to i_S(U)$, $ind_g : i_S(U) \to i_S(V)$,

$res_S(U, g, V) : a_S(V) \to a_S(U)$, $ind_S(V, g, U) : a_S(U) \to a_S(V)$.

Moreover, the following assertions hold:

(I) If $u \in U \leq G$, then

$$res_S(U, u, U) = ind_S(U, u, U) = Id_{a_S(U)}.$$

(II) If $U \leq G$, $g \in G$, then

$$res_S(U, g, gUg^{-1}) \circ ind_S(gUg^{-1}, g, U) = Id_{a_S(U)}$$

and

$$\mathrm{ind}_S(gUg^{-1}, g, U) \circ \mathrm{res}_S(U, g, gUg^{-1}) = \mathrm{Id}_{a_S(gUg^{-1})} .$$

(III) If $U, V, W \leq G$, $g, h \in G$, $gUg^{-1} \subseteq V$ and $hVh^{-1} \subseteq W$ (thus $hgU(hg)^{-1} \subseteq W$), then

$$\mathrm{res}_S(U, g, V) \circ \mathrm{res}_S(V, h, W) = \mathrm{res}_S(U, gh, W)$$

and

$$\mathrm{ind}_S(W, h, V) \circ \mathrm{ind}_S(V, g, U) = \mathrm{ind}_S(W, gh, U).$$

(IV) If $U, V \leq W \leq G$, then

$$\mathrm{res}_S(U, 1, W) \circ \mathrm{ind}_S(W, 1, V) = \sum_{VwU \subseteq W} \mathrm{ind}_S(U, 1, w^{-1}Vw \cap U) \circ \mathrm{res}_S(w^{-1}Vw \cap U, w, U)$$

where the sum is taken over all double cosets $VwU \subseteq W$.

(V) If $U, V \leq G$, $g \in G$ and $gUg^{-1} \subseteq V$ and if $x, y \in a_S(V)$, $z \in a_S(U)$ and $t \in k_S(U)$, then we have (with $\mathrm{res}_S = \mathrm{res}_S(U, g, V)$, $\mathrm{ind}_S = \mathrm{ind}_S(V, g, U)$, $\mathrm{ind} = \mathrm{ind}_g : k_S(U) \to k_S(V)$)

$$\mathrm{res}_S(xy) = \mathrm{res}_S(x) \cdot \mathrm{res}_S(y),$$

$$x \cdot \mathrm{ind}_S(z) = \mathrm{ind}_S(\mathrm{res}_S(x) \cdot z),$$

$$x \cdot \mathrm{ind}(t) = \mathrm{ind}(\mathrm{res}_S(x) \cdot t).$$

Moreover, we have $\mathrm{res}_S(1_{a_S(V)}) = 1_{a_S(U)}$, $1_{a_S(V)}$ acts as identity on $k_S(V)$ and the Cartan map $c_S(-) : k_S(-) \to a_S(-)$ as well as its components $k_S(-) \hookrightarrow a(-)$, $a(-) \twoheadrightarrow a_S(-)$ commute with both restriction and induction.

Remark 1: Since for $S = G/G = \bullet$, the trivial G-set, we have $k_\bullet(U) = a(U) = a_\bullet(U)$ and $\mathrm{res}_\bullet = \mathrm{res}$, $\mathrm{ind}_\bullet = \mathrm{ind}$ the assertions (I)-(IV) hold particularly for "res" and "ind", considered as maps defined on $a(-)$ (which is well known) as well as on $k_S(-)$ or $i_S(-)$.

Remark 2: Proposition 1.5 just says that $a(-)$, $k_S(-)$, $i_S(-)$ and $a_S(-)$ can be considered a G-functor in the sense of Green [7]. We will come back to this point of view in §3 and exploit it in §§4 and 5.

Proof: We have to prove only the very first part of Proposition 1.5 since the rest is known to hold for "res" and "ind" on a(-) and thus must hold on $a_S(-)$ as well. But by Proposition 1.3 we know that $\varphi_g : U \to V : u \mapsto gug^{-1}$ induces the maps as said once we know that restricting $S(= S|_V)$ to U via φ_g leads to a U-set S', which is isomorphic to $S(= S|_U)$. But U acts on S' via $u \circ s = gug^{-1}s$ $(u \in U,\ s \in S' = S)$, thus a U-isomorphism $S \tilde{\to} S'$ is given by $s \mapsto gs$ (thus $us \mapsto g(us) = (gug^{-1})gs = u \circ gs!$).

Finally we will need the following result which is based on some kind of "Maschke"-argument (see [5] or [1], §62/63):

Proposition 1.6: If for some p-Sylow-subgroup $U_p \leq U\ (\leq G)$ one has $S^{U_p} \neq \phi$, then the Cartan-homomorphism $k_S(U) \to a_S(U)$ as well as its components $k_S(U) \hookrightarrow a(U)$, $a(U) \twoheadrightarrow a_S(U)$ are isomorphisms.

Proof: It is enough to show that $k_S(U) \tilde{\to} a(U)$ is surjective, i.e., that any U-module N is S-projective. So let $s \in S^{U_p} \neq \phi$ and consider the U-orbit $Us = S' \subseteq S$ of s in S. Because $U_p \leq U_s = \{u \in U \mid us = s\}$ we have $|S'| = |Us| = (U : U_s) \not\equiv 0(p)$, thus $N \to N[S] : n \mapsto |S'|^{-1} \sum_{s' \in S'} ns'$ is right inverse to $\varphi_N : N[S] \to N$, so N is isomorphic to a direct summand of N[S] and thus S-projective.

§2. Some Results of T. Y. Lam and I. Reiner.

The most basic question concerning the structure of relative Grothendieck rings posed by I. Reiner and T. Y. Lam is whether or not $a_S(G)$ is a (torsion-)free $\mathbb{Z}$-module. In a considerable number of different cases they were able to show that $a_S(G)$ is in fact free, but in general this is still a rather interesting conjecture.

A second question is whether or not $a_S(G)$ is finitely generated as an abelian group. Here Reiner and Lam gave a complete answer: $a_S(G)$ is finitely generated if and only if any p-group in $\mathfrak{U}(S)$ $(= \{U \leq G \mid S^U \neq \phi\}!)$ is cyclic. In this case one can also compute the rank of $a_S(G)$.

Another question is to ask for subgroups $H \leq G$ such that the restriction map $a_S(G) \to a_S(H)$ is injective. They proved that for $H \trianglelefteq G$, $S = G/H$, $(G : H)$ a power of p and Ω algebraically closed the restriction map

$$a_S(G) \to a_S(H) \cong a(H)$$

is indeed injective. They also considered the image of this map and proved it to be "as large as possible"; more precisely G acts by conjugation on H, thus on $a(H)$ and the image of $a_S(G) \to a(H)$ is just the G-invariant part $a(H)^G$ of $a(H)$, i.e., we have an isomorphism

$$a_S(G) \xrightarrow{\sim} a(H)^G.$$

An even more delicate result is the "Excision Theorem". Let $H \leq G$, $K \trianglelefteq G$ and consider the factor groups $\overline{G} = G/K$, $\overline{H} = HK/K$. Then considering $\overline{G}$-modules as G-modules we get an inflation map $\inf : a(\overline{G}) \to a(G)$ which induces (for $S = G/H$, $\overline{S} = \overline{G}/\overline{H}$) a map

$$\inf_S : a_{\overline{S}}(\overline{G}) \to a_S(G).$$

Reiner and Lam looked for reasonable conditions on H and K which imply that $\inf_S$ is an isomorphism. It is more or less obvious that K has to be a p-group and that $H \cap K = 1$ must hold. It was an open question whether these conditions are already sufficient. In accordance with a conjecture stated by Lam and Reiner ([15], p. 161) we will answer this question in the negative in §4.

The positive result proved by Lam and Reiner is as follows: Assume $H \cap K = 1$, $|K| = p^\alpha$ and assume that there exist elements $g_1 = 1, g_2, \dots, g_r \in G$ ($r = (G : HK)$) with $G = \bigcup_{i=1}^{r} g_i HK$ and $\bigcup_{i=1}^{r} Hg_i \subseteq \bigcup_{i=1}^{r} g_i H$. Then $\inf_S : a_{\bar{S}}(\bar{G}) \to a_S(G)$ is an isomorphism.

We will discuss this result later on thoroughly. Let us just remark right now that the somehow opaque coset condition implies

(*) If $H_1, H_2 \leq G$ are subconjugate to H in G, i.e., if there exist $x_i \in G$ with $x_i H_i x_i^{-1} \leq H$ ($i = 1, 2$) and if $H_1 K \subseteq H_2 K$, then there exists $k \in K$ with $kH_1k^{-1} \subseteq H_2$.

<u>Proof:</u> W.l.o.g. we may assume $x_2 = 1$, i.e., $H_2 \leq H$. Then $H \cap K = 1$ implies $g_j hH_2 K \cap \bigcup_{i=1}^{r} g_i H = g_j hH_2$ for any $j \in \{1, \dots, r\}$, $h \in H$. Now let us write x_1 in the form $x_1 = g_j hk$ ($j \in \{1, \dots, r\}$, $h \in H$, $k \in K$). Then one gets $kH_1k^{-1} \subseteq H_2 K \cap kx_1^{-1} Hx_1 k = H_2 K \cap h^{-1} g_j^{-1} Hg_j h \subseteq h^{-1} g_j^{-1} (g_j hH_2 K \cup \bigcup_{i=1}^{r} g_i H) = h^{-1} g_j^{-1} (g_j hH_2) = H_2$.

The excision theorem has many interesting consequences, for instance the product isomorphism theorem (cf. [15], Thm. 4.4) which generalizes the result that for $G = G_1 \times G_2$, $H_1 \leq G_1$, $S_1 = G_1/H_1$, $S_2 = G_2/1$, $S = S_1 \times S_2 = G/H_1$ and Ω a splitting field for G_2 and all of its subgroups one has an isomorphism

$$a_{S_1}(G_1) \otimes a_{S_2}(G_2) \xrightarrow{\sim} a_S(G).$$

§3. The Relative Grothendieck Ring as G-functor.

Let us first recall Green's definition of a G-functor (cf. [7]). So let δG be the "category of subgroups of G", i.e., the objects are the subgroups $U, V, \ldots \leq G$, the morphisms from U to V are tripels (V, g, U) $(g \in G)$ such that $gUg^{-1} \subseteq V$ and the composition of morphisms is given by $(W, h, V) \circ (V, g, U) = (W, hg, U)$. A G-functor X from δG to the category R-mod of left R-modules (R some commutative ring with $1 \in R$) is a pair (X_*, X^*) of a contravariant functor $X_* : \delta G \to$ R-mod and a covariant functor $X^* : \delta G \to$ R-mod which coincide on the objects

$$X_*(U) = X^*(U) =: X(U)$$

such that the following holds:

(G1) $X_*(U, u, U) = X^*(U, u, U) = \mathrm{Id}_{X(U)}$ for $u \in U$;

(G2) $X_*(gUg^{-1}, g, U)$ and $X^*(gUg^{-1}, g, U)$ are inverse isomorphisms between $X(gUg^{-1})$ and $X(U)$;

(G3) if $U, V \leq W \leq G$ and if $w_1, \ldots, w_t \in W$ is a complete system of representatives of the double cosets UwV $(w \in W)$, i.e., if $W = \bigcup_{i=1}^{t} Uw_iV$, then the composition
$$X_*(W, 1, U) \circ X^*(W, 1, V) : X(V) \to X(W) \to X(U)$$
coincides with
$$\sum_{i=1}^{t} X^*(U, w_i, w_i^{-1}Uw_i \cap V) \circ X_*(V, 1, w_i^{-1}Uw_i \cap V) : X(V) \to \bigoplus_{i=1}^{t} X(w_i^{-1}Uw_i \cap V) \to X(U).$$

To define a pairing of G-functors, let X, Y, Z be three G-functors from δG to R-mod. Then a pairing $X \times Y \to Z$ is a family of R-bilinear maps

$$\Theta_U : X(U) \times Y(U) \to Z(U) : (x, y) \mapsto x \cdot y$$

such that for any morphism $\alpha = (V, g, U)$ and any $x_U \in X(U)$, $x_V \in X(V)$, $y_U \in Y(U)$, $y_V \in Y(V)$ the following holds:

(P1) $$(x_V \cdot y_V)_* = (x_V)_* \cdot (y_V)_*$$

(with $\ldots_*$ indicating the image of ... under $Z_*(\alpha)$, $X_*(\alpha)$ and $Y_*(\alpha)$ respectively),

(P2) $$(x_U)^* \cdot y_V = (x_U \cdot (y_V)_*),$$

(P3) $$x_V \cdot (y_U)^* = ((x_V)_* \cdot y_U)^*$$

(with $\ldots^*$ defined correspondingly).

We now define a multiplicative G-functor with a unit to be a G-functor $X : \delta G \to \underline{\underline{R\text{-mod}}}$ together with a pairing $\Theta : X \times X \to X$, such that for any $U \leq G$ there exists a (uniquely determined!) element $1_{X(U)} = 1_U \in X(U)$ which acts as left and right unit on $X(U)$ via Θ and is preserved under X_*, i.e., we have $1_U \cdot x = x \cdot 1_U = x$ for any $x \in X$ (with $x \cdot x' =: \Theta_U(x, x')$ for $x, x' \in X(U)$) and $X_*(V, g, U)(1_V) = 1_U$ whenever $gUg^{-1} \subseteq V$.

The following result shows that even without any additional assumption the multiplication in such a multiplicative G-functor with a unit is not too far from being commutative and associative. It will be proved in an appendix to this paper:

Proposition 3.1: Let $X : \delta G \to \underline{\underline{R\text{-mod}}}$ be a multiplicative G-functor with a unit and let

$$\underline{X}(U) = \langle X^*(U, 1, V)(1_V) \mid V \leq U \rangle$$

denote the R-linear span of the elements $1_V^U = X^*(U, 1, V)(1_V)$ $(V \leq U)$. Then $\underline{X}(U)$ is a commutative and associative subring of $X(U)$ (with $1_U = 1_U^U$ as a unit) and $X_*(-)$ and $X^*(-)$ map $\underline{X}(\ldots)$ into $\underline{X}(\ldots)$.

Finally let us define modules over X, X being such a multiplicative G-functor with a unit, to be G-functors $Y : \delta G \to \underline{\underline{R\text{-mod}}}$ together with a pairing $\Theta' : X \times Y \to Y$ such that $1_U \in X(U)$ acts unitary on $Y(U)$ via Θ'. An X-module-homomorphism $\Xi : Y \to Y'$ of course is then defined to be a family of

R-linear maps

$$\Xi_U : Y(U) \to Y'(U) \quad (U \leq G)$$

such that Ξ_U is $X(U)$-linear (i.e., $\Xi_U(x \cdot y) = x \cdot \Xi_U(y)$ for $x \in X(U)$, $y \in Y(U)$) and Ξ is a natural transformation from Y_* to Y'_* as well as from Y^* to Y'^*.

Examples: By Proposition 1.5 it is obvious that $a(-)$, $k_S(-)$, $i_S(-)$ and $a_S(-)$ are G-functors (with $\ldots_*$ and $\ldots^*$ defined by "res" and "ind"), $a(-)$ and $a_S(-)$ are moreover multiplicative G-functors with a unit, $k_S(-)$ (as well as $a_S(-)$ itself) is an $a_S(-)$-module and the Cartan-map $c_S : k_S \to a_S$ is an $a_S(-)$-module-homomorphism.

In §4 we will rely on

Proposition 3.2: Let $X : \delta G \to \underline{\underline{R\text{-mod}}}$ be a multiplicative G-functor with a unit and let Y be an X-module. Let $\mathcal{C}$ be a subconjugately closed family of subgroups of G (i.e., $C, C' \leq G$, $C \in \mathcal{C}$, $g \in G$ and $gC'g^{-1} \leq C$ imply $C' \in \mathcal{C}$) and assume that

$$\sum_{C \in \mathcal{C}} X^*(G,1,C) : \sum_{C \in \mathcal{C}} X(C) \to X(G)$$

is surjective. Then the product

$$\prod_{C \in \mathcal{C}} Y_*(G,1,C) : Y(G) \to \prod_{C \in \mathcal{C}} Y(C)$$

is injective and its image is

$$\varprojlim_{C \in \mathcal{C}} Y(C) = \{(y_C)_{C \in \mathcal{C}} \in \prod_{C \in \mathcal{C}} Y(C) \mid Y_*(C,g,C')(y_C) = y_{C'}$$

$$\text{whenever } C, C' \in \mathcal{C},\ g \in G \text{ and } gC'g^{-1} \leq C\}.$$

Proof: The injectivity is well known (and it is for instance one of the main ingredients of Lam's thesis [10]): if

$$1_G = \sum_{C \in \mathcal{C}} x_C^*$$

(with $x_C \in X(C)$ and $x_C^* = X^*(G,1,C)(x_C)$) and if $y \in \mathrm{Ker}(Y(G) \to \prod_{C \in \mathcal{C}} Y(C))$,

then using (P2) we get:

$$y = 1_G \cdot y = (\sum_{C \in \mathcal{C}} x_C^*) \cdot y = \sum_{C \in \mathcal{C}} (x_C \cdot y|_C)^* = 0,$$

since $y|_C = Y_*(G,1,C)(y) = 0$ for any $C \in \mathcal{C}$.

It is easy to see that the image of $\prod_{C \in \mathcal{C}} Y_*(G,1,C) : Y(G) \to \prod_{C \in \mathcal{C}} Y(C)$ is contained in $\varprojlim_{C \in \mathcal{C}} Y(C)$, since $y \in Y(G)$, $C, C' \in \mathcal{C}$, $g \in G$ and $gC'g^{-1} \leq C$ together imply $Y_*(C,g,C')(y|_C) = Y_*(C,g,C')(Y_*(G,1,C)(y)) = Y_*(G,g,C')(y) = Y_*(G,1,C')(Y_*(G,g,G)(y)) \overset{(G1)}{=} Y_*(G,1,C')(y) = y|_{C'}$.

To show the opposite inclusion let $(y_C)_{C \in \mathcal{C}} \in \varprojlim_{C \in \mathcal{C}} Y(C)$ and consider $y = \sum_{C \in \mathcal{C}} (x_C y_C)^*$ (with $1_G = \Sigma x_C^*$ as above). We claim that $y|_D = y_D$ for any $D \in \mathcal{C}$: for any fixed $C \in \mathcal{C}$ one has

$$(x_C y_C)^*|_D \overset{(G3)}{=} \sum_{DgC \subseteq G} Y^*(D,g,g^{-1}Dg \cap C)((x_C y_C)|_{g^{-1}Dg \cap C})$$

$$\overset{(P1)}{=} \sum_{DgC} Y^*(D,g,g^{-1}Dg \cap C)(x_C|_{g^{-1}Dg \cap C} \cdot y_{g^{-1}Dg \cap C})$$

(since $y_C|_{g^{-1}Dg \cap C} = y_{g^{-1}Dg \cap C}$, because $(y_C) \in \varprojlim_{C \in \mathcal{C}} Y(C)$)

$$\overset{(P2)}{=} (\sum_{DgC \subseteq G} X^*(D,g,g^{-1}Dg \cap C)(x_C|_{g^{-1}Dg \cap C}))\ y_D$$ (again, since $(y_C) \in \varprojlim_{C \in \mathcal{C}} Y(C)$ and thus $Y_*(D,g,g^{-1}Dg \cap C)(y_D) = y_{g^{-1}Dg \cap C}$) $\overset{(G3)}{=} x_C^*|_D \cdot y_D$

and thus $y|_D = \sum_{C \in \mathcal{C}} (x_C y_C)^*|_D = \sum_{C \in \mathcal{C}} x_C^*|_D \cdot y_D = 1_G|_D \cdot y_D = 1_D \cdot y_D = y_D$.

<u>Remark</u>: A more transparent proof can be given by associating to any G-functor

Y a "Mackey-functor" $\hat{Y}$ in the sense of [4] defined on the category $\hat{G}$ of G-sets by

$$\hat{Y}(S) = (\bigoplus_{s \in S} Y(G_s))^G .$$

Here $\bigoplus_{s \in S} Y(G_s)$ is considered as an RG-module by defining

$g \cdot y_s = Y_*(gG_sg^{-1}, g, G_s)(y_s)$ for $g \in G$, $y_s \in Y(G_s) \subseteq \bigoplus_{s \in S} Y(G_s)$ and $(\ldots)^G$ denotes the G-invariant part of $(\ldots)$.

For any G-map $\varphi : S \to T$ we have a "restriction map": $\varphi_* : \hat{Y}(T) \to \hat{Y}(S)$ derived from $Y_*(G_{\varphi(s)}, 1, G_s)$ $(s \in S)$ and an "induction map" $\varphi^* : \hat{Y}(S) \to \hat{Y}(T)$, derived from $Y^*(G_{\varphi(s)}, 1, G_s)$ $(s \in S)$. The "Mackey-axiom" (G3) can then be rephrased in the following way: for any pull back diagram in $\hat{G}$

$$\begin{array}{ccc} S & \xrightarrow{\Phi} & S_2 \\ {\scriptstyle\Psi}\downarrow & & \downarrow{\scriptstyle\psi} \\ S_1 & \xrightarrow[\varphi]{} & T \end{array}$$

the diagram

$$\begin{array}{ccc} \hat{Y}(S) & \xrightarrow{\Phi^*} & \hat{Y}(S_2) \\ {\scriptstyle\Psi_*}\downarrow & & \downarrow{\scriptstyle\psi_*} \\ \hat{Y}(S_1) & \xrightarrow[\varphi^*]{} & \hat{Y}(T) \end{array}$$

commutes.

The above proposition turns out to be a special case of the following theorem: if $\bullet = G/G$ denotes the trivial G-set and $\varphi : S \to \bullet$ the obvious unique G-map from S onto $\bullet$, and if $\varphi^* : \hat{X}(S) \to \hat{X}(\bullet)$ is surjective, then $\hat{Y}_*$ applied to the semisimplicial complex of G-sets

$$\bullet \leftarrow S \leftleftarrows S \times S \leftleftarrows S \times S \times S \ldots$$

gives rise to an exact sequence

$$\hat{Y}(\bullet) \to \hat{Y}(S) \to \hat{Y}(S^2) \to \dots ,$$

the morphisms being the alternating sums of $\hat{Y}_*(\dots)$ applied to the various projection maps $S^n \to S^{n-1}$. More generally, the cohomology groups of this complex, considered as $\hat{X}(\bullet)$-modules, are annihilated by $\mathrm{Im}(\varphi_* : \hat{X}(S) \to \hat{X}(\bullet))$.

Let us further remark that for $Y = a_S(-)$ the values $\hat{Y}(T)$ have a rather natural interpretation as the Grothendieck ring of ΩG-bundles over T relative to sequences of such bundles which split, when restricted to $T \times S$. This interpretation offers particularly an easy and conceptual proof of the Mackey subgroup theorem (G3) in the pull-back-diagram-form stated above.

For details of this version of the theory see [4] (cf. also §5, step 5, below).

As an application of Proposition 3.2 let us just mention

Corollary: Let X, Y and $\mathcal{C}$ be as in Proposition 3.2. Let Y' be another X-module and $\Xi : Y \to Y'$ an X-module-homomorphism. If $\Xi_C : Y(C) \xrightarrow{\sim} Y'(C)$ is an isomorphism for any $C \in \mathcal{C}$, then $\Xi_G : Y(G) \to Y'(G)$ is an isomorphism.

Proof: We have a commutative diagram

$$\begin{array}{ccc} Y(G) & \xrightarrow{\Xi_G} & Y'(G) \\ \wr\downarrow & & \wr\downarrow \\ \varprojlim_{C \in \mathcal{C}} Y(C) & \xrightarrow{\sim} & \varprojlim_{C \in \mathcal{C}} Y'(C), \end{array}$$

thus Ξ_G must be an isomorphism.

The following result will be needed in §5 and proved in the appendix:

Proposition 3.3: Let R be a subring of $\mathbb{Q}$ and let $X : \delta G \to \underline{\underline{R\text{-mod}}}$ be a multiplicative G-functor with a unit. Let $\mathcal{C}$ be a subconjugately closed family of subgroups of G and $\mathcal{G} = \{H \leq G \mid \text{ex. } C \trianglelefteq H \text{ with } C \in \mathcal{C} \text{ and } H/C \text{ a } q\text{-group for some prime } q \text{ with } qR \neq R\}$.

Assume that

$$\sum_{C \in \mathcal{C}} \mathbb{Q} \otimes X(C) \to \mathbb{Q} \otimes X(G)$$

is surjective and that any torsion element in $\underline{X}(G)$ is nilpotent. Then $\sum_{H \in \mathcal{H}} X(H) \to X(G)$ is surjective.

Let us now consider the G-functor $a_S(-)$. As already remarked, $a_S(-)$ is multiplicative and has a unit. Let $\mathcal{C}(S) = \{C \leq G \mid$ the p-Sylow-subgroup C_p is normal in C, C/C_p is cyclic and $S^{C_p} \neq \phi\}$ and let $\mathcal{H}(S) = \{H \leq G \mid$ there exists $C \trianglelefteq H$ with $C \in \mathcal{C}(S)$ and H/C a q-group for some prime $q \neq p\}$. Let $a'_S(U) = \mathbb{Z}(\frac{1}{p}) \otimes a_S(U)$ and $A_S(U) = \mathbb{Q} \otimes a_S(U) = \mathbb{Q} \otimes a_S(U)$ (as well as $a'(U) = \mathbb{Z}(\frac{1}{p}) \otimes a(U)$, $A(U) = \mathbb{Q} \otimes a(U)$ of course). Then we can state the following induction theorem which will be applied in §4 via Proposition 3.2 and its corollary.

<u>Theorem</u>: The sums of the induction maps

$$\sum_{C \in \mathcal{C}(S)} A_S(C) \twoheadrightarrow A_S(G)$$

and

$$\sum_{H \in \mathcal{H}(S)} a'_S(H) \twoheadrightarrow a'_S(G)$$

are surjective.

The proof will be given in §5.

§4. Applications.

The first and perhaps most fundamental application of the above theorem is toward the Cartan map. Since $S^{H_p} \neq \phi$ for $H \in \mathcal{H}_p(S)$ we know by Proposition 1.6 that the Cartan map $k_S(H) \to a_S(H)$ as well as its components $k_S(H) \hookrightarrow a(H)$, $a(H) \twoheadrightarrow a_S(H)$ are isomorphisms. Thus the corollary of Proposition 3.2 implies immediately:

Proposition 4.1: The Cartan map

$$k_S(G) \to a_S(G)$$

induces an isomorphism

$$k_S'(G) \xrightarrow{\sim} a_S'(G).$$

We mention the following consequences, which follow from Proposition 4.1 and the more or less obvious fact, that $k_S(G)$ is $\mathbb{Z}$-free with basis the isomorphism classes of indecomposable S-projective G-modules.

Corollary 1: $a_S'(G)$ is $\mathbb{Z}(\frac{1}{p})$-free.

Corollary 2: $a_S(G)$ has at most p-torsion.

Corollary 3: The Cartan map $k_S(G) \to a_S(G)$ is injective and its cokernel is a p-torsion group.

Corollary 4: There exists a p-power p^α such that $p^\alpha \cdot a_S(G) = \{p^\alpha \cdot x \mid x \in a_S(G)\}$ is $\mathbb{Z}$-free, particularly $p^\alpha \cdot x = 0$ for all $x \in \mathrm{Tor}(a_S(G))$.

Proof: The image of the Cartan map $k_S(G) \hookrightarrow a_S(G)$ is an ideal in $a_S(G)$ and by Corollary 3 there exists p^α with $p^\alpha \cdot 1_G \in k_S(G)$, thus $p^\alpha(a_S(G)) \subseteq k_S(G)$.

Conjecture: In Corollary 4 one can always choose $p^\alpha = |G|_p$, the p-part of the order $|G|$ of G.

Corollary 5: $a_S(G)$ is $\mathbb{Z}$-free if and only if it is torsion free.

Remark: This is not clear a priori, since $a_S(G)$ is generally not finitely generated.

Corollary 6: $a_S(G)$ has finite rank if and only if any p-group U in $\mathfrak{U}(S)$ is cyclic. If this is the case and if $s(H)$ denotes for any finite group H the number of simple H-modules, then

$$\operatorname{rk} a_S(G) = \sum_{U \in \mathfrak{U}(S),\, U \text{ p-group}}{}' \varphi(|U|) \cdot s(N_G(U))$$

(with $\varphi(\dots)$ the Euler function and $N_G(U) = \{g \in G \mid gUg^{-1} = U\}$), where the sum Σ' is taken over all conjugacy classes of such subgroups.

Particularly for $H \trianglelefteq G$ a normal cyclic p-subgroup of G and $S = G/H$ we get (cf. [11])

$$\operatorname{rk} a_S(G) = |H| \cdot s(G).$$

Proof: Since $a'_S(G) \cong k'_S(G)$ and thus $\operatorname{rk} a_S(G) = \operatorname{rk} k_S(G)$ we may replace $a_S(G)$ by $k_S(G)$ in Corollary 6, but then everything is well known.

Now we want to apply Proposition 3.2 directly with $X = Y = a'_S(-)$ (or $A_S(-)$). We get immediately (using $a_S(H) = a(H)$ for $H \in \mathfrak{H}(S)$):

Proposition 4.2: Restriction defines isomorphisms

$$a'_S(G) \xrightarrow{\sim} \varprojlim_{H \in \mathfrak{H}(S)} a'(H),$$

$$A_S(G) \xrightarrow{\sim} \varprojlim_{C \in \mathfrak{C}(S)} A(C).$$

Corollary 1: The product of the restriction maps

$$a'_S(G) \xrightarrow{\sim} \prod_{C \in \mathcal{C}(S)} a'(C)$$

is injective.

Proof: Since $a'_S(G)$ is free, it imbeds into $A_S(G)$.

Corollary 2: If $H \trianglelefteq G$, $(G : H)$ a power of p and $G_s \leq H$ for $s \in S$, then restriction defines an isomorphism

$$a'_S(G) \xrightarrow{\sim} a'_S(H)^G.$$

Proof: This is easy to derive from Proposition 4.2, since any subgroup in $\mathcal{G}(S)$ is contained in H. A rather trivial proof can be derived from the fact that both compositions

$$a_S(G) \xrightarrow{res_S} a_S(H)^G \xrightarrow{ind_S} a_S(G)$$

and

$$a_S(H)^G \xrightarrow{ind_S} a_S(G) \xrightarrow{res_S} a_S(H)^G$$

are just multiplication by $(G : H) = p^\beta$, as already observed (in case $S = G/H$) by Reiner and Lam.

Corollary 3: If G is a p-group, then

$$a'_S(G) \xrightarrow{\sim} \varprojlim_{C \in \mathfrak{N}(S)} a'(C).$$

Proof: In this case we have obviously $\mathcal{G}(S) = \mathfrak{N}(S)$.

Corollary 4 (cf. [14], §2): If G is a p-group and $S = G/H$ for some cyclic subgroup H, then

$$a'_S(G) \xrightarrow{\sim} a'(H)$$

is an isomorphism.

Proof: Since $H \in \mathfrak{U}(G/H) = \{C \leq G \mid \text{there exists } g \in G \text{ with } gCg^{-1} \subseteq H\}$ there is an obvious projection map $\pi: \varprojlim_{C \in \mathfrak{U}(G/H)} a'(C) \subseteq \prod_C a'(C) \to a'(H)$. To show that this is an isomorphism we construct an inverse map by observing that for a cyclic p-group C the automorphism group $\mathrm{Aut}(C)$ acts trivially on $a(C)$ (since any indecomposable C-module is characterized by its dimension, for instance) and thus

$$a'(H) \to \varprojlim_{C \lesssim H} a'(C) : x \mapsto (\mathrm{res}(C, g, H)(x))_{C \lesssim H}$$

(with $g \in G$ any element with $gCg^{-1} \subseteq H$) is a well-defined map which then must be the inverse of π.

Corollary 5: $\mathrm{Ke}(a_S(G) \to \prod_{C \in \mathcal{C}(S)} a(C)) = \mathrm{Tor}\, a_S(G)$.

Corollary 6: Two S-projective G-modules P, Q are isomorphic if and only if their restrictions $P|_C$ and $Q|_C$ are isomorphic for any $C \in \mathcal{C}(S)$.

Proof: $k_S(G)$ is free and maps injectively into $a_S(G)$, thus intersects $\mathrm{Ke}(a_S(G) \to \prod_C a(C)) = \mathrm{Tor}\, a_S(G)$ only trivially. So $k_S(G) \to \prod_{C \in \mathcal{C}(S)} a(C)$ is injective, i.e., $P|_C \cong Q|_C$ for every $C \in \mathcal{C}(S)$ implies $[P]-[Q] = 0$ in $k_S(G)$. But this in turn implies $P \cong Q$, since the Krull-Schmidt-Theorem holds for G-modules.

Corollary 7: For any S-projective G-module P there exist C-modules M'_C, M''_C $(C \in \mathcal{C}(S))$ and some $n \in \mathbb{N}$ with

$$\underbrace{P \oplus \ldots \oplus P}_{n \text{ times}} \oplus \bigoplus_{C \in \mathcal{C}(S)} \Omega G \underset{\Omega C}{\otimes} M'_C \cong \bigoplus_{C \in \mathcal{C}(S)} \Omega G \underset{\Omega C}{\otimes} M''_C$$

Proof: Use the induction theorem to choose some $n \in \mathbb{N}$, such that there exist elements $y_C \in a(C)$ $(C \in \mathcal{C}(S))$ with $\sum_{C \in \mathcal{C}(S)} \mathrm{ind}_S(y_C) = n \cdot 1_{a_S(G)}$. Let $y_C = [N'_C]-[N''_C]$; then multiplying $[P] \in k_S(G)$ with $n \cdot 1_{a_S(G)}$ yields $n \cdot [P] = [P] \cdot \Sigma\, \mathrm{ind}_S(y_C) = \Sigma\, \mathrm{ind}_S([P|_C] \cdot y_C) = \Sigma[\Omega G \underset{\Omega C}{\otimes} (P|_C \otimes N'_C)] - \Sigma[\Omega G \underset{\Omega C}{\otimes} (P|_C \otimes N''_C)]$. Thus putting $M'_C = P|_C \otimes N'_C$, $M''_C = P|_C \otimes N''_C$ and using the Krull-Schmidt-Theorem once more one gets the result.

Remark: Actually the proof of the last two statements follows closely the work of R. G. Swan mentioned in the introduction and looking for such a proof was the starting point for trying to establish an induction theorem for relative Grothendieck rings.

Now let us consider the excision theorem. Let $K \trianglelefteq G$ and put $\bar{G} = G/K$, $\bar{S} = K\backslash S$ the $\bar{G}$-set of K-orbits in S. Considering $\bar{G}$-modules as G-modules defines ring homomorphisms

$$\mathrm{inf} : a_{\bar{S}}(\bar{G}) \to a_S(G),$$

$$\mathrm{inf}' : a'_{\bar{S}}(\bar{G}) \to a'_S(G),$$

$$\mathrm{Inf} : A_{\bar{S}}(\bar{G}) \to A_S(G).$$

Our first result is

Proposition 4.3: The following statements are equivalent:

(i) $\mathrm{inf}' : a'_{\bar{S}}(\bar{G}) \to a'_S(G)$ is injective.

(ii) $\mathrm{Inf} : A_{\bar{S}}(\bar{G}) \to A_S(G)$ is injective.

(iii) For any $\bar{C} \leq \bar{G}$ with $\bar{C} \in \mathcal{C}(\bar{S})$ there exists a $C \in \mathcal{C}(S)$ with $\bar{C} = CK/K$.

<u>Proof:</u> Since $a'_{\bar{S}}(\bar{G})$ is torsion free (Corollary 1 to Proposition 4.1) the first two conditions are obviously equivalent. To prove the equivalence of the last two conditions let us observe at first that $CK/K \in \mathcal{C}(\bar{S})$ holds whenever $C \in \mathcal{C}(S)$. Thus, considering CK/K-modules as C-modules for any $C \in \mathcal{C}(S)$ we get a map $A(CK/K) \to A(C)$ which is well known to be injective and fits into a commutative diagram

$$\begin{array}{ccccc} A_{\bar{S}}(\bar{G}) & \tilde{\hookrightarrow} & \varprojlim_{\bar{C} \in \mathcal{C}(\bar{S})} A(\bar{C}) & \subseteq & \prod_{\bar{C} \in \mathcal{C}(\bar{S})} A(\bar{C}) \\ \downarrow & & \vdots & & \downarrow \\ A_S(G) & \tilde{\hookrightarrow} & \varprojlim_{C \in \mathcal{C}(S)} A(C) & \subseteq & \prod_{C \in \mathcal{C}(S)} A(C) \end{array}$$

(the middle vertical arrow $\vdots$ exists for instance by the commutativity of this diagram and Proposition 4.2). Since the last vertical arrow must be injective, too, if (iii) holds, this proves already "(iii) $\Rightarrow$ (ii)".

To prove the opposite implication we need the following two facts:

<u>Lemma 1:</u> For any subgroup $\bar{C} \leq \bar{G}$ there exist $\bar{G}$-sets $\bar{T}_1$ and $\bar{T}_2$ with $|\bar{T}_1^{\bar{C}}| \neq |\bar{T}_2^{\bar{C}}|$, but $\bar{T}_1|_{\bar{H}} \cong \bar{T}_2|_{\bar{H}}$ for any $\bar{H} \leq \bar{G}$, which does not contain a conjugate of $\bar{C}$.

<u>Proof:</u> This should be well known. For the convenience of the reader we recall a simple proof (in which we may omit the upper bars of course). So let $\mathfrak{R} = \{U_1 = 1, U_2, \dots, U_r = C, \dots, U_t = G\}$ be a complete system of representatives of conjugacy classes of subgroups of G and assume $|U_i| \leq |U_j|$ for $i \leq j$. Put

$$a_{i,j} = |\{G/U_i^{\;U_j}\}| = |\{gU_i \in G/U_i \,|\, U_j gU_i = gU_i\}|.$$

Then it is obvious that the matrix $A_G = A = (a_{ij})$ is triangular with diagonal entries $a_{i,i} = (N_G(U_i) : U_i) > 0$. Thus A is nonsingular and so we can find a vector $(n_1, n_2, \ldots, n_t)$ with $n_i \in \mathbb{Z}$ and

$$\sum_{i=1}^{t} n_i a_{i,j} = \begin{cases} 0 & U_j \neq C \text{ (i.e., } j \neq r), \\ \neq 0 & U_j = C \text{ (i.e., } j = r). \end{cases}$$

Let S_i denote the disjoint union of $|n_i|$ copies of G/U_i and put

$$T_1 = \dot{\bigcup_{n_i > 0}} S_i, \quad T_2 = \dot{\bigcup_{n_i < 0}} S_i .$$

One has

$$|T_1^{U_j}| - |T_2^{U_j}| = \sum_{i=1}^{t} n_i a_{ij}$$

and thus $|T_1^C| \neq |T_2^C|$, but $|T_1^{H'}| = |T_2^{H'}|$ for any subgroup $H' \leq H$, whenever $H \leq G$ does not contain a conjugate of C (using $|T^U| = |T^{gUg^{-1}}|$ for $U \leq G$, $g \in G$, T a G-set). Thus by a well-known theorem of Burnside's one has $T_1|_H \cong T_2|_H$ for any such $H \leq G$ (cf. [0], p. 238--by the way the theorem follows more or less immediately from the fact that any H-set is isomorphic to a disjoint union of H-sets of the type H/V_i with V_i ranging over some complete system of representatives of conjugacy classes of subgroups of H and the nonsingularity of $A_H = (|H/V_i^{V_j}|)_{i,j}$).

<u>Lemma</u> 2: If $\bar{C} \in \mathfrak{C}(\bar{S})$ and if $\bar{T}_1, \bar{T}_2$ are two $\bar{G}$-sets with $|\bar{T}_1^{\bar{C}}| \neq |\bar{T}_2^{\bar{C}}|$, then $\Omega[\bar{T}_1]|_{\bar{C}} \not\cong \Omega[\bar{T}_2]|_{\bar{C}}$.

<u>Proof</u>: (Again we omit the upper bars.) It is sufficient to construct an additive map $\psi : a(C) \to \mathbb{C}$ with $\psi([\Omega[T]]) = |T^C|$ for any C-set T. By the Krull-Schmidt-Theorem it is enough to define $\psi([M])$ for any indecomposable C-module M. Let C_p denote the normal p-Sylow-subgroup of C and let $g \in C$ be a p-regular element such that $C = \langle g \rangle \cdot C_p$. Choose some imbedding of the group

of roots of unity of some algebraic closure of Ω into $\mathbb{C}^*$ and define Brauer characters $\chi_M(g) \in \mathbb{C}$ accordingly. Then

$$\psi([M]) = \begin{cases} 0 & M \text{ indec., } C_p \text{ acts nontrivially on } M \\ \chi_M(g) & M \text{ indec., } C_p \text{ acts trivially on } M \end{cases}$$

defines the sought for additive map, because $\psi([\Omega[C/U]]) \neq 0$ if and only if $U = C$, since C_p acts trivially on an indecomposable summand of $\Omega[C/U]$ if and only if $C_p \leq U$, in which case C_p acts trivially on $\Omega[C/U]$ and one has $\psi([\Omega[C/U]]) = \chi_{\Omega[C/U]}(g) = 0$ unless $U = C$.

It is easy now to prove that (ii) implies (iii): Let $\bar{C} \in \mathcal{C}(\bar{S})$ and choose $\bar{T}_1$ and $\bar{T}_2$ accordingly to Lemma 1. Consider $x = [\Omega[\bar{T}_1]]-[\Omega[\bar{T}_2]] \in A_{\bar{S}}(\bar{G})$. Because the restriction $x|_{\bar{C}}$ in $A_{\bar{S}}(\bar{C}) = A(\bar{C})$ is nonzero by Lemma 2, one has $x \neq 0$. Thus the injectivity of the first vertical arrow (that's our hypothesis) and of the lower horizontal arrow in the above diagram imply the existence of some $C \in \mathcal{C}(S)$ such that $\Omega[\bar{T}_1]|_C \not\cong \Omega[\bar{T}_2]|_C$. Thus $\Omega[\bar{T}_1]$ and $\Omega[\bar{T}_2]$ are not isomorphic when considered as CK/K-modules. By Lemma 1 this implies that CK/K must contain some conjugate $\bar{g}\bar{C}\bar{g}^{-1}$ of $\bar{C}$. Let $g \in G$ be a pre-image of $\bar{g} \in \bar{G}$ and consider $C' = \{g^{-1}cg \mid c \in C$ and $cK \in \bar{g}\bar{C}\bar{g}^{-1}\}$. Because C' is subconjugate to C one has $C' \in \mathcal{C}(S)$ and because $\bar{g}\bar{C}\bar{g}^{-1} \subseteq CK/K$ one has $C'K/K = \bar{C}$. q. e. d.

<u>Corollary</u> 1 (cf. [15], p. 161): The coset condition in the excision theorem of Lam and Reiner cannot be waived (i. e., not without some kind of compensation).

<u>Proof</u>: Let G be the semidirect product of $G_p = Z_p \times Z_p = \{(u_1, u_2) \mid u_i \in Z_p\} \trianglelefteq G (Z_p = \mathbb{Z}/_p\mathbb{Z},\ p \neq 2)$ and $Z_2 = \{1, \alpha\}$, which acts on G_p by $(u_1, u_2)^\alpha = (u_1, -u_2)$. Put $K = \{(0,u) \mid u \in Z_p\} \trianglelefteq G$, $H = \{(u,u) \mid u \in Z_p\}$ and $S = G/H$. Then we have $\bar{G} \in \mathcal{C}(\bar{S})$, but $\mathcal{C}(S) = \{H, H^\alpha, 1, Z_2^g \ (g \in G)\}$ and thus

$CK/K \neq \bar{G}$ for all $C \in \mathcal{C}(S)$.

Corollary 2: If $\bar{G}$ is a p-group, then

$$\text{inf}' : a'_{\bar{S}}(\bar{G}) \to a'_S(G)$$

is always injective.

Proof: One has $\mathcal{C}(\bar{S}) = \mathfrak{N}(\bar{S}) = \{\bar{C} \leq \bar{G} \mid \bar{S}^{\bar{C}} \neq \phi\}$. Pick $s \in \bar{s} = Ks \in \bar{S}^{\bar{C}}$ and let $C' \leq G$ denote the full pre-image of $\bar{C} \leq \bar{G}$ in G. $C'Ks = Ks$ implies $C'_s K = C'K$, thus with C denoting the p-Sylow-subgroup of C'_s one gets $CK/K = (C'_s K/K)_p = (C'K/K)_p = (\bar{C})_p = \bar{C}$ as well as $C \in \mathcal{C}(S)$.

This last result can be generalized to

Corollary 3: For any finite group U let $U^{(p)}$ denote the subgroup generated by all p-regular elements in U. If for any p-group D in $\mathcal{C}(S)$ the inclusion $N_G(DK)^{(p)} \subseteq N_G(D) \cdot K$ holds, then $\text{inf}' : a'_{\bar{S}}(\bar{G}) \to a'_S(G)$ is injective.

Proof: Let $\bar{C} \in \mathcal{C}(\bar{S})$ and choose some p-regular element $g \in G$ such that $\bar{g} = gK \in \bar{G}$ together with $\bar{C}_p \trianglelefteq \bar{C}$ generates $\bar{C}$ ($\bar{C}_p$ of course is the normal p-Sylow-subgroup of $\bar{C}$). Since $\bar{C}_p \in \mathfrak{N}(\bar{S})$ there exists (by the argument used in the proof of Corollary 2) a p-group $D \in \mathfrak{N}(S)$ with $DK/K = \bar{C}_p$. Since $gK \in N_{\bar{G}}(\bar{C}_p)$ one has $g \in N_G(DK)$, thus $g \in N_G(DK)^{(p)} \subseteq N_G(D) \cdot K$: $g = n \cdot k$ with $n \in N_G(D)$, $k \in K$. Since $\bar{n} = nK = gK = \bar{g}$ is p-regular we may as well replace n by its p-regular part, again denoted by n. Then $C = D \cdot \langle n \rangle$ is contained in $\mathcal{C}(S)$ and $CK/K = \langle n \rangle \cdot DK/K = \langle \bar{n} \rangle \cdot \bar{C}_p = \langle \bar{g} \rangle \cdot \bar{C}_p = \bar{C}$. q.e.d.

Let us now look for reasonable conditions which imply that "inf' " or "Inf" are isomorphisms. Obviously, K must be a p-group which acts freely on S, since otherwise K would contain a nontrivial subgroup $C \in \mathcal{C}(S)$ and there could be no element $x \in A_{\bar{S}}(\bar{G})$ with $\mathrm{Inf}(x)|_C = [\Omega G]|_C$, thus neither inf' nor Inf could be surjective.

Moreover, one may of course state that inf' (or Inf) is an isomorphism if and only if the map

$$\varprojlim \mathrm{inf}' : \varprojlim_{\bar{H} \in \mathfrak{H}(\bar{S})} a'(\bar{H}) \to \varprojlim_{H \in \mathfrak{H}(S)} a'(H)$$

$$\left(\text{or } \varprojlim \mathrm{Inf} : \varprojlim_{\bar{C} \in \mathfrak{C}(\bar{S})} A(\bar{C}) \to \varprojlim_{C \in \mathfrak{C}(S)} A(C)\right)$$

induced by $H \to HK/K$ (or $C \to CK/K$ respectively) is an isomorphism. Even if this condition doesn't look too easy to handle, one may use it to prove

<u>Proposition</u> 4.4: Let $K \trianglelefteq G$ be a p-group which acts freely on S and put $\bar{G} = G/K$, $\bar{S} = K \backslash S$. If for any two p-groups $D_1, D_2 \in \mathfrak{U}(S)$ with $D_1K \subseteq D_2K$ there exists $k \in K$ with $kD_1k^{-1} \subseteq D_2$,[1)] then

$$\mathrm{inf}' : a'_{\bar{S}}(\bar{G}) \to a'_S(G)$$

is an isomorphism.

<u>Proof</u>: Since $\varprojlim \mathrm{inf}'$ is well defined anyhow, it is enough to verify

(a) any $\bar{H} \in \mathfrak{H}(\bar{S})$ is of the form HK/K for some $H \in \mathfrak{H}(S)$;

(b) for any $F, H \in \mathfrak{H}(S)$ with $FK \subseteq HK$ there exists $k \in K$ with $kFk^{-1} \subseteq H$,

to conclude that $\varprojlim \mathrm{inf}'$ is an isomorphism.

(a) Let $\bar{H} \in \mathfrak{H}(\bar{S})$ and let H' be the full pre-image of $\bar{H} \leq \bar{G}$ in G. Since K is a p-group, H' has a normal p-Sylow-subgroup H'_p containing K and $H'_p/K = \bar{H}_p \in \mathfrak{U}(\bar{S})$. Thus there exists a p-group $D \in \mathfrak{U}(S)$ with $H'_p = DK$ (choose $s \in \bar{s} = k \cdot s \in \bar{S}^{\bar{H}_p}$ and put $D = (H'_p)_s$). Since $H' \subseteq N_G(H'_p) = N_G(DK) =$

---

1) cf. (*) in §2.

$N_G(D) \cdot K(gDKg^{-1} = DK$ implies $kgDg^{-1}k^{-1} = D$ for some $k \in K$ by our hypothesis!) one has $H' = H''K$ with $H'' = N_G(D) \cap H'$.

Let E be a p-complement of the normal p-Sylow-subgroup $H''_p \trianglelefteq H''$. Then $H = ED \in \mathfrak{H}(S)$ and $HK = EDK = EH'_p = EH''_pK = H''K = H'$, thus $HK/K = H'/K = \bar{H}$.

q.e.d.

(b) Let F_p and H_p denote the normal p-Sylow-subgroups of F, $H \in \mathfrak{H}(S)$ respectively, thus $F_p, H_p \in \mathfrak{A}(S)$ and $F_pK = (FK)_p \subseteq (HK)_p = H_pK$. So there exists $k \in K$ with $kF_pk^{-1} \subseteq H_p$ and w.l.o.g. we may assume $k = 1$, i.e., $F_p \subseteq H_p$.

Now let $F' \leq F$ denote a p-complement of $F_p \trianglelefteq F$ in F. There exists a p-complement $\bar{H}'$ of $\bar{H}_p = H_pK/K$ in $\bar{H} = HK/K \cong H$ containing $\bar{F}' = F'K/K$. Thus if we denote by H' the full pre-image of $\bar{H}'$ in H, then H' is a p-complement of $H_p \trianglelefteq H$ in H with $F'K \subseteq H'K$. From this one gets $F'K = (F'K \cap H')K$, i.e., F' and $F'K \cap H'$ both are p-complements of K in $F'K$. Moreover $F' \subseteq N_G(F_p)$ and $F'K \cap H' \subseteq N_G(H_p) \cap FK \subseteq N_G(H_p \cap FK) = N_G(F_p)$, thus F' and $F'K \cap H'$ are already p-complements of $K \cap N_G(F_p)$ in $F'K \cap N_G(F_p) = F' \cdot (K \cap N_G(F_p))$. So there exists $k \in K \cap N_G(F_p)$ with $kF'k^{-1} = F'K \cap H' \subseteq H'$, which implies $kFk^{-1} = (kF'k^{-1})(kF_pk^{-1}) = (F'K \cap H') \cdot F_p \subseteq H' \cdot H_p = H$.

q.e.d.

Remark. The condition stated in Proposition 4.4 is definitely not necessary: Let G be the nonabelian group of exponent p and order p^3 $(p \neq 2)$; let $K = Z(G)$ be the center of G; let $H \leq G$ be any subgroup of order p not equal to K and put $S = G/H$, $\bar{G} = G/K$, $\bar{H} = HK/K \cong H$ and $\bar{S} = K\backslash S \cong \bar{G}/\bar{H}$. Then the commutative diagram

$$\begin{array}{ccc} a'_{\bar{S}}(\bar{G}) & \xrightarrow{\sim} & a'(\bar{H}) \\ {\scriptstyle \inf'}\downarrow & & \downarrow\wr \\ a'_S(G) & \xrightarrow{\sim} & a'(H) \end{array}$$

shows that inf' is an isomorphism, whereas for $H \in \mathfrak{U}(S)$ one has $N_G(HK) = G \neq N_G(H)K = HK$.

So one may replace Proposition 4.4 by the following result, which follows more or less directly from the "$\varprojlim$"-description of the relative Grothendieck rings and also covers the just mentioned case, but might generally be harder to check.

<u>Proposition</u> 4.4': Let $K \trianglelefteq G$, $S, \bar{G}$ and $\bar{S}$ be as in Proposition 4.4 and assume that any $\bar{H} \in \mathcal{H}(\bar{S})$ $(\bar{C} \in \mathcal{C}(\bar{S}))$ is of the form $\bar{H} = HK/K$ for some $H \in \mathcal{H}(S)$ $(\bar{C} = CK/K$ for some $C \in \mathcal{C}(S))$. Moreover, assume that for any $H_1, H_2 \in \mathcal{H}(S)$ with $H_1K \subseteq H_2K$ $(C_1, C_2 \in \mathcal{C}(S)$ with $C_1K \subseteq C_2K)$ there exists some $g \in G$ with $gH_1g^{-1} \subseteq H_2$ $(gC_1g^{-1} \subseteq C_2)$ such that the diagram

$$\begin{array}{ccc} a'(\bar{H}_2) & \xrightarrow{\mathrm{res}(\bar{H}_1, 1, \bar{H}_2)} & a'(\bar{H}_1) \\ \downarrow\wr & & \downarrow\wr \\ a'(H_2) & \xrightarrow{\mathrm{res}(H_1, g, H_2)} & a'(H_1) \end{array}$$

$$\begin{array}{cccc} (\text{resp.} & A(\bar{C}_2) & \xrightarrow{\mathrm{res}(\bar{C}_1, \bar{1}, \bar{C}_2)} & A(\bar{C}_1)) \\ & \downarrow\wr & & \downarrow\wr \\ & A(C_2) & \xrightarrow{\mathrm{res}(C_1, g, C_2)} & A(C_1) \end{array}$$

commutes (with $\bar{H}_i = H_iK/K \cong H_i$, $\bar{C}_i = C_iK/K \cong C_i$).

Then

$$\inf' : a'_{\bar{S}}(\bar{G}) \to a'_S(G)$$

$$(\text{resp. } \operatorname{Inf} : A_{\bar{S}}(\bar{G}) \to A_S(G))$$

is an isomorphism.

I do not know whether the conditions given in Proposition 4.4' are also necessary. They are at least rather close to being necessary since for instance the surjectivity of $A_{\bar{S}}(\bar{G}) \to A_S(G)$ implies that the $\mathbb{Q}$-subalgebra $B_{\bar{S}}(\bar{G})$ of $A_{\bar{S}}(\bar{G})$ generated by the permutation representations $\Omega[\bar{T}]$ of $\bar{G}$ maps surjectively onto the correspondingly defined $\mathbb{Q}$-subalgebra $B_S(G) \subseteq A_S(G)$ which in turn is equivalent with the condition that any $C, C' \in \mathcal{C}(S)$ with $CK = C'K$ are conjugate in G (as can be deduced from the theory of Burnside rings, cf. [4]).

Let us now consider the product isomorphism theorem or rather its special form as stated at the end of §2, from which the general theorem can be deduced by using the excision theorem (cf. [15]). So let $G = G_1 \times G_2$ and $S = S_1 \times S_2$ with S_i a G_i-set. There is an obvious map $t : a(G_1) \otimes a(G_2) \to a(G) : (M_1, M_2) \mapsto M_1 \otimes M_2$ which is well known to be injective in general and surjective if, say, G_2 is p-regular and Ω is a splitting field for G_2. Moreover, it induces a map $t_{S_1,S_2} : a_{S_1}(G_1) \otimes a_{S_2}(G_2) \to a_S(G)$.

<u>Proposition</u> 4.5: t_{S_1,S_2} induces always an injective map

$$t'_{S_1,S_2} : a'_{S_1}(G_1) \otimes a'_{S_2}(G_2) \to a'_S(G).$$

Moreover, t'_{S_1,S_2} is an isomorphism if $\mathfrak{U}(S_2)$ contains no p-groups (except 1) and Ω is a splitting field for any p-regular subgroups of G_2.

Proof: Since the projections $H_i \leq G_i$ of any $H \in \mathcal{U}(S)$ are in $\mathcal{U}(S_i)$ one may identify $a'_S(G)$ with $\varprojlim_{(H_1, H_2) \in \mathcal{U}(S_1) \times \mathcal{U}(S_2)} a'(H_1 \times H_2)$ and thus derive the result from the commutative diagram

$$\begin{array}{ccc} a'_{S_1}(G_1) \otimes a'_{S_2}(G_2) & \longrightarrow & a'_S(G) \\ \downarrow \wr & & \downarrow \wr \\ \varprojlim_{H_1 \in \mathcal{U}(S_1)} a'(H_1) \otimes \varprojlim_{H_2 \in \mathcal{U}(S_2)} a'(H_2) & \rightarrow & \varprojlim_{(H_1, H_2) \in \mathcal{U}(S_1) \times \mathcal{U}(S_2)} a'(H_1 \times H_2). \end{array}$$

The stronger form of the product isomorphism theorem can now be derived by just following the proof given by Reiner and Lam in [15].

Finally I want to discuss a few obvious conjectures. The most basic one still is the conjecture of Reiner and Lam that $a_S(G)$ is $\mathbb{Z}$-free. By our results this is equivalent to either one of the following statements:

(i) $a_S(G)$ is torsion free;

(ii) $a_S(G) \to \prod_{C \in \mathcal{C}(S)} a(C)$ is injective.

It would imply the conjecture following Corollary 4 to Proposition 4.1 as well as the following statements:

(iii) If $H \trianglelefteq G$, $(G : H)$ a power of p and $G_s \leq H$ for $s \in S$, then $a_S(G) \to a_S(H)$ is injective (cf. [15], p. 162, (4.6) as well as [8], p. 101; actually the conjecture stated there can't hold unless "$a(E, H)$" is replaced by "$a_{G/H}(E)$" since H may be G-conjugate to some $H' \leq E$ without being E-conjugate).

(iv) If G is a p-group, then

$$a_S(G) \to \prod_{s \in S'} a(G_s)$$

(with S' a complete system of representatives of the G-orbits in S) is injective.

(v) If $K \trianglelefteq G$, $\bar{G} = G/K$ and $\bar{S} = K\backslash S$ then $\inf : a_{\bar{S}}(\bar{G}) \to a_S(G)$ is injective if and only if $\text{Inf} : A_{\bar{S}}(\bar{G}) \to A_S(G)$ is injective. Particularly $\inf : a_{\bar{S}}(\bar{G}) \to a_S(G)$ is always injective if G is a p-group.

(vi) $t_{S_1,S_2} : a_{S_1}(G_1) \otimes a_{S_2}(G_2) \to a_S(G)$ as defined just above Proposition 4.5 is always injective.

(vii) (cf. [8], p. 104) If $\tilde{\Omega}$ is a field containing Ω and if $\tilde{a}_S(G)$ denotes the Grothendieck ring of $\tilde{\Omega}G$-modules relative to S, then $a_S(G) \to \tilde{a}_S(G)$: $M \mapsto \tilde{\Omega} \otimes_{\Omega} M$ is injective.

It seems worthwhile to study these conjectures even if (i) is not true. Similarly one may conjecture that the hypothesis of Proposition 4.4 implies that even $\inf : a_{\bar{S}}(\bar{G}) \to a_S(G)$ is an isomorphism.

The induction methods we have employed in this paper won't help much to study these refined questions, so probably one would have to extend the more module theoretic methods used by Reiner and Lam.

Still Proposition 3.3 together with the first step of the proof of the induction theorem established in §5 and the surjectivity of $\sum_{C \in \mathcal{C}(S)} A_S(C) \to A_S(G)$ imply the surjectivity of

$$\sum_{H \in \mathfrak{H}_1(S)} a_S(H) \to a_S(G)$$

with $\mathfrak{H}_1(S) = \mathfrak{H}(S) \cup \{H \leq G \mid H^{(p)} \in \mathcal{C}(S)\}$ and thus the injectivity of

$$a_S(G) \to \prod_{H \in \mathfrak{H}_1(S)} a_S(H).$$

Thus $a_S(G)$ is torsion free if $a_S(H)$ is torsion free for any $H \leq G$ with $H^{(p)} \in \mathcal{C}(S)$. (This can be used for instance to prove that $a_S(A_4)$ and $a_T(S_4)$ are torsion free with $S = A_4/Z_2$, $T = S_4/Z_2$, Z_2 some subgroup of order 2 in A_4, and Ω a field of characteristic 2, cf. the discussion at the end of [13].) So after all the case of p-groups may already be the most difficult one and is definitely the first one, one should study in detail.

§5. Proof of the Induction Theorem.

We are now going to prove the induction theorem stated at the end of §3. We will proceed in several steps, freely using results and methods from [4].

1st Step: It is enough to prove the surjectivity of $\sum_{C \in \mathcal{C}(S)} A(C) \to A_S(G)$, since this implies the surjectivity of $\sum_{H \in \mathcal{H}(S)} a'(H) \to a'_S(G)$ by Proposition 3. 3 and by

Lemma 3: Any torsion element in $a_S(G)$ is nilpotent.

Proof: Using a lemma due to G. Segal (cf. [4], §9, Lemma 9.6) it is enough to show that $a_S(G)$ is a (not necessarily special) λ-ring[1]. But $a(G)$ is well known to be a λ-ring, whose λ-structure is induced from exterior powers of G-modules, i.e., $\lambda_i([M]) = [\Lambda^i(M)]$, $\Lambda^i(M)$ the i-th exterior power of the Ω-vector space M, considered as a G-module in the obvious way. So it remains to check that $\lambda_i(\chi_{\mathcal{E}})$ is contained in $i_S(G)$ for any S-split sequence $\mathcal{E}: 0 \to M' \to M \to M'' \to 0$. This can be done straight forwardly by using the fact that $\Lambda^i(M)$ has a natural filtration $0 = F_1(\Lambda^i(M)) \subseteq F_0(\Lambda^i(M)) \subseteq \ldots \subseteq F_i(\Lambda^i(M)) = \Lambda^i(M)$ defined by $F_j(\Lambda^i(M)) = \langle x_i \wedge \ldots \wedge x_i \mid x_i \in M$ and $x_i \notin M'$ for at most j indices "i"$\rangle$ and that this filtration is "S-split" (i.e., $0 \to F_{j-1}(\Lambda^i(M)) \to F_j(\Lambda^i(M)) \to F_j(\Lambda^i(M))/F_{j-1}(\Lambda^i(M)) \to 0$ is S-split for $j = 1, \ldots, i$), as can be seen

[1] For the convenience of the reader let us recall Segal's proof that any torsion element in any λ-ring is nilpotent: R is λ-ring if it is endowed with a homomorphism $\lambda_t : R \to R[[t]]$ (formal power series) with $\lambda_t(r+r') = \lambda_t(r) \cdot \lambda_t(r')$ and $\lambda_t(r) = 1+rt+\ldots$. Now assume $p^n \cdot r = 0$ for some $r \in R$. Then $1 = \lambda_t(p^n r) = \lambda_t(r)^{p^n} = (1+rt+\ldots)^{p^n} \equiv 1+r^{p^n} \cdot t^{p^n} + \ldots \pmod{p^R}$, thus there exists $r' \in R$ with $r^{p^n} = pr'$ which implies $r^{(p^n+1)\cdot n} = (r^{p^n+1})^n = (prr')^n = p^n \cdot r \cdot r^{n-1} \cdot r'^n = 0.$

most easily by using Proposition 1.2.

2nd Step: By [4], §9, Proposition 9.1, it is known that for any finite group H

$$\sum_{C \in \mathcal{C}(H/H)} A(C) \to A(H)$$

is surjective (see also the remark at the end of this section).

Thus by the transitivity of induction it is sufficient to show that $\sum_{H \leq G,\, H_p \in \mathfrak{U}(S)} A(H) \to A_S(G)$ is surjective, since obviously $\mathcal{C}(S) = \bigcup_{H \leq G,\, H_p \in \mathfrak{U}(S)} \mathcal{C}_H(H/H)$.

3rd Step: Let $\mathfrak{D}$ denote the class of all pairs (H, T), where H is a finite group, T an H-set and

$$\sum_{U \lneq H} A_T(U) \to A_T(H)$$

is <u>not</u> surjective. Transitivity of induction implies easily that

$$\sum_{H \leq G,\, (H, S|_H) \in \mathfrak{D}} A_S(H) \to A_S(G)$$

is surjective for any finite group G (since either (G, S) is in $\mathfrak{D}$, in which case there is nothing to prove, or $\sum_{U \lneq G} A_S(U) \to A_S(G)$ is surjective, in which case one is reduced to the groups U, which have a smaller order). So it remains to show

(***) $$(H, T) \in \mathfrak{D} \Rightarrow H_p \in \mathfrak{U}(T).$$

But this follows easily from the following facts:

(I) $(H, T) \in \mathfrak{D},\ U \leq H \Rightarrow (U, T|_U) \in \mathfrak{D}$,

(II) $(H, T_1 \dot{\cup} T_2) \in \mathfrak{D} \Rightarrow (H, T_1) \in \mathfrak{D}$ or $(H, T_2) \in \mathfrak{D}$,

(III) $(H, T) \in \mathfrak{D},\ K \trianglelefteq H \Rightarrow (H/K, K \backslash T) \in \mathfrak{D}$,

(IV) $(Z_p, Z_p/1) \notin \mathfrak{D}$.

Proof of (***): By (I) $(H, T) \in \mathfrak{D}$ implies $(H_p, T|_{H_p}) \in \mathfrak{D}$ and by (II) this in turn implies $(H_p, H_p/U) \in \mathfrak{D}$ for some $U \in \mathfrak{N}(T)$. If $U = H_p$, we are through. Otherwise there exists a maximal subgroup K with $U \leq K \trianglelefteq_{\neq} H_p$ and by (III) implies $(H_p/K, K\backslash H_p/U) \cong (Z_p, Z_p/1) \in \mathfrak{D}$, a contradiction to (IV).

4th Step: We still have to verify (I)-(IV). (IV) is obvious, since for any p-group H one has $A_{H/1}(H) \cong \mathbb{Q}$ and the induction-map never is zero. (III) follows obviously from the existence of an inflation map $a_{\bar{T}}(\bar{H}) \to a_T(H)$ (with $\bar{H} = H/K$, $\bar{T} = K\backslash T$), the commutativity of

$$\begin{array}{ccc} \sum\limits_{K \leq U \subsetneq H} A_{\bar{T}}(U/K) & \to & A_{\bar{T}}(\bar{H}) \\ \downarrow & & \downarrow \\ \sum\limits_{K \leq U \subsetneq H} A_T(U) & \to & A_T(H) \end{array}$$

and the fact that the lower horizontal arrow (as any induction map) is surjective if and only if $1_{A_T(H)}$ is in its image, since by the Frobenius reciprocity law (see §3, (P2/3)) the image is always an ideal.

To verify (II) let us first observe that for any two exact sequences $\mathcal{E}_j : 0 \to M_j' \to M_j \to M_j'' \to 0$ $(j = 1, 2)$ with $\mathcal{E}_j$ being T_j-split the tensor product $\mathcal{E}_1 \otimes \mathcal{E}_2 : 0 \to M_1' \otimes M_2' \to M_1' \otimes M_2 \oplus M_1 \otimes M_2' \to$

$\to M_1' \otimes M_2'' \oplus M_1 \otimes M_2 \oplus M_1'' \otimes M_2' \to M_1 \otimes M_2'' \oplus M_1'' \otimes M_2 \to M_1'' \otimes M_2'' \to 0$

splits over T_1 as well as over T_2 and thus over T, so its Euler-characteristic $\chi\mathcal{E}_1 \otimes \mathcal{E}_2 = \chi\mathcal{E}_1 \cdot \chi\mathcal{E}_2$ vanishes in $a_T(H)$, i.e., one has

$$i_{T_1}(H) \cdot i_{T_2}(H) \subseteq i_T(H).$$

Now let L denote the image of $\sum\limits_{U \subsetneq H} A(U) \to A(H)$--so L is an ideal in A(H). If $(H, T_j) \notin \mathfrak{D}$ $(j = 1, 2)$, then $L + \mathbb{Q} \otimes i_{T_j}(H) = A(H)$ for $j = 1, 2$ and thus one has

$A(H) = (L + \mathbb{Q} \otimes i_{T_1}(H))(L + \mathbb{Q} \otimes i_{T_2}(H)) = L + \mathbb{Q} \otimes i_{T_1}(H) \cdot i_{T_2}(H) \subseteq$ $L + \mathbb{Q} \otimes i_T(H)$, i.e., $(H, T) \notin \mathfrak{D}$, which proves (II).

(I) will be verified in the last step.

5th Step: Let $\bar{A}_T(H) = A_T(H)/\mathrm{Im}(\sum_{U \lneqq H} A_T(U) \to A_T(H))$, thus $\bar{A}_T(H) \neq 0$ if and only if $(H, T) \in \mathfrak{D}$. It is enough to define a ring homomorphism $\mu_T : \bar{A}_T(U) \to \bar{A}_T(H)$ (which maps 1_U onto 1_H), since the mere existence of such a ring homomorphism implies

$$(H, T) \in \mathfrak{D} \iff \bar{A}_T(H) \neq 0 \iff 1_H \neq 0_H \Rightarrow 1_U \neq 0_U \iff \bar{A}_T(U) \neq 0 \iff (U, T|_U) \in \mathfrak{D}.$$

To construct μ_T we use the technique of multiplicative induction which was used already in [4], §8 for similar purposes. So let N be a U-module and observe that the induced module $\Omega H \underset{\Omega U}{\otimes} N$ splits naturally into a direct sum of blocks $N_x = h \otimes N$ $(x = hU \in H/U)$ which are permuted by H. Thus the tensor product $\mu(N) = \bigotimes_{x \in H/U} N_x$ is endowed with a natural G-module structure. One has obviously

$$\mu(\Omega) \cong \Omega \text{ (}\Omega\text{ the trivial U-, resp. H-module)}$$

as well as

$$\mu(N' \underset{\Omega}{\otimes} N'') \cong \mu(N') \underset{\Omega}{\otimes} \mu(N''),$$

but of course μ is not additive. Still $\mu(N' \oplus N'') = \bigotimes_{x \in H/U} (N' \oplus N'')_x$ $= \bigoplus_{j=0}^{(H:U)} \mu_j(N', N'')$ with $\mu_j(N', N'') = \bigoplus_{X \subseteq H/U, |X| = j} (\bigotimes_{x \in X} N'_x \otimes \bigotimes_{y \notin X} N''_y)$ splits into a direct sum of H-modules $\mu_j(N', N'')$ with $\mu_0(N', N'') = \mu(N'')$, $\mu_{(H:U)}(N', N'') = \mu(N')$, whereas for $0 < j < (H:U)$ the H-module $\mu_j(N', N'')$ is a direct sum of blocks $\mu_j(N', N'')_X = \bigotimes_{x \in X} N'_x \otimes \bigotimes_{y \notin X} N''_y$, where X runs through the H-set of all subsets of H/U with precisely j elements and the blocks

$\mu_j(N', N'')_X$ are permuted by H according to the way H acts on the indices X. Thus any $\mu_j(N', N'')$ $(0 < j < (H:U))$ is a direct sum of modules which are induced from the various $H_X = \{h \in H \mid hX = X\}$ all of which are proper subgroups of H, since no $X \subseteq H/U$ with $\phi \neq X \neq H/U$ can be fixed by every $h \in H$. Thus μ induces a ring homomorphism

$$A(U) \to \bar{A}(H) = A(H)/\mathrm{Im}(\sum_{V \lneqq H} A(V) \to A(H)).$$

Using a very similar argument one can as well prove that this ring homomorphism factors through $A(U) \to \bar{A}(U) = A(U)/\mathrm{Im}(\sum_{W \lneqq U} A(W) \to A(U))$ and thus gives rise to a ring homomorphism (cf. [4], §8):

$$\mu : \bar{A}(U) \to \bar{A}(H).$$

It remains to be seen that μ induces a ring homomorphism

$$\mu_T : \bar{A}_T(U) \to \bar{A}_T(H).$$

So let $\mathcal{E} : 0 \to N' \to N \to N'' \to 0$ be a T-split sequence of U-modules. $\mu(N)$ has a natural filtration

$$\mathcal{F} : 0 = F_{-1}(\mu(N)) \subseteq F_0(\mu(N)) \subseteq \dots \subseteq F_{(H:U)}(\mu(N)) = \mu(N)$$

with $F_j(\mu(N)) = \langle \bigotimes_{x \in H/U} n_x \mid n_x \notin N'$ for at most j indices $x \in H/U \rangle$ and thus

$$F_j(\mu(N))/F_{j-1}(\mu(N)) \cong \mu_j(N'', N') \cong \mu_{(H:U)-j}(N', N'').$$

Once we can show that this filtration is T-split, we are done, since this implies

$$\mu(N) = \sum_{j=0}^{(G:H)} \mu_j(N', N'') = \mu(N' \oplus N'') \text{ in } A_T(H) \text{ and thus } \mu(\chi_{\mathcal{E}}) = 0 \text{ in } \bar{A}_T(H).$$

So let us show that there exists a right inverse

$$\Psi_j : \mu_j(N'', N')[T] \to F_j(\mu(N))[T]$$

to the obvious map

$$\lambda_j[T] : F_j(\mu(N))[T] \to \mu_j(N'', N')[T].$$

We need the following simple

Lemma 4: $\mathscr{C}: 0 \to N' \to N \xrightarrow{\lambda} N'' \to 0$ is a T-split sequence of U-modules, if and only if there exist Ω-linear maps $\psi_t : N'' \to N$ $(t \in T)$ all of which are right inverse to λ and satisfy $u(\psi_t(n'')) = \psi_{ut}(un'')$ for any $u \in U$, $t \in T$ and $n'' \in N''$.

Proof: If the ψ_t $(t \in T)$ are given, then

$$\psi : N''[T] \to N[T] : \Sigma n_t''t \mapsto \Sigma \psi_t(n_t'') \cdot t$$

is a U-homomorphism which is right inverse to $\lambda[T]$:

$$\psi(u(\Sigma n_t''t)) = \psi(\Sigma un_t''ut) = \Sigma\psi_{ut}(un_t'')ut = \Sigma u\psi_t(n_t'')ut = u\psi(\Sigma n_t''t),$$

$$\lambda[T] \circ \psi(\Sigma n_t''t) = \lambda[T](\Sigma\psi_t(n_t'')t) = \Sigma\lambda \circ \psi_t(n_t'')t = \Sigma n_t''t.$$

On the other hand, if $\psi : N''[T] \to N[T]$ is right inverse to $\lambda[T]$, then

$$\psi_t : N'' \to N : n'' \mapsto \psi_t(n'') = \text{coeff. of } t \text{ in } \psi(n''t)$$

is right inverse to λ and the ψ_t satisfy $u\psi_t(n'') = u(\text{coeff. of } t \text{ in } \psi(n''t)) =$ coeff. of ut in $u\psi(n''t) =$ coeff. of ut in $\psi(un'' \cdot ut) = \psi_{ut}(un'')$.

Now to construct Ψ_j let $\psi_t : N'' \to N$ $(t \in T)$ be given as in Lemma 4. We want to construct well-defined maps $\psi_{t,x} : N_x'' \to N_x$ $(t \in T)$ for any $x \in H/U$. So let $h \in x = hU$, put $N_x'' = h \otimes N''$, $N_x = h \otimes N''$ and define

$$\psi_{t,h} : N_x'' = h \otimes N'' \to N_x = h \otimes N : h \otimes n'' \mapsto h \otimes \psi_{h^{-1}t}(n'').$$

Actually, $\psi_{t,h}$ does not depend on the choice of $h \in x$, since $\psi_{t,hu}(h \otimes n'') = \psi_{t,hu}(hu \otimes u^{-1}n'') = hu \otimes \psi_{u^{-1}h^{-1}t}(u^{-1}n'') = hu \otimes u^{-1}\psi_{h^{-1}t}(n'') = \psi_{t,h}(h \otimes n'')$. Thus we may write $\psi_{t,x}$ instead of $\psi_{t,h}$ if $x = hU \in H/U$. The $\psi_{t,x}$ satisfy the equation

$$g\psi_{t,x}(n''_x) = \psi_{gt,gx}(gn''_x) \quad (g \in H,\ t \in T,\ x \in H/U,\ n''_x \in N''_x)$$

and are right inverse to

$$\lambda_x : N_x = h \otimes N \to N''_x = h \otimes N'' : h \otimes n \mapsto h \otimes \lambda(n)$$

(which again does not depend on the choice of $h \in x$).

Now we can define

$$\Psi_j : \mu_j(N'', N')[T] \to F_j(\mu(N))$$

by

$$\Psi_j((\bigotimes_{x \in X} n''_x \otimes \bigotimes_{y \notin X} n'_y)t) = \bigotimes_{x \in X} \psi_{t,x}(n''_x) \otimes \bigotimes_{y \notin X} n'_y)t$$

and it is straightforward to check that Ψ_j is a G-homomorphism and right inverse to $\lambda_j[T]$. q. e. d.

One could give a perhaps less technical proof by introducing the concept of equivariant ΩH-bundles ("H-bundles" for short) which is implicit in the proof of (I) anyhow. For any H-set S let $\underline{\underline{S}}$ denote the category, whose objects are the elements $s \in S$ with morphisms $[s, s']_{\underline{\underline{S}}} = \{(s', h, s) | h \in H,\ s' = hs\}$, composed by $(s'', h', s') \circ (s', h, s) = (s'', h'h, s)$. A covariant functor $\zeta : \underline{\underline{S}} \to \underline{\underline{\Omega\text{-mod}}}$ from $\underline{\underline{S}}$ into the category of finite dimensional Ω-vector spaces is called an H-bundle over S. It consists of an indexed family ζ_s $(s \in S)$ of vector spaces--the fibers--and an indexed family of maps $\zeta_{(s', h, s)} : \zeta_s \to \zeta_{hs} = \zeta_{s'}$, which compose properly. Let $[\underline{\underline{S}}, \underline{\underline{\Omega\text{-mod}}}] = \widetilde{S}$ denote the category of all such functors. $\widetilde{S}$ is a nice abelian category endowed with an internal tensor product which is defined fiberwise: $(\zeta \otimes \zeta')_s = \zeta_s \otimes_\Omega \zeta'_s$. Thus one may form the Grothendieck ring $K_H(S)$ of isomorphism classes of $\widetilde{S}$.

Any H-map $\varphi : S' \to S$ between two H-sets S', S gives rise to a functor $\underline{\underline{\varphi}} : \underline{\underline{S'}} \to \underline{\underline{S}}$ and thus to a functor $\widetilde{\varphi} : \widetilde{S} \to \widetilde{S}'$, the restriction functor, which induces a ring homomorphism $\mathcal{G}_* : K_H(S) \to K_H(S')$. Moreover $\mathcal{G}$ induces two functors from $\widetilde{S'}$ to $\widetilde{S}$, one given by

$$\mathcal{G}^{\oplus} : \widetilde{S'} \to \widetilde{S} : \zeta' \mapsto \mathcal{G}^{\oplus}(\zeta')$$

with $\mathcal{G}^{\oplus}(\zeta')_s = \bigoplus_{s' \in \mathcal{G}^{-1}(s)} \zeta'_{s'}$, the other one given by:

$$\mathcal{G}^{\otimes} : \widetilde{S'} \to \widetilde{S} : \zeta' \mapsto \mathcal{G}^{\otimes}(\zeta')$$

with $\mathcal{G}^{\otimes}(\zeta')_s = \bigotimes_{s' \in \mathcal{G}^{-1}(s)} \zeta'_{s'}$.

For any pull back diagram

$$\begin{array}{ccc} S_0 & \xrightarrow{\Phi} & S'' \\ \Psi \downarrow & & \downarrow \psi \\ S' & \xrightarrow[\mathcal{G}]{} & S \end{array}$$

one has two diagrams commuting up to canonical equivalences:

$$\begin{array}{ccc} \widetilde{S}_0 & \xrightarrow{\Phi^{\oplus}} & \widetilde{S''} \\ \widetilde{\Psi} \Big| & & \Big| \widetilde{\psi} \\ \widetilde{S'} & \xrightarrow[\mathcal{G}^{\oplus}]{} & \widetilde{S} \end{array}$$

and

$$\begin{array}{ccc} \widetilde{S}_0 & \xrightarrow{\Phi^{\otimes}} & \widetilde{S''} \\ \widetilde{\Psi} \Big| & & \Big| \widetilde{\psi} \\ \widetilde{S'} & \xrightarrow[\mathcal{G}^{\otimes}]{} & \widetilde{S} \end{array} .$$

In the particular case $S = H/U$ one has canonical equivalences

$$\widetilde{S} \rightleftarrows \underline{\underline{\text{U-mod}}},$$

the upper arrow given by $\zeta \mapsto \zeta_U$, ζ_U the fiber over $U \in H/U = S$, considered as U-module, the lower arrow given by $N \mapsto H \times_U N$ with

$(H \times_U N)_s = N_s = N_{hu} = h \otimes N$ $(s = hU \in H/U)$, the block in $\Omega H \underset{\Omega U}{\otimes} N$ corresponding to $s = hU$. Thus $K_H(H/U) \cong a(U)$.

Using this identification one gets commutative diagrams with respect to the H-map $\varphi : H/U \to H/H$

$$\begin{array}{ccc} \widetilde{H/H} & \xrightarrow{\sim} & \underline{\underline{H\text{-mod}}} \\ \tilde{\varphi}\downarrow & & \downarrow \text{res} \\ \widetilde{H/U} & \xrightarrow{\sim} & \underline{\underline{U\text{-mod}}} \end{array} ,$$

$$\begin{array}{ccc} \widetilde{H/U} & \xrightarrow{\sim} & \underline{\underline{U\text{-mod}}} \\ \varphi^{\oplus}\downarrow & & \downarrow \text{ind} \\ \widetilde{H/H} & \xrightarrow{\sim} & \underline{\underline{H\text{-mod}}} \end{array}$$

and

$$\begin{array}{ccc} \widetilde{H/U} & \xrightarrow{\sim} & \underline{U\text{-mod}} \\ \varphi^{\otimes}\downarrow & & \downarrow \mu \\ \widetilde{H/H} & \xrightarrow{\sim} & \underline{H\text{-mod}} \end{array} .$$

Moreover, by Lemma 4 it is obvious that a sequence $0 \to M' \to M \to M'' \to 0$ of H-modules--i.e., H-bundles over H/H--is T-split if and only if the sequence restricted with respect to $\varphi_T : T \to H/H$ is split. More generally, a sequence of U-modules $0 \to N' \to N \to N'' \to 0$ is T-split if and only if the restriction of the sequence $0 \to H \times_U N' \to H \times_U N \to H \times_U N'' \to 0$ of H-bundles over H/U to $T \times H/U$ with respect to the projection $T \times H/U \to H/U$ is split (cf. the construction of $\psi_{t,x}$). Thus we may define relative Grothendieck rings $K_H(S;T)$ of H-bundles over S with respect to exact sequences of such bundles which split over $T \times S$, and obtain

$$K_H(H/U, T) \cong a_T(U).$$

Now to prove that the filtration $0 = F_{-1}(\mu(N)) \subseteq F_0(\mu(N)) \subseteq \dots \subseteq F_{(H:U)}(\mu(N)) = \mu(N)$

is T-split, we replace the T-split sequence $0 \to N' \to N \to N'' \to 0$ of U-modules by the corresponding sequence

$$\mathscr{G} : 0 \to \zeta' \to \zeta \to \zeta'' \to 0$$

of H-bundles over H/U, which splits over $T \times H/U$, and consider the commutative diagram

$$\begin{array}{ccc} \widetilde{H/U \times T} & \xrightarrow{\Phi^{\otimes}} & \widetilde{T} \\ \widetilde{\Psi} \uparrow & & \uparrow \widetilde{\psi} \\ \widetilde{H/U} & \xrightarrow[\mathscr{G}^{\otimes}]{} & \widetilde{H/H} \end{array},$$

induced from the pull back diagram

$$\begin{array}{ccc} H/U \times T & \xrightarrow{\Phi} & T \\ \Psi \downarrow & & \downarrow \psi \\ H/U & \xrightarrow[\mathscr{G}]{} & H/H \end{array}.$$

The result follows from the following observations:

If $\mathscr{G} : S' \to S$ is an H-map and $0 \to \zeta' \to \zeta \to \zeta'' \to 0$ an exact sequence of H-bundles over S', then $\mathscr{G}^{\otimes}(\zeta)$ is endowed with a natural filtration $\{F_j(\mathscr{G}^{\otimes}(\zeta))\}_{j=-1,0,\ldots}$ which is split, whenever $0 \to \zeta' \to \zeta \to \zeta'' \to 0$ is split (since $F_j(\mathscr{G}^{\otimes}(\zeta)) = \bigoplus_{\ell \leq j} \mathscr{G}_j^{\otimes}(\zeta'', \zeta')$, using an obvious definition of $\mathscr{G}_j^{\otimes}(\ldots)$), and which behaves naturally with respect to restriction, i.e., for any H-set T the diagram

$$\begin{array}{ccc} T \times S' & \xrightarrow{\Phi} & T \times S \\ \Psi \downarrow & & \downarrow \psi \\ S' & \xrightarrow[\mathscr{G}]{} & S \end{array}$$

induces a natural isomorphism

$$F_j(\Phi^{\otimes}(\widetilde{\Psi}(\zeta))) \cong \widetilde{\psi}(F_j(\mathscr{G}^{\otimes}(\zeta))).$$

Remark: Let us finally remark that concerning the 2nd step, it is not difficult to prove the quoted result once (I) and (III) are established together with

(V) $(Z_q \times Z_q, \bullet) \notin \mathfrak{B}$ for $q \neq p$,

(VI) $(Z_q \rtimes Z_\ell, \bullet) \notin \mathfrak{B}$ for $q \neq p$, ℓ any prime with $Z_\ell \leq \mathrm{Aut}(Z_q)$,

($\bullet$ the trivial one-point-set),

what may be done by straightforward computations (cf. [4], §9). (I), (III) and (V) imply that any q-Sylow-subgroup of any H with $(H, \bullet) \in \mathfrak{B}$ must be cyclic $(q \neq p)$ and then (I), (III) and (VI) together with Frobenius's transfer theorem imply that H must be q-nilpotent $(q \neq p)$. Thus H_p, being the intersection of all normal q-complements $(q \neq p)$, must be normal and H/H_p, being q-nilpotent for any $q \mid |H/H_p|$ and thus nilpotent with cyclic q-Sylow-subgroups, must be cyclic.

Appendix, concerning multiplicative G-functors.

We still have to prove Propositions 3.1 and 3.3. So let us start by recalling the definition of the Burnside functor. For any $U \leq G$ let $\hat{U}$ denote the category of U-sets and let $\Omega(U)$ denote the Grothendieck ring of isomorphism classes in $\hat{U}$ with sum defined by the disjoint union and product defined by the cartesian product with diagonal group action. Thus as an additive group $\Omega(U)$ is the free abelian group generated by the isomorphism classes U/W of transitive U-sets. For any $U, V \leq G$ and $g \in G$ with $gUg^{-1} \leq V$ one has an obvious restriction map $res(U,g,V) : \Omega(V) \to \Omega(W)$ as well as an induction map $ind(V,g,U) : \Omega(U) \to \Omega(V) : S \mapsto V \underset{U}{\times} S = \{\overline{(v,s)} \in V \underset{U}{\times} S \mid v \in V,\ s \in S$ and $\overline{(v,s)} = \overline{(v',s')}$ if and only if there exists $u \in U$ with $v' = v\,gug^{-1},\ s' = u^{-1}s\}$, thus $ind(V,g,U)(U/W) \cong V/gWg^{-1}$.

So $U \rightsquigarrow \Omega(U)$ may be considered as a multiplicative G-functor $\Omega(-)$ with a unit and a commutative and associative multiplication. $\Omega(-)$ acts on any G-functor X by

$$\Omega(U) \times X(U) \to X(U)$$

$$(U/W, x) \mapsto ind(U,1,W) \circ res(W,1,U)(x)$$

and it is easy to check that this defines a pairing of G-functors, which makes $X(U)$ a module over the commutative ring $\Omega(U)$. Particularly if X is a multiplicative G-functor with a unit, then

$$\Omega(U) \to X(U) : S \mapsto S \cdot 1_{X(U)}$$

is a multiplicative homomorphism which maps $\Omega(-)$ onto $\underline{X}$. This proves Proposition 3.1 (of course one could also check the statements of Proposition 3.1 directly). Now to prove Proposition 3.3 we need

Lemma 5: Let R, $\mathcal{C}$ and $\mathcal{G}$ be as in Proposition 3.3.

Let $K_\Omega(\mathcal{C}) = \mathrm{Ke}(\Omega(G) \xrightarrow{\text{res}} \prod_{C \in \mathcal{C}} \Omega(C))$ and $I_\Omega(\mathcal{H}) = \mathrm{Im}(\sum_{H \in \mathcal{H}} \Omega(H) \xrightarrow{\text{ind}} \Omega(G))$. Then $R \otimes K_\Omega(\mathcal{C}) + R \otimes I_\Omega(\mathcal{H}) = R \otimes \Omega(G)$.

Sketch of Proof: $L = R \otimes K_\Omega(\mathcal{C}) + R \otimes I_\Omega(\mathcal{H})$ definitely is an ideal in $R \otimes \Omega(G)$. Thus if it is not equal to $R \otimes \Omega(G)$ there must be a maximal ideal $\mathcal{Y}$ in $R \otimes \Omega(G)$ which contains L. But by Burnside's Theorem quoted in the proof of Lemma 1 the ring homomorphism $\Omega(G) \to \prod_{U \leq G} \mathbb{Z} : S \mapsto (|S^U|)_{U \leq G}$ is injective. Thus by the Cohen-Seidenberg "going-up-theorem" $\mathcal{Y}$ must be of the form $\mathcal{Y} = R \otimes \mathcal{Y}(U,q)$ with $\mathcal{Y}(u,q) = \{S - T \in \Omega(G) \mid |S^U| \equiv |T^U| \, (q)\}$ for some q with $qR \neq R$. Since $R \otimes K_\Omega(\mathcal{C}) \subseteq \mathcal{Y}$ we may again use the going-up-theorem to conclude that U may be chosen as a subgroup in $\mathcal{C}$: $U = C \in \mathcal{C}$. Now consider $C_1 = C^{(q)} \in \mathcal{C}$ and let C_2 equal the pre-image in $N_G(C_1) \leq G$ of some q-Sylow-subgroup in $N_G(C_1)/C_1$. Then $C_2^{(q)} = C_1 \in \mathcal{C}$, thus $C_2 \in \mathcal{H}$ and therefore $G/C_2 \in I_\Omega(\mathcal{H}) \subseteq \mathcal{Y}(C,q)$ in contradiction to $|G/C_2{}^C| \equiv |G/C_2{}^{C_1}| \equiv |G/C_2{}^{C_2}| = (N_G(C_2) : C_2) \not\equiv 0(q)$ by the choice of C_2.

Now consider the map $\varepsilon : \Omega(G) \to X(G)$ constructed above as well as

$$\varepsilon_R : R \otimes \Omega(G) \twoheadrightarrow R \otimes \underline{X}(G) = \underline{X}(G)$$

and

$$\varepsilon_{\mathbb{Q}} : \mathbb{Q} \otimes \Omega(G) \to \mathbb{Q} \otimes \underline{X}(G) \subseteq \mathbb{Q} \otimes X(G).$$

Since $\sum_{C \in \mathcal{C}} \mathbb{Q} \otimes X(C) \to \mathbb{Q} \otimes X(G)$ is surjective

$$\mathbb{Q} \otimes X(G) \to \prod_{C \in \mathcal{C}} \mathbb{Q} \otimes X(C)$$

must be injective and thus the image of $R \otimes K_\Omega(\mathcal{C})$ in $\underline{X}(G)$ is annihilated by $\mathbb{Q}$. So by our assumption any element in $\varepsilon_R(R \otimes K_\Omega(\mathcal{C}))$ is nilpotent. Use Lemma 5 to write $1_{R \otimes \Omega(G)} = a+b$ with $a \in R \otimes K_\Omega(\mathcal{C})$, $b \in R \otimes I_\Omega(\mathcal{H})$. Apply ε_R to get

$$1_{\underline{X}(G)} = \varepsilon_R(1_{R \otimes \Omega(G)}) = \varepsilon_R(a) + \varepsilon_R(b).$$

But $\varepsilon_R(a)$ is nilpotent, thus $\varepsilon_R(b)$ must be a unit, contained in the ideal $\mathrm{im}(\sum_{H \in \mathfrak{H}} X(H) \to X(G))$, so this ideal must be all of $X(G)$, i.e., $\sum_{H \in \mathfrak{H}} X(H) \to X(G)$ must be surjective. q.e.d.

REFERENCES

[0] W. Burnside, "Theory of groups of finite order", 2d ed., Cambridge Univ. Press, Cambridge, 1911.

[1] C. Curtis and I. Reiner, "Representation theory of finite groups and associative algebras", Interscience, New York, 1962.

[2] A. Dress, On relative Grothendieck rings, Bull. Amer. Math. Soc. 75(1969), 955-958.

[3] ________, Vertices of integral representations, Math. Z. 114(1970), 159-169.

[4] ________, Contributions to the theory of induced representations, in "Algebraic K-Theory II, Battelle Institute Conference 1972", Springer Lecture Notes, 342, (1973), 183-240.

[5] W. Gaschütz, Über den Fundamentalsatz von Maschke zur Danstellungstheorie der endlichen Gruppen, Math. Z. 56(1952), 376-387.

[6] J. A. Green, The modular representation algebra of a finite group, Ill. J. Math. 6(1962), 607-619.

[7] ___________, Axiomatic representation theory for finite groups, J. Pure and Appl. Algebra 1 (1971), 41-77.

[8] W. H. Gustafson, The theory of relative Grothendieck rings, Proceedings of the Conference on Orders, Group Rings and Related Topics (Ohio State Univ., Columbus, Ohio, 1972), Lecture Notes in Mathamatics, Vol. 353, pp. 95-112, Springer, Berlin, 1973.

[9] D. G. Higman, Induced and produced modules, Canadian J. Math. 7(1955), 490-508.

[10] T. Y. Lam, Induction theorems for Grothendieck groups and Whitehead groups of finite groups, Ann. Sci. École Norm. Sup. 4e série 1(1968), 99-148.

[11] __________ and I. Reiner, Relative Grothendieck groups, J. Algebra 11(1969), 213-242.

[12] __________ and __________, Reduction theorems for relative Grothendieck rings, Trans. Amer. Math. Soc. 142(1969), 421-435.

[13] __________ and __________, Finite generation of Grothendieck rings relative to cyclic subgroups, Proc. Amer. Math. Soc. 23(1969), 481-489.

[14] __________ and __________, Restriction maps on relative Grothendieck rings, J. Algebra 14(1970), 260-298.

[15] __________ and __________, An excision theorem for Grothendieck rings, Math. Z. 115(1970), 153-164.

[16] __________ , __________ and D. Wigner, Restrictions of representations over fields of characteristic p, Proc. Symposia Pure Math. 21(1971), 99-106.

[17] R. G. Swan, Induced representations and projective modules, Ann. of Math. (2)71(1960), 552-578.

[18] __________, The Grothendieck ring of a finite group, Topology 2(1963), 85-110.

FINITE REPRESENTATION TYPE IS OPEN

Peter Gabriel

Let V be a vector space of finite dimension n over an algebraically closed field k. The bilinear maps $V \times V \to V$ form a vector space $\mathrm{Hom}_k(V \otimes V, V)$ of dimension n^3, which we do consider here together with its natural structure of an algebraic variety over k. Clearly, the associative maps form a Zariski-closed subset S of $\mathrm{Hom}_k(V \otimes V, V)$. A further simple investigation shows that the associative algebra-structures with 1 on V form a Zariski-open affine subset of S, which we denote by Alg_V or Alg_n. The purpose of this paper is to show that the algebra-structures (with 1) of finite representation type also form an open subset of Alg_n. In other words, if A is an associative algebra with 1, A is of finite representation type iff some polynomials $P_1, \ldots, P_r$ do not vanish on the constant structures of A. This answers to a question of M. Auslander.

In §2 we write down explicitely some routine properties of the algebraic set Alg_n. Similar properties of the varieties of modules will be needed in §3. They are given in §1. In §3 we show that, for any natural number r, the algebra structures on V, for which there are only finitely many isomorphism classes of modules of dimension $\leq r$, form an open subset S_r of Alg_n. Our main statement then follows from the fact that, for large r, S_r is equal to the set of all algebra structures of finite representation type. This follows from the conjecture-theorem of Brauer-Thrall-Nazarova-Roiter ! The last paragraph is devoted to a thorough investigation of Alg_n for $n = 4$.

We denote schemes by "script letters" as $\underline{\underline{X}}$. The corresponding roman letter X then stands for the set $\underline{\underline{X}}(k)$ of rational points of $\underline{\underline{X}}$.

§1. The module varietes.

1.1. Let A be an associative algebra with 1 over the algebraically closed field k. For the sake of simplicity we suppose A to be of finite dimension n over k. If W is a k-vector space of dimension r, the bilinear maps $A \times W \to W$ form a vector space $\mathrm{Hom}_k(A \otimes W, W)$ of a dimension $r^2 n$, whereas the A-module structures on W form a Zariski-closed subset, which we denote by Mod_r^A. Clearly Mod_r^A is the set of rationel points of an algebraic scheme $\underline{\underline{\mathrm{Mod}}}_r^A$ over k, which may be described in the functorial point of view as follows: for any commutative k-algebra R, we have

$$\underline{\underline{\mathrm{Mod}}}_r^A(R) = \{R \otimes_k A\text{-module structures on } R \otimes_k W\}$$

The consideration of the scheme $\underline{\underline{\mathrm{Mod}}}_r^A$ does not need any special commentary. For previous use of it we refer to M. Artin [1], Procesi [9] or Voigt [10] and recall some of their observations. The linear group $\underline{\underline{GL}}(W)$ obviously operates on $\underline{\underline{\mathrm{Mod}}}_r^A$ by means of the formula

$$(g\varphi)(a \otimes w) = g(\varphi(a \otimes g^{-1} w)), \text{ where } g \in \underline{\underline{GL}}(W), \varphi \in \mathrm{Hom}_k(A \otimes W, W).$$

The orbits of $GL(W) = \underline{\underline{GL}}(W)(k)$ in $\mathrm{Mod}_r^A = \underline{\underline{\mathrm{Mod}}}_r^A(k)$ are the isomorphism classes of A-modules with k-dimension r. If $M \in \mathrm{Mod}_r^A$ is an A-module with underlying vector space W, we denote by T_M the Zariski-tangent space of $\underline{\underline{\mathrm{Mod}}}_r^A$ at the point M, by T_M^o the Zariski-tangent space at M of the orbit of $\underline{\underline{GL}}(W)$ through M. We then have the following

<u>Proposition</u> (Voigt) : <u>For any A-module M with underlying vector space W, there is a canonical isomorphism</u>

$$T_M / T_M^o \xrightarrow{\sim} \mathrm{Ext}_A^1(M, M) .$$

<u>Sketch of the proof</u>. Let $k[\varepsilon] = k \oplus k\varepsilon$ be the algebra of dual numbers, $\varepsilon^2 = 0$. By definition, a tangent vector to

$\underline{\underline{\mathrm{Mod}}}_n^A$ at M is a $k[\varepsilon]\otimes_k$ A-module structure on $k[\varepsilon]\otimes_k W$ which gives rise to M by reduction mod ε [DG.II, 4, 3.3]. Such a $k[\varepsilon]\otimes_k$ A-module structure induces an extension of A-modules

$$0 \longrightarrow M \xrightarrow{i} k[\varepsilon]\otimes_k W \xrightarrow{p} M \longrightarrow 0 ,$$

where $i(x) = \varepsilon\otimes x$ and $p((a+b\varepsilon)\otimes w) = aw$. Thus we get a map $T_M \to \mathrm{Ext}_A^1(M, M)$. Now the isotropy group of $M \in \mathrm{Mod}_r^A$ in $\underline{\underline{GL}}(W)$ is the group of automorphisms $\underline{\underline{\mathrm{Aut}}}\ M$ of the A-module M. This group is smooth, being an open subscheme of the vector space $\mathrm{Hom}_A(M, M)$. Hence the canonical map $\underline{\underline{GL}}(W) \to \underline{\underline{GL}}(W)\cdot M$, $g \mapsto g\cdot M$ is smooth, so that the induced tangent map $\mathrm{End}(W) = \mathrm{Lie}\ \underline{\underline{GL}}(W) \to T_M^o$ is surjective. The image of an endomorphism $D \in \mathrm{End}(W)$ under this map is the $k[\varepsilon]\otimes_k$ A-module structure on $k[\varepsilon]\otimes_k W$ given by

$$(1\otimes a).(1\otimes w) = 1\otimes aw + \varepsilon\otimes(Da-aD)w \quad \text{if} \quad a \in A,\ w \in W .$$

The end of the proof is now routine.

1.2 Corollary: The orbit of $M \in \mathrm{Mod}_r^A$ under $\underline{\underline{GL}}(W)$ is an open subscheme of $\underline{\underline{\mathrm{Mod}}}_r^A$ iff $\mathrm{Ext}_A^1(M, M) = 0$.

Proof. Clearly $T_M = T_M^o$ if the orbit $\underline{\underline{GL}}(W)\cdot M$ is open in $\underline{\underline{\mathrm{Mod}}}_r^A$. Reversely, suppose that $T_M = T_M^o$ and let $d = [T_M^o : k]$. As $\underline{\underline{GL}}(W)\cdot M$ is smooth, we get $d = \dim(\underline{\underline{GL}}(W).M) \leq \dim_M(\underline{\underline{\mathrm{Mod}}}_r^A)$ = local dimension of $\underline{\underline{\mathrm{Mod}}}_r^A$ at M. The equality $T_M = T_M^o$ therefore implies that $\dim T_M = \dim_M(\underline{\underline{\mathrm{Mod}}}_r^A) = d$, so that $\underline{\underline{\mathrm{Mod}}}_r^A$ is smooth at M of the same local dimension as $\underline{\underline{GL}}(W).M$. Hence $\underline{\underline{\mathrm{Mod}}}_r^A$ and $\underline{\underline{GL}}(W)\ M$ coincide in a neighbourhood of M.

1.3 Corollary (Artin, Voigt) : The orbit of $M \in \mathrm{Mod}_r^A$ under $\underline{\underline{GL}}(W)$ is closed in $\underline{\underline{\mathrm{Mod}}}_r^A$ iff M is a semi-simple module.

Proof. Let $\underline{r}$ be the radical of A. Clearly $\underline{\underline{\mathrm{Mod}}}_r^{A/\underline{r}}$ is a closed subscheme of $\underline{\underline{\mathrm{Mod}}}_r^A$. By 1.2 the orbits of $\underline{\underline{GL}}(W)$ in $\underline{\underline{\mathrm{Mod}}}_r^{A/\underline{r}}$ are all open and hence closed. This shows that the orbit of M is closed whenever M is a semi-simple A-module.

Now consider any module structure M on W and denote by N a semi-simple module structure on W having the same Jordan-Hölder factors as M. We shall show by induction on the height of M that N lies in the Zariski-closure of the orbit of M. The <u>height</u> of M is defined inductively as follows: height (0) = 0 , height (semi-simple module $\neq 0$) = 1 , height (M) = 1 + height (M/S) if S is the <u>socle</u> of M , that is the largest semi-simple submodule. In order to prove our statement, choose any supplementary subspace U of S in W . The scalar multiplication $\underset{1}{\cdot}$ in M is such that

$$a \underset{1}{\cdot} u = au + \sigma(a,u)$$

with $au \in U$ and $\sigma(a,u) \in S$ if $a \in A$ and $u \in U$ .For any $t \in k \setminus 0$ we get a new module structure M_t on W by setting

$$a \underset{t}{\cdot} u = au + t\sigma(a,u) \quad \text{and} \quad a \underset{t}{\cdot} s = a \underset{1}{\cdot} s$$

if $a \in A$, $u \in U$ and $s \in S$. Of course, this new structure is related to the old one by the vector space isomorphism $S \oplus U \longrightarrow S \oplus U$, $s+u \mapsto ts+u$. Hence the new structure lies in the orbit of M . However, when t tends to 0 , M_t tends to a structure M_0 , which is isomorphic to $S \oplus M/S$. Applying our induction hypotesis to M/S we may "degenerate" M/S to a semi-simple module structure, thus getting our statement.

1.4. <u>Corollary</u>: <u>Every connected component of $\underline{\underline{Mod}}_r^A$ contains exactly one closed orbit</u>. <u>Two module structures on W belong to the same connected component of $\underline{\underline{Mod}}_r^A$ iff they have the same Jordan-Hölder factors</u>.

This follows directly from corollary 1.3. and from its proof.

1.5. Corollary: *Let S, T be two simple A-modules such that $[S:k]+[T:k]=r$. Then the connected component of $S\oplus T$ in Mod_r^A has at most 2 irreducible components. It has two components iff $S\neq T$ and $\mathrm{Ext}_A^1(S,T)\neq 0\neq \mathrm{Ext}_A^1(T,S)$.*

Proof. Identify S and T with two supplementary subspaces of W and consider the module structures M_σ on W such that

$$a\underset{\sigma}{\cdot} s = as \qquad \text{and} \qquad a\underset{\sigma}{\cdot} t = at + \sigma(a,t)$$

if $a\in A$, $s\in S$ and $t\in T$. Here we write as and at for the original scalar multiplication within S and T, and σ stands for a bilinear map $A\times T\to S$ such that

$$\sigma(1,t) = 0 \quad \text{and} \quad \sigma(ba,t) = b\sigma(a,t) + \sigma(b,at)$$

These maps σ form a vector space $Z_{T,S}$. Denote by $E_{T,S}$ the image of the map

$$GL(W)\times Z_{T,S} \longrightarrow \mathrm{Mod}_r^A, \quad (g,\sigma)\longrightarrow g\cdot M_\sigma$$

Clearly $E_{T,S}$ is an irreducible subset of the connected component $C_{T,S}$ of $S\oplus T$ in Mod_r^A. It is also closed, because the module structures having a submodule structure isomorphic to S form a closed subset of Mod_r^A (following 1.3. a submodule structure isomophic to S cannot be degenerated in anything else but S). Hence $C_{T,S}$ is the union of two irreducible closed subsets $E_{T,S}$ and $E_{S,T}$. The condition $E_{T,S}\subset E_{S,T}$ clearly means that either $T=S$ or $\mathrm{Ext}_A^1(T,S)=0$.

1.6. Example: $A=k[x]/(x^n)$. Denote by I_s the A-module $I_s=k[x]/(x^s)$, $s=1,2,\ldots,n$. If M and N are two A-module structures of dimension r, we write $M\to N$ whenever N is a specialization (or "degeneration") of M, which means that N belongs to the Zariski-closure of the orbit of M in Mod_r^A . This gives rise to an ordering of the orbits in Mod_r^A . Associating with any A-module

$$M = I_1^{r_1} \oplus \ldots \oplus I_n^{r_n} \quad , \quad r_1 + 2r_1 + \ldots + nr_n = r$$

the natural numbers

$$t_i = r_1 + 2r_2 + \ldots + (i-1)r_{i-1} + i(r_i + r_{i+1} + \ldots + r_n)$$
$$= \left[{}_iM : k\right] , \text{ with } {}_iM = \{m \in M \mid x^i m = 0\} \text{ and } 1 \leqq i \leqq n-1 ,$$

we get an order-isomorphism of the set of orbits onto the subset of the product $\mathbb{N}^{n-1}$ formed by all $t = (t_1, \ldots, t_{n-1})$ such that $2t_1 \geqq t_2$, $2t_2 \geqq t_3 + t_1$, $2t_3 \geqq t_4 + t_2$, ... , $2t_{n-2} \geqq t_{n-1} + t_{n-3}$, $2t_{n-1} \geqq r + t_{n-2}$ and $r \geqq t_{n-1}$.
In particular , $\underline{\underline{\mathrm{Mod}}}_r^A$ is irreducible and all module structures of dimension r are specialisations of $I_n^q \oplus I_\rho$, with $r = qn + \rho$. The scheme $\underline{\underline{\mathrm{Mod}}}_r^A$ is generically reduced iff $\mathrm{Ext}_A^1(I_n^q \oplus I_\rho , I_n^q \oplus I_\rho) = \mathrm{Ext}_A^1(I_\rho, I_\rho) = 0$, which takes places iff $\rho = n$, that is iff r is a multiple of n . In case $n = 3$ and $r = 8$ for instance we get the following points of $\mathbb{N}^2$

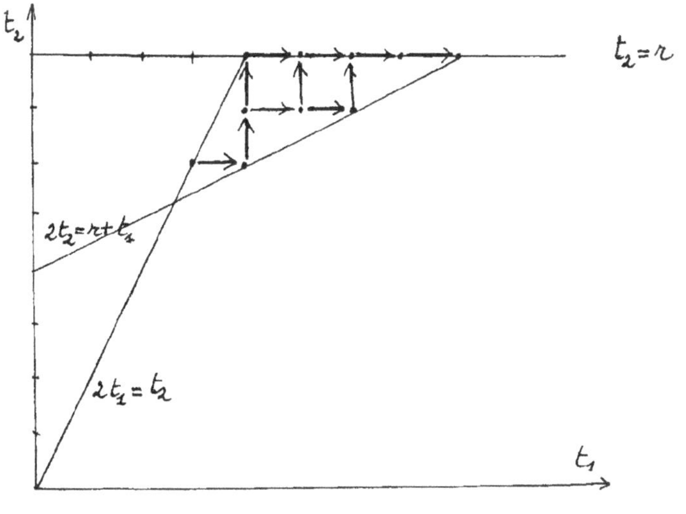

These points are attached to the following module structures

$$\begin{array}{ccccccccc}
 & & I_2^4 & \longrightarrow & I_2^3\oplus I_1^2 & \longrightarrow & I_2^2\oplus I_1^4 & \longrightarrow & I_2\oplus I_1^6 \longrightarrow I_1^8 \\
 & & \uparrow & & \uparrow & & \uparrow & & \\
 & & I_3\oplus I_2^2\oplus I_1 & \longrightarrow & I_3\oplus I_2\oplus I_1^3 & \longrightarrow & I_3\oplus I_1^5 & & \\
 & & \uparrow & & & & & & \\
I_3^2\oplus I_2 & \longrightarrow & I_3^2\oplus I_1^2 & & & & & &
\end{array}$$

1.7. Example: $A = k[x,y] / (x^2,xy,y^2)$. Denote by S the only simple A-module, by I_λ the quotient-module $A/A(y-\lambda x)$ by I_∞ the quotient A/Ax, by $\forall$ the dual of A(=injective hull of S).

Then $\underline{\underline{\mathrm{Mod}}}_3^A$ has 2 irreducible components of dimension 6. These are the closures of the open orbits of A and $\forall$. Defining an order on the orbits as in 1.6, we get for $r = 3$ the following diagram

$$\begin{array}{ccc}
A & & \forall \\
 & \searrow \quad \swarrow & \\
 & I_\lambda \oplus S, \ \lambda \in k \cup \infty & \\
 & \downarrow & \\
 & S^3 &
\end{array}$$

2. The varieties of algebras

2.1. Let us now come back to the notations of the introduction. Clearly, the associative bilinear maps $V \times V \longrightarrow V$ are the rational points of an affine algebraic scheme $\mathcal{S}_n$ over k, which may be defined by the formula

$$\mathcal{S}_n(R) = \{\text{associative } R\text{-algebra structures on } R \otimes_k V\},$$

where R runs through the commutative k-algebras.

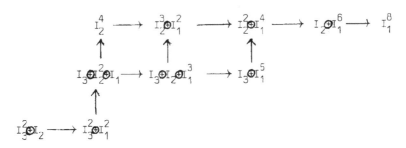

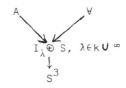

Lemma. The associative algebra structures with 1 form an affine open subscheme $\underline{\underline{Alg}}_n$ of $\mathcal{S}_n$.

Proof. Consider the pairs $(\varphi, x) \in \mathrm{Hom}_k(V \otimes V, V) \times V$, where φ is an associative bilinear map and x a unit element for the multiplication φ . Clearly, the set of all these pairs is Zariski-closed in $\mathrm{Hom}_k(V \otimes V, V) \times V$. From a functorial point of view, there is an affine algebraic scheme $\mathcal{C}_n$ over k , whose R-points are the pairs (φ, x) , where φ is an associative R-algebra structure on $R \otimes_k V$ and x a unit element. The first projection $(\varphi, x) \mapsto \varphi$ gives a morphism $\mu : \mathcal{C}_n \to \mathcal{S}_n$. It is sufficient to show that this is an open enbedding. But $\mathcal{C}_n(R) \to \mathcal{S}_n(R)$ is clearly an injection for any R. In particular μ is radicial (EGA, I 3.5.4) . It is etale by DG, I, §4, 4.7. Our statement hence follows from SGA1, I, 5.1.

2.2. The linear group $\underline{\underline{GL}}(V)$ clearly operates on $\underline{\underline{Alg}}_n$ by means of the formula

$$(g\varphi)(x,y) = g\varphi(g^{-1}x,\ g^{-1}y) ,$$

where $g \in \underline{\underline{GL}}(V)(R)$, $\varphi \in \underline{\underline{Alg}}_n(R)$ and $x,y \in R \otimes_k V$.

Proposition. Let $A \in \mathrm{Alg}_n$ be a k-algebra structure on V . The orbit of A in $\underline{\underline{Alg}}_n$ under $\underline{\underline{GL}}(V)$ is closed iff $A \xrightarrow{\sim} k \oplus U$ with $U^2 = 0$.

Proof. Clearly $\underline{\underline{Alg}}_n$ contains at least one closed orbit. It is therefore sufficient to show that any algebra structure A with 1 may be degenerated into $k \oplus U$. For this purpose, take a basis $l_1, l_2, \ldots, l_n$ such that $l_1 = 1$. Let $e_i e_j = \sum_k c_{ij}^h e_h$, where the coefficients $c_{ij}^h \in k$ are the structure constants. Taking the new basis $f_1 = e_1, f_2 = te_2, f_3 = te_3, \ldots, f_n = te_n$ with $t \in k\setminus 0$, the new structure constants d_{ij}^h are such that $d_{ij}^1 = t^2 c_{ij}^1$ and $d_{ij}^h = tc_{ij}^h$ for $i,j,h \neq 1$. These structure constants define an algebra structure A_t on V , which tends towards $k \oplus U$, when t tends to 0.

As a consequence, we see that the scheme $\underline{\underline{\mathrm{Alg}}}_n$ is connected. It should be one of the main tasks of associative algebra to determine for every n the number of irreducible components of $\underline{\underline{\mathrm{Alg}}}_n$.

2.3. Examples: a) When $n=2$, $\underline{\underline{\mathrm{Alg}}}_n$ has only one irreducible component, containing an open orbit, that of $k \times k$, and a closed orbit, that of $k[x]/(x^2)$.

b) When $n=3$, $\underline{\underline{\mathrm{Alg}}}_n$ has 2 irreducible components, namely the closure of the orbit of $k \times k \times k$ and the closure of the orbit of the algebra $\begin{pmatrix} k & k \\ 0 & k \end{pmatrix}$ of triangular matrices. Writing again $A \to B$ if the algebra structure B may be obtained by degenerating the structure A, we get the following diagram of 3-dimensional algebras

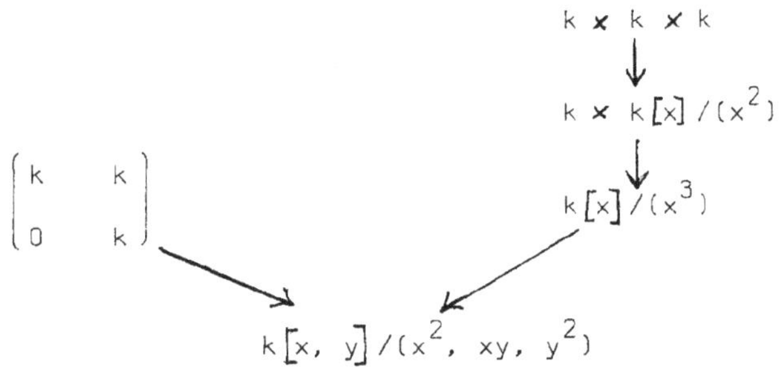

The description of $\underline{\underline{\mathrm{Alg}}}_n$ is given at the end of this paper.

c) For $n=2$, $\mathcal{S}_2$ has 3 irreducible components and 7 orbits under the action of $\underline{\underline{\mathrm{GL}}}(V)$. The diagram of 2-dimensional associative algebras with or without 1 is as follows

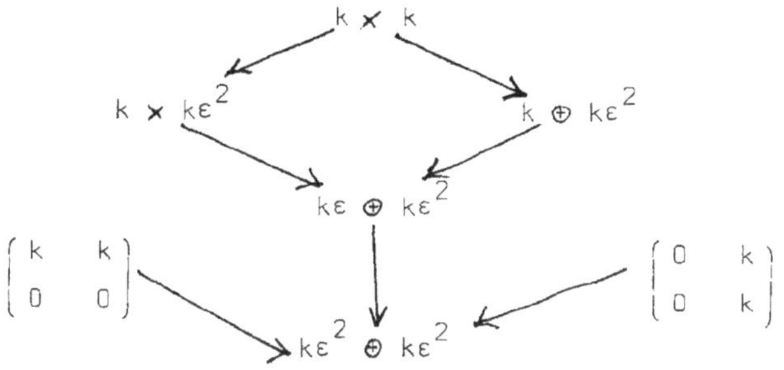

where $\varepsilon = x \bmod x^3 \in k[x]/(x^3)$. In this picture $k\varepsilon \oplus k\varepsilon^2$ for instance is considered as a subalgebra of $k[x]/(x^3)$, $\begin{pmatrix} k & k \\ 0 & 0 \end{pmatrix}$ as a subalgebra of $\begin{pmatrix} k & k \\ 0 & k \end{pmatrix}$, $k \times k\varepsilon^2$ stands for the product algebra structure ...

2.4. Whenever $A \in \mathrm{Alg}_n$, we denote by T_A the Zariski-tangent space of $\underline{\underline{\mathrm{Alg}}}_n$ at A , by T_A^o the Zariski-tangent space at A of the orbit of A under $\underline{GL}(V)$.

<u>Proposition</u>. <u>For any</u> $A \in \mathrm{Alg}_n$, <u>there is a canonical isomorphism</u>

$$T_A/T_A^o \xrightarrow{\sim} H^2(A, A) ,$$

<u>where</u> $H^i(A, M)$ <u>stands for the</u> i-th <u>Hochschild cohomology group of an</u> A-<u>bimodule</u> M .

<u>Sketch of the proof</u>. It is similar to that of proposition 1.1. Let $k[\varepsilon] = k + k\varepsilon$ be the algebra of dual numbers. By definition [DG, II, 4, 3.3.] , a tangent vector to Alg_n at A is a $k[\varepsilon]$ -algebra structure on $k[\varepsilon] \otimes_k V$ which induces A by reduction mod ε. Such a structure obviously gives rise to an algebra extension

$$0 \longrightarrow A \xrightarrow{i} k[\varepsilon] \otimes_k V \xrightarrow{p} A \longrightarrow 0 ,$$

where $i(x) = \varepsilon \otimes x$ and $p((a + b\varepsilon) \otimes v) = av$. The image of A is an ideal of square radical 0 bearing the natural A-bimodule structure. As is well known, extensions of this type are classified by $H^2(A, A)$, so that we may finish our sketch as in 1.1.

2.5. Corollary: The orbit of $A \in \underline{\underline{Alg}}_n$ under $\underline{\underline{GL}}(V)$ is an open subscheme of $\underline{\underline{Alg}}_n$ iff $H^2(A, A) = 0$.
Compare with 1.2.

2.6. Corollary: The orbit of $A \in Alg_n$ under $\underline{\underline{GL}}(V)$ is an open subscheme of $\underline{\underline{Alg}}_n$ whenever the global dimension of A is ≤ 1.

In example 2.3.b) this happens to be right for $A = k \times k \times k$ and $A = \begin{pmatrix} k & k \\ 0 & k \end{pmatrix}$.

Proof. We prove more generally that (finite dimensional) algebras A of global (homological) dimension ≤ 1 also have Hochschild dimension ≤ 1. Equivalently we shall prove that, whenever B is a finite dimensional algebra and I an ideal of B with square 0, then any algebra homomorphism $A \to B/I$ may be lifted to B. This follows at once from the structure of algebras of global dimension ≤ 1 (see for instance [5]) : set r = radical of A, $K = A/r$, $w = r/r^2$. Then A is of global dimension ≤ 1 iff $A \tilde{\to} \otimes_K \omega$ = tensor algebra of the K-bimodule ω. Now, whenever $\varphi : \otimes_K \omega \to B/I$ is an algebra homomorphism, we can first lift $\varphi|K$ to an algebra-homomorphism $K \to B$, and then further $\varphi|\omega$ to a K-bimodule homomorphism $w \to B$. Both "liftings", then extend uniquely to a lifting of φ :

2.7. Proposition: The following subsets of Alg_n are Zariski-closed:

$$\{A \in Alg_n \mid [\text{Rad } A : k] \geq s\}$$
$$\{A \in Alg_n \mid [\text{Cent } A : k] \geq s\}$$

$\{A \in \underline{\underline{Alg}}_n \mid$ number of blocks of $A \leq s\}$
$\{A \in \underline{\underline{Alg}}_n \mid A$ is basic, i.e. $A/\mathrm{Rad}A \cong k^t$ for some $t\}$.

Proof. Consider for any natural number $m \leq n$ the grassmarian $\underline{\underline{Gr}}_{m,n-m}$ formed by the m-dimensional subspaces of $V \cong k^n$. Consider in the product scheme $\underline{\underline{Alg}}_n \times \underline{\underline{Gr}}_{m,n-m}$ the pairs (A,I) where I is an m-dimensional nilpotent two-sided ideal of A . These pairs give rise to a closed subscheme $\underline{\underline{\mathcal{Y}}}_m$ of $\underline{\underline{Alg}}_n \times \underline{\underline{Gr}}_{m,n-m}$. As $\underline{\underline{Gr}}_{m,n-m}$ is complete, the projection F_m of $\underline{\underline{\mathcal{Y}}}_m(k)$ into $\underline{\underline{Alg}}_n(k)$ is Zariski-closed. Hence our first statement follows from the equality

$$\{ A \in \underline{\underline{Alg}}_n \mid [\mathrm{Rad}\, A : k] \geq s\} = \bigcup_{s \leq m \leq n} F_m .$$

A similar proof works for the second subset of our proposition.

In order to investigate the third subset consider the scheme $\underline{\underline{E}}$, whose R-points are the pairs (A,e) , where A is an R-algebra structure on $R \otimes_k V$ and e a central idempotent of A . In case $R = R'/J$ with $J^2 = 0$ and $A =$ $A = A'/A'J$ with $A' \in \underline{\underline{Alg}}_n(R')$, it is easily seen that e may be lifted in a unique way to a central idempotent e' of A' . By DG, I, §4, 4.7 this means that the morphism $\underline{\underline{E}} \to \underline{\underline{Alg}}_n$, $(A,e) \mapsto A$ is etale. Hence the cardinality of the fibres can only decrease by specialization.

As for the last subset, just replace the scheme $\underline{\underline{\mathcal{Y}}}_m$ above by the closed subscheme of those pairs (A,I) such that A/I is commutative: the proof then goes through.

2.8. <u>Proposition: Let</u> B <u>be a finite-dimensional semi-simple k-algebra and</u> M <u>a finite dimensional</u> B-<u>bimodule. Then the following subsets of</u> Alg_n <u>are locally closed:</u>

$\{A \in \mathrm{Alg}_n \mid A/\mathrm{Rad}\, k \cong B\}$
$\{A \in \mathrm{Alg}_n \mid A$ is of species (B,M), i.e. $(A/\mathrm{rad}\, A,$
$\mathrm{Rad}\, A/(\mathrm{Rad}\, A)^2) \cong (B,M)\}$

Proof. Set $[B : k] = n-m$. For the first statement it is sufficient to show that the pairs (A,I) with $A/I \xrightarrow{\sim} B$ form an open subset T of $\underline{\underline{\mathfrak{L}}}_m(k)$ (2.7). Indeed, the projection $\underline{\underline{\mathfrak{L}}}_m(k) \to F_m$ is closed and T is nothing else but the inverse image of the first subset of our proposition. Hence this subset is open within F_m.

In order to study T consider any $(n-m)$-dimensional subspace U of V . Clearly, the pairs (A,I) such that I is a supplementary subspace of U in V give rise to an open subscheme $\underline{\underline{\mathfrak{L}}}_m^U$ of $\underline{\underline{\mathfrak{L}}}_m$. Giving to U the algebra structure of A/I , we get a morphism $\underline{\underline{\mathfrak{L}}}_m^U \to \underline{\underline{\mathrm{Alg}}}_U$. By 2.6 the algebra structures isomorphic to B form an open subset of $\underline{\underline{\mathrm{Alg}}}_U$. The inverse image of this subset in $\underline{\underline{\mathfrak{L}}}_m^U(k)$ is $T \cap \underline{\underline{\mathfrak{L}}}_m^U(k)$. Hence T is open in $\underline{\underline{\mathfrak{L}}}_m(k)$, being covered by open subsets of $\underline{\underline{\mathfrak{L}}}_m(k)$.

The investigation of the second subset is similar, but in some way more delicate. We shall not need it and hence omit the proof.

2.9. Proposition: The subset H of Alg_n formed by algebras of global homological dimension $\leq s$ is constructible, i.e. is a finite union of locally closed subsets.

We do not know wether H is open or not.

Proof. Let F be any irreducible closed subset of Alg_n such that $F \cap H$ is dense in F . We have to prove that $F \cap H$ contains a subset which is open within F [DG, I, § 3, 3.2]. We only sketch this proof. Let $\underline{\underline{F}}$ be the reduced closed subscheme of $\underline{\underline{\mathrm{Alg}}}_n$ having F as set of rational points. Let R be the algebra of functions on $\underline{\underline{F}}$.

There is by definition a canonical bijection between the k-algebra structures $A \in F$ and the algebra-homomorphismes $R \to k$. Moreover, $R \otimes_k V$ bears a canonical R-algebra structure B with the property that any $A \in F$ is identified with the induced structure on $V \tilde{\to} k \otimes_{\varphi_A} R \otimes_k V$, if $\varphi_A : R \to k$ stands for the homomorphism attached to A .

Let K be the field of fractions of R and $B' = K \otimes_R B$. Let

$$(*) \quad P'_{s+1} \to P'_s \to \dots \to P'_1 \to B' \to B'/\mathrm{Rad}\, B' \to 0$$

be the minimal projective resolution of $B'/\mathrm{Rad}\, B'$. Set $I = B \cap \mathrm{Rad}\, B'$. We can prove by means of the usual techniques of commutative algebra the existence of a sequence of B_t-modules

$$(**) \quad P_{s+1} \to P_s \to \dots \to P_1 \to B_t \to B_t/I_t \to 0$$

for some $0 \neq t \in R$, which induces $(*)$ by localization from R_t to K . Moreover, t can be chosen in such a way that, whenever $A \in F$ and $t(A) \neq 0$, we have $\mathrm{Rad}\, A \tilde{\to} k \otimes_{\varphi_A} I$ and

$$k \otimes_{\varphi_A} P_{s+1} \to k \otimes_{\varphi_A} P_s \to \dots \to k \otimes_{\varphi_A} P_1 \to A \to A/\mathrm{Rad} \to 0$$

is a minimal projective resolution of length $s+1$ (in fact $(**)$ will split !).

Now, by hypothesis, A is of global dimension $\leq s$ on a dense subset of F . Hence $k \otimes_{\varphi_A} P_{s+1} = 0$ on a dense subset of the open subset $F_t = \{A \in F \mid t(A) \neq 0\}$ of F . This implies that $k \otimes_{\varphi_A} P_{s+1} = 0$ all over F_t .

3. Algebras with finitely many representations of a given dimension.

3.1. For any $A \in \mathrm{Alg}_n$ and any $r \in \mathbb{N}$, we denote by $\nu_A(r)$ the number of isomorphism classes of A-modules of dimension r . The main aim of the present paragraph is to prove the following.

Proposition: The set $\{A \quad \mathrm{Alg}_n \mid \nu_A(r) < +\infty\}$ is open in Alg_n .

The proof of this statement rests on a study of the scheme $\underline{\underline{\mathrm{Algmod}}}_{n,r}$, whose R-points are the pairs (A,M) formed by an R-algebra structure A on $R \otimes_k V \xrightarrow{\sim} R^n$ and an A-module structure on $R \otimes_k W \xrightarrow{\sim} R^r$. Clearly, the fibres of the morphism

$$p : \underline{\underline{\mathrm{Algmod}}}_{n,r} \to \underline{\underline{\mathrm{Alg}}}_n , \quad (A,M) \mapsto A$$

are the schemes $\underline{\underline{\mathrm{Mod}}}_r^A$ of § 1, on which the group $\underline{\underline{\mathrm{GL}}}(W)$ is operating. We shall prove the following statement, which is equivalent to our proposition: whenever F is a closed irreducible subset of Alg_n and $A \in F$ a point whose fiber $p^{-1}(A)$ contain s only finitely many GL(W) - orbits, then there is an non empty open subset $U \subset F$ such that any fiber over U contains only finitely many orbits. The proof of this statement rests on the following lemmas.

3.2. Lemma. Let F be a closed irreducible subset of Alg_n and $A \in F$. Then any closed irreducible subset F' of $\mathrm{Modalg}_{n,r}$, which dominates F and is stable under GL(W), cuts $p^{-1}(A)$.

Proof. Let $\underline{\underline{F}}$ and $\underline{\underline{F}}'$ be the reduced closed subschemes having F and F' as sets of rationel points. Let K be a large algebraically closed field extension of k and $B \in \underline{\underline{F}}(K)$ a generic geometric point of $\underline{\underline{F}}$. By 1.3, there is a pair $(B,N) \in \underline{\underline{F}}'(K)$ over B such that N is semi-simple. Let $N = B/I_1 \oplus \ldots \oplus B/I_s$ be a decomposition of N into a direct sum of simple B-modules of dimensions $r_1, \ldots, r_s$. Now consider within

$$\underline{\underline{\mathrm{Alg}}}_n \times \underline{\underline{\mathrm{Gr}}}_{n-r_1,r_1} \ldots \times \underline{\underline{\mathrm{Gr}}}_{n-r_s,r_s}$$

the k-closure $\underline{\underline{G}}$ of $(B,I_1, \ldots, I_s)$. Clearly, $\underline{\underline{G}}$ dominates $\underline{\underline{F}}$ and is proper over $\underline{\underline{F}}$. Hence $\underline{\underline{G}}$ contains a point $(A,J_1, \ldots, J_s)$ over A.

Now choose for every i a vector subspace U_i of V

such that $V = U_i \oplus J_i$ and $K \otimes_k V = K \otimes_k U_i \oplus I_i$. The points $(C, L_1, \ldots, L_s) \in \underline{\underline{F}}(R)$ such that $R \otimes_k V = R \otimes_k U_i \oplus L_i$ for every i form an open irreducible subscheme $\underline{\underline{G}}'$ of $\underline{\underline{G}}$ containing both $(A, J_1, \ldots, J_s)$ and $(B, I_1, \ldots, I_s)$. Identifying C/L_i with $R \otimes_k U_i$ and $U_1 \oplus \ldots \oplus U_s$ with W, we then get a morphism

$$\underline{\underline{G}}' \to \underline{\underline{\mathrm{Modalg}}}_{n,r},\ (C, L_1, \ldots, L_1) \mapsto C/L_1 \oplus \ldots \oplus C/L_s .$$

By definition of $\underline{\underline{G}}$ this morphism has to factorize through $\underline{\underline{F}}'$. Hence F' cuts $p^{-1}(A)$ at $(A, A/J_1 \oplus \ldots \oplus A/J_s)$.

3.3. Lemma. Let F be a closed irreducible subset of Alg_n and A a point of F such that $p^{-1}(A)$ contains only finitely many $GL(W)$-orbits. Then every $GL(W)$-stable closed irreducible subset F' of $\mathrm{Modalg}_{n,r}$, which dominates F, contains a non-empty $GL(W)$-stable open subset U' such that the number of $GL(W)$-orbits within $U' \cap p^{-1}(B)$ is finite for every $B \in F$.

Proof. By 3.2. F' cuts $p^{-1}(A)$ in some point (A,M). As $p^{-1}(A)$ contains only finitely many orbits, we may suppose that the orbit $G_{A,M}$ of (A,M) is open within $F' \cap p^{-1}(A)$. We then have

$$\dim G_{A,M} = \dim_{(A,M)} p^{-1}(A) \text{ (=local dimension of the fibre).}$$

Now it is well known in algebraic geometrie that the local dimension of the fibres is an upper semicontinuous function, whereas the dimension of the orbits is lower semicontinuous. Hence the subset

$$U' = \{(B,N) \mid \dim G_{B,N} \geq \dim G_{A,M} \text{ and } \dim_{(B,N)} p^{-1}(B) \leq \leq \dim_{(A,M)} p^{-1}(A)\}$$

is open in F', being the intersection of two open subsets. Within U' we get

$$\dim_{(B,N)} p^{-1}(B) \leq \dim_{(A,M)} p^{-1}(A) \simeq \dim G_{A,M} \leq \dim G_{B,M} \leq$$

$\leq \dim_{(B,N)} p^{-1}(B)$, so that $\dim G_{B,N} = \dim_{(B,N)} p^{-1}(B)$, which means that every orbit is open it its fibre. Hence there can be only finitely many orbits in $U' \cap p^{-1}(B)$.

3.4. Now we are in a position to prove proposition 3.1., or equivalently the statement following this proposition. In fact, we shall prove more generally by noetherian induction that, for any GL(W)-stable closed subset F'_1 of $\mathrm{Modalg}_{n,r}$ lying over F , there is an open subset $U \subset F$ such that the number of GL(W)-orbits in $F'_1 \cap p^{-1}(u)$ is finite for every $u \in U$. The statement of 3.1. is just the special case $F'_1 = p^{-1}(F)$.

The assertion to be proved here is obvious, when F'_1 does not dominate F , hence when F'_1 is small. In case F'_1 dominates F , take any irreducible component F' of F'_1 dominating F . Then take U' as in 3.3. We may suppose that U' is open in F'_1 (otherwise replace U' by $U' \setminus (F'_1 \setminus F')$) . Now applying our induction hypothesis to $F'_1 \setminus U'$ we get an open subset $U \subset F$ such that the number of GL(W)-orbits in $(F'_1 \setminus U') \cap p^{-1}(u)$ is finite for every $u \in U$. Clearly this U works.

4. The main statement.

4.1. Lemma. *The set*

$$S_\infty = \{A \in \mathrm{Alg}_n \mid A \text{ is of finite representation type}\}$$

is a countable union of constructible subsets of Alg_n .

Proof. The detailed proof of this lemma is routine work for algebraic geometers and tedious for everybody. We only outline two possible proofs.

In order to ascertain wether a given algebra A is of finite representation type or not, we may use the algorithm given in [6], starting with the family of functors consisting only in the forgetful functor, which attaches to any A-module its underlying group and is represented by A itself. Arguing as in 2.9. we may prove that the algebras A for which the algorithm of [6] stops at the step i form a constructible subset C_i of Alg_n. Hence $S_\infty = \bigcup_i C_i$.

Another proof is obtained by applying Auslander's construction of algebras of finite representation type [2]. Let us say that a finite-dimensional k-algebra Γ is an <u>Auslander algebra</u> if it is basic, if its global dimension is ≤ 2 and if in the minimal injective resolution $0 \to \Gamma_\Gamma \to I_0 \to I_1 \cdots$ the injectives I_0 and I_1 are also projective. Now consider the scheme $\underline{\underline{Bimod}}_{n,r,s}$ whose R-points are the triples (A,M,Γ), where A is an R-algebra structure on $R \otimes_k V \xrightarrow{\sim} R^n$, Γ an A-algebra structure on R^s and M an A-Γ-bimodule structure on $R \otimes_k W \xrightarrow{\sim} R^r$. Arguing as in 2.9. we may prove that the triples $(A,M,\Gamma) \in \underline{\underline{Bimod}}_{n,r,s}(k)$ with the properties a)-d) given below form a constructible subset $C(r,s)$ of $\underline{\underline{Bimod}}_{n,r,s}(k)$: a) Γ is an Auslander algebra; b) as a Γ-module M is both projective and injective; c) any indecomposable projective and injective Γ-module is a direct summand of Γ; d) A is identified with $End_\Gamma(M)$. The image of $C(r,s)$ under the first projection is therefore a constructible subset of Alg_n, and S_∞ is the union of these images, when r and s vary.

4.2. <u>Theorem</u>: <u>The set</u>

$$S_\infty = \{A \in Alg_n \mid A \text{ is of finite representation type}\}$$

<u>is open in</u> Alg_n. <u>In fact, we also have</u>

$$S_\infty = \{A \in Alg_n \mid A \text{ has only finitely many representations of dimension } \leq r\}$$

<u>if</u> r <u>is large enough</u>.

Proof. Let S_r be the set of all $A \in Alg_n$ having only finitely many representations of dimension $\leq r$. By the conjecture of Brauer-Thrall and the theorem of Nazarova-Roiter [8] we know that $S_\infty = \bigcap_{r\in \mathbb{N}} S_r$. In order to prove that $S_\infty = S_r$ for large r we may suppose that the ground field k is not countable (otherwise replace k by an algebraically closed extension K and notice that a k-algebra A is of finite representation type iff $K \otimes_k A$ is so). By 4.1 we also know that S_∞ is the union of increasing sequence of constructible subsets S'_s , so that finally

$$\bigcup_{s\in \mathbb{N}} S'_s = S_\infty = \bigcap_{r\in \mathbb{N}} S_r .$$

Our theorem follows therefore from the

Lemma. Let $\underline{X}$ be an algebraic scheme over a not countable algebraically closed field k . Let $S_1, S_2 \ldots$ be a decreasing sequence of open subsets of $X = \underline{X}(k)$ and $S'_1, S'_2, \ldots$ an increasing sequence of constructible subsets of X . In case

$$\bigcap_s S'_s = \bigcup_r S_r$$

we must have $S'_s = S_r$ for some s, r .

Proof. By noetherian induction. We may suppose the lemma proven for all closed proper subschemes of $\underline{X}$. Moreover, the lemma is true if it works for every irreducible component of X . Hence we may assume X irreducible and $S_r \neq \phi$ for any r . As X is the union of the closures $\overline{S'_s}$ and $X \setminus S_r$, and as $X \setminus S_r \neq X$, we must have $\overline{S'_s} = X$ for some s. Hence S'_s contains an open subset U of X . Now applying our induction hypothesis to $X \setminus U$, $S'_t \setminus U$ and $S_r \setminus U$ we get the wanted result.

5. Algebras of dimension 4.

We give the results without proof. The case of algebras of dimension 5 will be given elsewhere.

<u>Proposition</u>. <u>The scheme</u> Alg_4 <u>has 5 irreducible components</u>.

We now give the list of these five components together with the algebra structures contained in them and the dimensions of the orbits of these structures.

a) <u>The irreducible component of commutative algebras</u>.

1)	$k \times k \times k \times k$	represented by the symbol		Orbit dimension	15
2)	$k \times k \times k[x]/(x^2)$	"		"	14
3)	$k[x]/(x^2) \times k[y]/(y^2)$	"		"	13
4)	$k \times k[x]/(x^3)$	"		"	13
5)	$k[x]/(x^4)$	"		"	12
6)	$k \times k[x,y]/(x,y)^2$	"		"	11
7)	$k[x,y]/(x^2,y^2)$	"		"	11
8)	$k[x,y]/(x^3,xy,y^2)$	"		"	10
9)	$k[x,y,z]/(x,y,z)^2$	"		"	6

b) <u>The irreducible component of matrix algebras</u>.

10)	$\begin{pmatrix} k & k \\ k & k \end{pmatrix} = M_2$	"	M	"	12
11)	$\left\{ \begin{pmatrix} a & 0 & 0 & 0 \\ 0 & a & 0 & d \\ c & 0 & b & 0 \\ 0 & 0 & 0 & b \end{pmatrix} \;\; a,b,c,d \;\; k \right\}$	"		"	11
12)	Λk^2 = exterior algebra of k^2	"		"	9
9)	$k[x,y,z]/(x,y,z)^2$	"		"	6

c) <u>The irreducible component of triangular algebras</u>.

13)	$k \times \begin{pmatrix} k & k \\ 0 & k \end{pmatrix}$	represented by	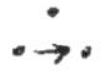	Dimension orbit	13
14)	$\left\{ \begin{pmatrix} a & 0 & 0 \\ c & a & 0 \\ d & 0 & b \end{pmatrix} \middle\vert \; a,b,c,d, \;\; k \right\}$	"	G·→	"	12

15)	$\left\{\begin{pmatrix} a & c & d \\ 0 & a & 0 \\ 0 & 0 & b \end{pmatrix} \middle\vert\, a,b,c,d, \in k\right\}$	"	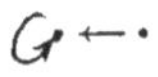	"	12
16)	$k\{x,y\}/(x^2,y^2,yx)$	"		"	11
8)	$k[x,y]/(x^3,xy,y^2)$	"		"	10
9)	$k[x,y,z]/(x,y,z)^2$	"		"	6

d) <u>The irreducible component of Kronecker algebras</u>

17)	$\left\{\begin{pmatrix} a & 0 & 0 \\ 0 & a & 0 \\ c & d & b \end{pmatrix} \middle\vert\, a,b,c,d, \in k\right\}$	"		"	9
9)	$k[x,y,z]/(x,y,z)^2$	"		"	6

e) <u>The irreducible component of the continuous series</u>

18)	$k\{x,y\}/(x^2,y^2,yx,-\lambda xy)$ for $\lambda \neq -1$	"		"	11
19)	$k\{x,y\}/(y^2,x^2+yx,xy+yx)$	"	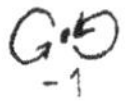	"	11
12)	Λk^2	"		"	9
8)	$k[x,y]/(x^3,xy,y^2)$	"		"	10
9)	$k[x,y,z]/(x,y,z)^2$	"		"	6

As usual $k\{x,y\}$ stands for the free associative algebra with 1 generated by x and y.

The diagram of algebras of dimension 4

(A ⟶ B means that the structure B may be obtained by degenerating the structure A)

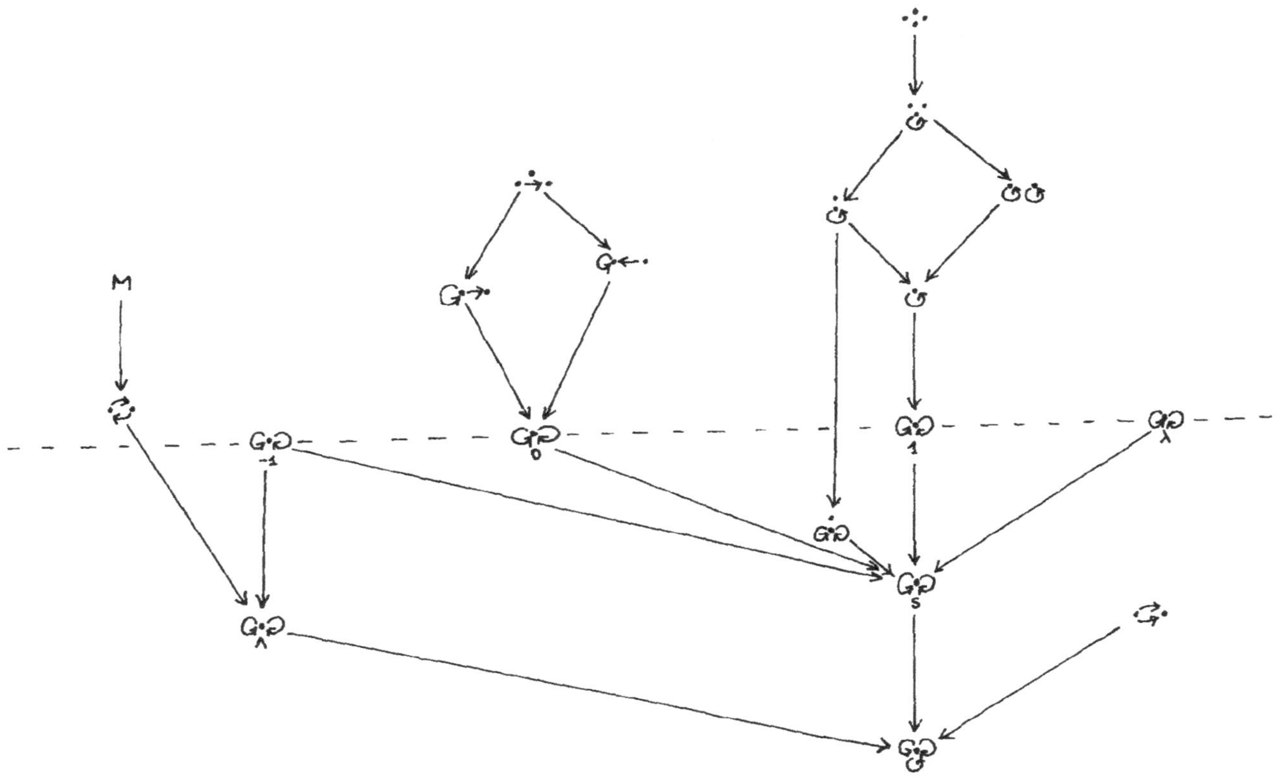

Literature

1 M.Artin, On Azumaya algebras and finite dimensional representations of rings, J.Algebra 11 (1969), p. 532-563.

2 M.Auslander, Representation dimension of artin algebras, Queen Mary College, Mathematics Notes, 1971.

3 M.Demazure - P.Gabriel, Groupes algébriques linéaires, North-Holland Publishing Company 1970, quoted DG

4 J.Dieudonne - A.Grothendieck, Eléments de Géometrie Algébriques, Presses Universitaires, Paris 1960, quoted EGA.

5 P.Gabriel, Indecomposable representations II, Symposia math. Ist. Naz. di Alta Mat., Vol. XI, 1973.

6 P.Gabriel, Représentations indécomposables, Séminaire Bourbaki, 26e année, 1973/74, no 444.

7 A.Grothendieck, Séminaire de Géometrie Algébrique I, Springer Lecture Notes, quoted SGA.

8 L.A.Nazarova - A.V.Roiter : "Categorical matricial problems and the conjecture of Brauer - Thrall", preprint Institute of Mathematics of the Academy of Sciences of Ukraine, Kiev 1974.

9 Procesi Finite dimensional representations of Algebras Israel Journal of Mathematics, 1975

10 D.Voigt, Induzierte Darstellungen in der Theorie der endlichen algebraischen Gruppen, Habilitationsschrift, Bonn 1974.

Peter Gabriel
Mathematisches Institut der Universität
Freiestrasse 36
8032 Zürich, Switzerland

After the redaction of the preceding I have been kindly informed by F.J. Flanigan of his own work on related topics. I therefore include a list of his most recent publications.

J.D. Donald and F.J. Flanigan
Deformations of Algebra Modules J. of Alg. 31, 1974 p.245-256

J.D. Donald and F.J. Flanigan
A Deformation-Theoretic Version of Maschke's Theorem for Modular Group Algebras: The commutative case
J. of Alg., 29, 1974, p.98-102.

F.J. Flanigan
Which Algebras Deform into a Total Matrix Algebra ?
J. of Alg., 29, 1974, p.103-112.

SIMPLE COHERENT FUNCTORS

Laurent Gruson

Notation: for a ring A , A^{op} is the opposite ring of A , $Mod(A)$ is the category of left A-modules, $mod(A)$ is the category of finitely presented left A-modules, $C(A)$ is the category of coherent functors from $mod(A)$ to Ab (cf. [1],[2]

In [1] , M. Auslander proves that for an Artin algebra A , every non-zero object in $C(A)$ admits a simple subobject and a simple quotient. This existence theorem does not hold for Artin rings. The present paper is devoted to a small exploration in this direction.

First, I recall some preperties of $C(A)$:

Lemma 1 - $C(A)$ <u>is the free abelian category over</u> A^{op} .

This statement, proved in [3] , means that for any object M in an abelian category C and any homomorphism $A^{op} \longrightarrow end_C(M)$, there is a functor $f : C(A) \longrightarrow$ unique up to unique isomorphism, exact and such that $f((A,)) = M$.

For instance, there is an equivalence $C(A) \longrightarrow C(A^{op})^{op}$ preserving the identity functor. This equivalence transforms the representable functor $(M,)$ into the tensor-product $M\otimes_A$, so that one gets the following lemma:

Lemma 2 - $C(A)$ <u>has sufficiently many projectives, which are the functors</u> $(M,)$ $(M \in mod(A^{op}))$; $C(A)$ <u>has sufficiently many injectives, which are the functors</u> $\otimes_A N$ $(N \in mod(A))$.

Assume from now that A is left artinian; then $C(A)$ has projective covers and injective envelopes [1] . Given two indecomposable objects $M \in mod(A^{op})$, $N \in mod(A$ call M and N <u>correspondent</u> iff there is a simple object S of $C(A)$ with injecti envelope $\otimes_A N$ and with projective cover $(M,)$.

Lemma 3 - <u>If</u> A <u>is an algebra, this correspondance is induced by the usual duality between</u> $mod(A^{op})$ <u>and</u> $mod(A)$.

The usual duality is $hom_Z(,I)$ for $Z = center(A)$ and $I =$ the injective envelope of $Z/rad(Z)$. The existence theorem of [1] implies that if M is an indecomposable object of $mod(A^{op})$, if N is the dual of M and if x is a non-zero element in the socle of the $\text{end}_A(N)$ -module $M \otimes_A N$, the image of the morphism $x : (M, \ ) \longrightarrow \otimes_A N$ is a simple object of $C(A)$; lemma 3 follows.

In general the above correspondence is not bijective, for instance, if A is not right artinian, there is a simple object S of $\text{mod}(A)$ such that $\otimes_A S$ has no simple subobject in $C(A)$. I shall describe now a modification of the above argument which works under weaker assumptions.

Let M be a finite A -module, $C_M = \text{end}_A(M)$; consider the condition

(*) M is a C_M -module of finite length.

Note incidentally (without proof) the

Lemma 4 - M <u>satisfies</u> (*) <u>iff for any indecomposable direct summand</u> F <u>of</u> M <u>and any set</u> I , <u>there is a set</u> J <u>such that</u> $F^I \simeq F^{(J)}$.

If (*) is satisfied, C_M is left artinian and $(X, \otimes_A M)$ is a finite C_M -module for every object X in $C(A)$. An object X in $C(A)$ is called <u>neglegible</u> iff $(X, \otimes_A M) = 0$; each object X in $C(A)$ has a greatest neglegible subobject, which is the (finitely defined) intersection of the kernels of the morphisms from X to $\otimes_A M$. Assume now that M is a <u>generator</u> of $\text{mod}(A)$ satisfying (*): one then calculates the greatest neglegible subobject of $\otimes_A N$ as follows: Take an exact sequence $M^q \xrightarrow{e} M^p \longrightarrow N \longrightarrow 0$ and put $S = \text{coker}\ \text{hom}_A(e,M) \in \text{mod}(C_M)$; the required subobject is the monomorphism $\text{ext}^1_{C_M}(S, \otimes_A M) \longrightarrow \otimes_A N$, as proved in [2] .

Proposition 1 - <u>Let</u> M <u>be a generator of</u> $\text{mod}(A)$ <u>satisfying</u> (*). <u>The following conditions are equivalent</u>:

(i) <u>every object in</u> $C(A)$ <u>has a greatest neglegible quotient</u>

(ii) $\text{mod}(C_M)$ <u>has a finite cogenerator</u>

(iii) <u>for every</u> $P \in \text{mod}(A^{op})$, $f(P)$ <u>is a finitely generated</u> C_f <u>-module</u> (<u>where</u> f <u>is the coherent functor</u> $\text{ext}^1_{C_M}(C_M/\text{rad}(C_M), \otimes_A M)$ <u>and</u> C_f <u>is the endomorphism ring of</u> f)

(iv) <u>the object</u> $\otimes_A M$ <u>in</u> $C(A)$ <u>is the injective envelope of its socle</u>.

(iv) is simply (i) applied to the coherent functor $\text{hom}_{C_M}(C_M/\text{rad}(C_M), \otimes_A M)$. Assuming (iv), one gets (ii) by remarking that if $(P, \ )$ is the projective cover of the socle of $\otimes_A M$, then $P \otimes_A M$ is an injective cogenerator of $\text{mod}(C_M)$. Assume (ii):

then, for every finite C_M -module S, $\mathrm{ext}^1_{C_M}(C_M/\mathrm{rad}(C_M),S)$ is finite over C_M, a fortiori (iii) is true. Assume (iii): to prove (i), it is sufficient to prove that each projective $(P_o, \ )$ in $C(A)$ has a greatest neglegible quotient. For any P in $\mathrm{mod}(A^{op})$, denote by $Z(P)$ the intersection of the kernels of the morphisms from $(P,$ to f (by (iii) this intersection is finitely defined). By induction on i, define a morphism $u_i : P_i \longrightarrow P_{i+1}$ such that $(P_{i+1}, \ )$ is a projective cover of $Z(P_i)$. Let n be the smallest integer such that $\mathrm{rad}(C_M)^n = 0$. I claim that the cokernel of $(u_{n-1}u_{n-2}\dots u_o, \ ) : (P_n, \ ) \longrightarrow (P_o, \ )$ is the greatest neglegible quotient of $(P_o, \ )$ Since M is a generator of $\mathrm{mod}(A)$, it is sufficient to show that for any finite C_M module S, the functor $\mathrm{ext}^1_{C_M}(S, \otimes_A M)$ kills $u_{n-1}u_{n-2}\dots u_o$. Put $S_i = (\mathrm{rad}(C_M))^{n-i}$. then, by the very definition of u_i, one builds by induction on i a commutative diagram

$$\begin{array}{ccccc}
(P_{i+1}, \ ) & \xrightarrow{(u_i, \ )} & (P_i, \ ) & & \\
\downarrow & & \downarrow & & \\
\mathrm{ext}^1_{C_M}(S/S_{i+1}, \otimes_A M) & \longrightarrow & \mathrm{ext}^1_{C_M}(S/S_i, \otimes_A M) & \longrightarrow & \mathrm{ext}^1_{C_M}(S_{i+1}/S_i, \otimes_A M)
\end{array}$$

(using the exactness of the last row and the semi-simplicity of S_{i+1}/S_i). For $i = n-$ this gives the desired assertion.

Corollary - <u>If every object of</u> $\mathrm{mod}(A)$ <u>satisfies</u> (*), <u>then every non-zero object of</u> $C(A)$ <u>has a simple subobject.</u>

It is sufficient to verify condition (iii) of prop.1 for every generator M of $\mathrm{mod}(A)$: let $d : P \longrightarrow C_M$ a projective cover of $\mathrm{rad}(C_M)$ in $\mathrm{mod}(C_M)$; put $N = \mathrm{coker}\ \mathrm{hom}_{C_M}(d,M)$; the condition follows from the fact that N satisfies (*) .

To end this paper I want to give (without proof) some little information about a problem discussed by M. Auslander in [1] . Recall from [1] that the following conditions on an Artin ring A :

(i) A is of finite representation type

(ii) every object in $C(A)$ has finite length

(iii) every object in $C(A)$ is noetherian

are connected by the implications (i)$\Leftrightarrow$(ii)$\Rightarrow$(iii) and are equivalent when A is an algebra. Standard categorical arguments show that (iii) means: every A -module is a direct sum of finite A -modules. The problem is whether conditions (i) and (iii) are always equivalent: of course it would be sufficient to show that under condition (iii), every non-zero object of $C(A)$ has a simple subobject (this was the motivation for prop.1 above); in fact, I do not know whether (iii) implies A right artinian.

For an Artin ring satisfying (iii) one can prove the following:

(a) for M and N in $\mathrm{mod}(A)$, (M,N) is a $(\mathrm{end}_A(M))^{op}$-module of finite length and $\mathrm{mod}(\mathrm{end}_A(M)^{op})$ has a cogenerator (cf. condition (*));

(b) the Krull-dimension of the noetherian category $C(A)$ is 0 or $u+2$ for some ordinal u; in the second case there is a finite A-module M, finitely presented over $\mathrm{end}_A(M)$, and such that $\mathrm{end}_A(M)$ is coherent and not artinian;

(c) there is a sequence (M_n) of finite A-modules, such that M_p and M_q have no common direct summand $(p \neq q)$, and that for any n and any indecomposable A-module F, not a direct summand of M_p $(p < n)$, there is a monomorphism from F to some power of M_n.

I have been unable to decide if $C(A)$ can be of Krull-dimension 2 !

<u>References:</u>

[1] M. Auslander, Representation theory of Artin algebras I, II, III (to be published in Communications in Algebra).

[2] M. Auslander and M. Bridger, Stable module theory, Memoir A.M.S. n° 94.

[3] P. Freyd, Proceedings of the conference on categorical algebra, La Jolla, 1965.

U.E.R. de mathématiques

Université de Lille I

B.P. 36

59650 Villeneuve d'Ascq, France.

INDECOMPOSABLE MODULES WITH CYCLIC VERTEX

Wolfgang Hamernik

1. Introduction

By a result of D.G. Higman (cf. [1], (64.1)) the group algebra FG of a finite group G over a field F of characteristic $p \neq 0$ has just a finite number of (isomorphism classes of) indecomposable modules if and only if a Sylow p-subgroup of G is cyclic. It is well known that this assertion remains true for blocks $B \leftrightarrow e$ of FG, if one replaces "Sylow p-subgroup of G" by "defect group of the block $B \leftrightarrow e$".

Bearing in mind that the vertex of an indecomposable module in a block $B \leftrightarrow e$ of FG is always contained in a defect group of the block the statements above are special cases of the

Theorem

Let D be a p-subgroup of the finite group G and let $B \leftrightarrow e$ be a block of FG having a defect group containing D. Then the following statements hold:

a) $B \leftrightarrow e$ contains (at least) one indecomposable FG-module with vertex D.

b) D is cyclic if and only if $B \leftrightarrow e$ contains only a finite number of indecomposable FG-modules with vertex D.

c) If D is not cyclic, then $B \leftrightarrow e$ contains indecomposable FG-modules with vertex D of arbitrarily large dimension.

Concerning our notation we are considering finitely generated unitary right modules. Saying a block or group algebra has an infinite (or just a finite) number of indecomposable modules tacitely means that these are mutually non-isomorphic. By a block $B \leftrightarrow e$ of FG we understand a pair consisting of the block ideal B and the corresponding block idempotent e. If $H \leq G$ and U is an FH-module, we denote by U^G the induced module $U \otimes_{FH} FG$; for an FG-module V the FH-module $V|_H$ is obtained from V restricting the operators. For the elementary properties of defect groups we refer to [5] and for the theory of vertices to [3].

2. Vertices and the Green correspondence

Before we enter into the proof of the theorem we state some auxiliary facts (and reminders).

2.1 Lemma

Let D be a finite p-group which is not cyclic. Then FD has an infinite number of indecomposable modules with vertex D of arbitrarily large dimension.

Proof

Given an indecomposable FD-module X with vertex D_X we know by Green's theorem ([3], Theorem 9) that $|D : D_X|$ divides $\dim_F X$. The construction of Heller and Reiner (cf. [1], (64.3)) yields that FD has an indecomposable module of dimension $2n+1$ for every positive integer n. Since there are infinitely many odd integers not divisible by p the assertion is proved.

2.2 Lemma

Let H be a normal subgroup of G and let U be an indecomposable FH-module with vertex D which is normal in G. Then D is the vertex of every indecomposable component V of U^G.

Proof

Denoting by Γ a transversal of H in G we have

$$(U^G)|_H \cong \sum_{g\in\Gamma} \oplus (U \otimes g) .$$

Hence by the Krull-Schmidt theorem there are $g_1, \ldots, g_s \in G$ such that

$$V|_H \cong (U \otimes g_1) \oplus (U \otimes g_2) \oplus \ldots \oplus (U \otimes g_s).$$

Clearly the vertex of $U \otimes g_i$ is D for every $i \in \{1,\ldots,s\}$. If D_1 is a vertex of V, then by ([3], Theorem 6)

$$D \leq D_1$$

bearing in mind that D is normal in G. This proves the lemma since V is D-projective.

In the proof of the theorem we will make use of the Green correspondence ([4]); we recall the notations and statements:

For some arbitrary (but fixed) p-subgroup D of G (which is not normal in G) consider a subgroup H ($\neq G$) of G containing $N_G(D)$ and define the sets

$$\mathfrak{X} = \{D \cap D^g \mid g \in G \setminus H\}$$

$$\mathfrak{Y} = \{H \cap D^g \mid g \in G \setminus H\}$$

$$\mathfrak{A} = \{D' \leq D \mid \text{there is no } Z \in \mathfrak{X} \text{ with } D' \leq Z\} .$$

Green showed in [4]

2.3 Proposition (cf. [2], p. 167).

For $D_1 \leq D$ the following hold:

a) If X is an indecomposable FG-module with vertex D_1, then there is an indecomposable component Y of $X|_H$ with vertex D_1 such that X is a component of Y^G and $X|_H = Y \oplus Y'$ where Y' is $\mathfrak{Y}$-projective.

b) If Y is an indecomposable FH-module with vertex D_1, then there is an indecomposable component X of Y^G with vertex D_1 such that Y is a component of $X|_H$ and $Y^G = X \oplus X'$ where X' is $\mathfrak{X}$-projective.

2.4 Green correspondence ([4], Theorem 2)

There is a vertex preserving one-to-one correspondence between the set of isomorphism classes of indecomposable FG-modules V with vertex in $\mathfrak{A}$ and the set of isomorphism classes of indecomposable FH-modules U with vertex in $\mathfrak{A}$. U and V correspond if and only if V is a component of U^G (or equivalently U is a component of $V|_H$). Moreover

$$V|_H = U \oplus U' \text{ where } U' \text{ is } \mathfrak{Y}\text{-projective,}$$
$$U^G = V \oplus V' \text{ where } V' \text{ is } \mathfrak{X}\text{-projective.}$$

3. Proof of the theorem

(i) Consider first the case that D is normal in G.

The group algebra FD has a non-empty family of indecomposable modules A_i, $i \in I$, with vertex D and by 2.1 we know that in case D is not cyclic there are A_i of arbitrarily large dimension.

Let M be a simple FG-module in the block $B \leftrightarrow e$ of FG. M can be considered as a simple $F^{G}/_{D}$-module, i.e. up to isomorphism $M_{FG} \leq (F^{G}/_{D})_{FG}$. For every $i \in I$ all composition factors of A_i are isomorphic to the unique simple FD-module F_D. Since FG is a free (hence flat) FD-module, $F_D \otimes_{FD} FG \cong F^{G}/_{D}$ appears as a factor in $(A_i)^G$ for all $i \in I$. By the Krull-Schmidt theorem every $(A_i)^G$ $(i \in I)$ has an indecomposable component V_i belonging to the block $B \leftrightarrow e$, as M is a composition factor of $(A_i)^G$. Applying 2.2 we conclude that each V_i has vertex D. This proves a) under our assumption $D \trianglelefteq G$.

Suppose D is not cyclic. By the Krull-Schmidt theorem there are elements $g_{i1}, \dots, g_{in_i} \in G$ such that

$$(V_i)|_D = (A_i \otimes g_{i1}) \oplus \dots \oplus (A_i \otimes g_{in_i})$$

for every $i \in I$. Since there are A_i of arbitrarily high dimension there are V_i of arbitrarily high dimension also. In particular, $B \leftrightarrow e$ has infinitely many indecomposable modules with vertex D of arbitrarily large dimension.

(ii) Now let D be arbitrary. Denoting by $\sigma = \sigma_D$ the Brauer homomorphism with respect to D we have $\sigma(e) \neq 0$ since a defect group $\hat{D}$ of $B \leftrightarrow e$ contains D. This yields that

$$\sigma(e) = f_1 + f_2 + \dots + f_t$$

with block idempotents $f_1, \dots, f_t$ of FH, where $H = N_G(D)$.

Since D is normal in H we know that D is contained in every defect group of the blocks $b_j \leftrightarrow f_j$ $(j = 1,\dots,t)$ of FH. So the hypothesis of part (i) of the proof holds for all blocks $b_j \leftrightarrow f_j$ of FH. In particular, there is a non-empty family U_k, $k \in \Lambda$, of indecomposable FH-modules with vertex D satisfying

$$U_k\, \sigma(e) \neq 0 \quad (k \in \Lambda),$$

and under the assumption that D is not cyclic there are U_k of arbitrarily large dimension.

Consider the Green correspondence with respect to D between the indecomposable FG-modules and the indecomposable FH-modules. For all $k \in \Lambda$ let V_k be the Green correspondent of U_k. We are going to show that $V_k e = V_k$ for every $k \in \Lambda$.

In fact, by Conlon's theorem (cf. [2], p. 185) there is an indecomposable module W_k satisfying $W_k e = W_k$ such that

(*) U_k is a component of $(W_k)|_H$ and
W_k is a component of $(U_k)^G$.

By 2.3 b) there is (up to isomorphisms) exactly one component of $(U_k)^G$ (call it X_k) with vertex D. Suppose $W_k \not\cong X_k$. Then some vertex D_1 of W_k is properly contained in D and from 2.3 a) we conclude that either U_k is $\mathfrak{Y}$-projective or else has vertex $D_1 \neq D$. At any rate D is not a vertex of U_k, contradicting our assumption (U_k to be $\mathfrak{Y}$-projective would mean that there exist $g \in G \setminus H$, $h \in H$ such that $D^h \leq H \cap D^g$, i.e. $D \leq H \cap D^{gh^{-1}} = D^{gh^{-1}} \cap D \in \mathfrak{X}$, a contradiction).

Hence $W_k \cong X_k$ and in view of (*) and 2.4 we find that $W_k \cong V_k$ is the Green correspondent of U_k $(k \in \Lambda)$.

Since the Green correspondence is bijective and vertex preserving a), c), and the sufficiency of the condition in b) are proved.

(iii) Suppose now that D is cyclic. Then clearly FD has just a finite number of indecomposable modules and hence FG has just a finite number of D-projective indecomposable modules. This proves the theorem.

References

1. C.W. Curtis, I. Reiner, Representation theory of finite groups and associative algebras. Interscience, New York, 1962.

2. W. Feit, Representations of finite groups. Lecture notes, Yale University, 1970.

3. J.A. Green, On the indecomposable representations of a finite group. Math. Z. 70 (1958), 430 - 445.

4. ———, A transfer theorem for modular representations. J. Algebra 1 (1964), 73 - 84.

5. G. Michler, Blocks and centers of group algebras, in: Lectures on rings and modules, Springer Lecture Notes 246, 430 - 552 (Berlin, Heidelberg, New York, 1972).

Mathematisches Institut
der Universität

D 63 Giessen
Arndtstr. 2

W. Germany

UNIQUE DECOMPOSITION OF LATTICES OVER ORDERS

H. Jacobinski

Let o be a Dedekind ring with quotient field k and A a separable finite dimensional algebra over k and R an o-order in A (i.e. a subring of A with $1 \in R$ and $kR = A$ which at the same time is finitely generated as an o-module). An R-lattice is a finitely generated left R-module, which is torsion-free considered as an o-module. For a prime p of o let o_p be the localisation and $o_p^\times$ the completion of o at p, and put $R_p = o_p \otimes R$ and $R_p^\times = o_p^\times \otimes R$.

We say that the lattices over an order have unique decomposition (or that the Krull-Schmidt theorem holds) if in every decomposition $M = \bigoplus U_i$ into indecomposable lattices U_i, these U_i are uniquely determined up to order and isomorphism. It is well-known that the Krull-Schmidt theorem holds for $R_p^\times$ - lattices but that it does not hold in general for an arbitrary order R even if o is supposed to be local (cf. [5] , [6]). We will in the following give necessary and sufficient conditions on R in order that R-lattices have unique decomposition (Theorems 1,2 and 3). Finally we apply the general result to the case of lattices over the group ring oG of a finite group G (Theorem 4).

Orders over a local Dedekind ring.

We start by assuming that o is a local Dedekind ring with prime ideal p. Let $o^\times$, $R^\times$ be the completions. An arbitrary $R^\times$-lattice X will in general not be the completion of an R-lattice M. It is well-known that this is the case if and only if $k^{\times} \otimes X$ is the completion of an A-module. We define the complement $C(X)$ to be the minimal $A^{\times}$-module such that $k^{\times} \otimes X \oplus C(X)$ is the completion of an A-module. Clearly X determines $C(X)$ up to an isomorphism.

Now choose a maximal order O that contains R. Then $O^{\times}$ is a maximal order too. Two $O^{\times}$-lattices V and V' are isomorphic if and only if they span isomorphic $A^{\times}$-modules. Thus there is a unique $O^{\times}$-lattice $c(X)$ that spans $C(X)$. Then $X \oplus c(X)$ is an $R^{\times}$-lattice and it is the completion of an R-lattice M. Clearly, M is uniquely determined by X.

We need one more definition. Let $\Lambda(R)$ denote the set of all R-lattices, that do not have an O-lattice as a direct factor. $\Lambda(R)$ is well-defined even if R-lattices do not have unique decomposition; only in this case it may not be closed under direct sums. Similarly, we define $\Lambda(R^{\times})$ as the set of those $R^{\times}$- lattices, that do not have an $O^{\times}$-lattice as direct factor.

Proposition 1. R-lattices have unique decomposition if and only if $c(X \oplus Y) \cong c(x) \oplus c(Y)$ for all $X, Y \in \Lambda(R^{\times})$.

Proof. Put $X \oplus c(X) = M^{\times}$ and $Y \oplus c(Y) = N^{\times}$ where M and N are R-lattices. Moreover, since $X, Y \in \Lambda(R^{\times})$, M and N are in $\Lambda(R)$. If R-lattices have unique decomposition, this implies that $M \oplus N \in \Lambda(R)$. Now $c(X \oplus Y)$ is always a direct factor of $c(X) \oplus c(Y)$, in fact we have

$$c(X) \oplus c(Y) \cong c(X \oplus Y) \oplus V^{\times}$$

where V is an O-lattice. But $V^{\times}$ is a direct factor of $(M \oplus N)^{\times}$ and so V is a direct factor of $M \oplus N$. Since $M \oplus N \in \Lambda(R)$, this implies that V=0.

To show the converse, let U be an indecomposable R-lattice. Then $U^{\times}$ can be decomposable. Put $U^{\times} = X \oplus V$ with $X \in \Lambda(R^{\times})$ and V an $O^{\times}$-lattice. Clearly $c(X)$ must be a direct factor of V and since U is indecomposable, we have $V = c(X)$ provided $X \neq 0$. Now assume that X has a proper decomposition $X = X_1 \oplus X_2$ with $X_1 \neq 0, X$. But then $X_1 \oplus c(X_1)$ is the completion of an R-lattice U_1 and since $c(X) = c(X_1) \oplus c(X_2)$ we see that U_1 must be a direct factor of U, which is impossible. Thus we have shown that an indecomposable R-lattice U is either an O-lattice or $U^{\times} = X \oplus c(X)$ where X is an indecomposable $R^{\times}$-lattice. If we now consider an arbitrary R-lattice M we see that its decomposition into indecomposable lattices is completely determined by the decomposition of $M^{\times}$ into indecomposable lattices and so the decomposition of M is unique. This completes the proof.

Using the above additivity condition we can determine all orders with unique decomposition. We first define two types of $o^{\times}$-orders which we call σ-orders. Let first B be a simple $k^{\times}$-algebra, which is a ring of matrices of even degree 2r over a skew-field. Let Γ be a maximal order in B and $J(\Gamma)$ its radical. Then $\Gamma/J(\Gamma)$ is a ring of matrices of degree 2r over a skew-field F. We say that the $o^{\times}$-order S is a σ-order in B if $J(\Gamma) \subset S$ and if $S/J(\Gamma)$ is a ring of matrices of degree r over a skew-field K such that $(K:F) = 2$. Secondly, let $B = B_1 \oplus B_2$ be a direct sum of two simple $k^{\times}$-algebras and let $\Gamma = \Gamma_1 + \Gamma_2$ be a maximal order in B. Then S is a σ-order in B if S does not contain Γ_1 or Γ_2 but its projection on B_i equals Γ_i for $i=1,2$. Obviously such an order exists only if the algebras $\Gamma_1/J(\Gamma_1)$ and $\Gamma_2/J(\Gamma_2)$ are isomorphic.

We now return to our order R in the algebra A. Let B be a simple algebra occurring in a decomposition of A and e its unit element. Let

$$B^{\times} = B_1 \oplus \dots \oplus B_t$$
$$e = \varepsilon_1 + \dots + \varepsilon_t$$

be the decomposition of $B^{\times}$ into simple algebras and ε_i their unit elements. If l is a simple B-module and L_i a simple B_i-module we have

$$l^{\times} = \nu_1 L_1 \oplus \dots \oplus \nu_t L_t .$$

The number ν_i we call the splitting index of B_i. We now can state

Theorem 1. Let o be a local Dedekind ring and R an o-order in a separable algebra A. Then R-lattices have unique decomposition if and only if for every simple algebra B one of the following conditions is satisfied:

1) R contains a maximal order of B

2) $R^{\times}$ contains a maximal order of $B_2 \oplus \dots \oplus B_t$ and either

(i) B_1 has splitting index $\nu_1 = 1$, or

(ii) $\varepsilon_1 \in R^{\times}$, B_1 has splitting index $\nu_1 = 2$ and $\varepsilon_1 R^{\times}$ is a σ-order in B_1.

3) $R^{\times}$ contains a maximal order of $B_3 \oplus \dots \oplus B_t$, $\varepsilon_1 + \varepsilon_2 \in R^{\times}$, B_1 and B_2 have splitting index $\nu_1 = \nu_2 = 1$ and $(\varepsilon_1 + \varepsilon_2)R^{\times}$ is a σ-order in $B_1 \oplus B_2$.

Orders over semi-local Dedekind rings.

Now assume o to be semi- local with prime ideals $p_1, \dots, p_t$. Let us first remark that unique decomposition of R-lattices in this case cannot be trivially reduced to the localizations. In fact it is quite possible that R_{p_i}- lattices have unique decomposition for all i, but R-lattices do not have unique decomposition.

Since o is semi-local, every R-lattice M is uniquely determined, up to an isomorphism, by the set M_{p_i} , $i = 1,\dots,t$. Suppose that for every p_i there is a decomposition $M_{p_i} = X^i \oplus Y^i$ of M_{p_i} as an R_{p_i} -lattice. If all the A-modules kX^i are isomorphic, there is a decomposition $M = X \oplus Y$ such that $X^i \cong X_{p_i}$. Using this remark it is easy to show

Lemma. Let U be the set of those primes p of o for which R_p is not a maximal order and let $\bar{o}$ be the ring of quotients $\frac{a}{b}$, $a,b \in o$, such that b is not divisible by any prime of U. Put $\bar{R} = \bar{o} \otimes R$. Then R-lattices and $\bar{R}$-lattices have unique decomposition simultaneously.

In particular if U contains only one single prime p, i.e. R_q is a maximal order for all primes $q \neq p$, then R-lattices have unique decomposition if and only if R_p-lattices have. We now state

Theorem 2. Let o be a semi-local Dedekind ring and R an o-order in a separable algebra. Then R-lattices have unique decomposition if and only if

1) R has a two-sided decomposition

$$R = R^1 \oplus \dots \oplus R^t$$

such that for each i, R^i_q is a maximal order for all $q \neq p_i$.

2) R^i-lattices have unique decomposition for $i = 1,\dots,t$.

The second condition can be decided by means of Theorem 1, since R^i-lattices and $R^i_{p_i}$ -lattices have unique decomposition simultaneously.

For later use we mention the following

Corollary. Let e be a primitive central idempotent of A. If

R-lattices have unique decomposition, then there is a power p^a of a prime ideal p in o such that $p^a e \subset R$.

This follows at once from condition 2) since e must belong to one of the algebras kR^i.

<u>Orders over global Dedekind rings.</u>

Now let o be an arbitrary Dedekind ring whose quotient field k is a global field and R an o-order in a separable algebra. Recall that two R-lattices M and N belong to the same genus, notation $M \sim N$, if $M_p \cong N_p$ for all primes p. The sum of two genera $\Gamma_1 + \Gamma_2$ is defined in the natural way as the genus containing $M_1 \oplus M_2$ where $M_i \in \Gamma_i$. Then it is possible to talk about unique decomposition of genera. In terms of the lattices this means that if $M = \bigoplus_{i=1}^{s} V_i = \bigoplus_{i=1}^{t} W_i$ with V_i, W_i indecomposable, then $s = t$ and after rearranging, $V_i \sim W_i$ for all i.

The localisations R_p are maximal orders for all primes p except for finitely many. Let U be a finite non-empty set of primes containing all these. As before, let $\bar{o}$ be the ring of quotients $\frac{a}{b}$, $a,b \in o$ with b not divisible by any prime of U and put $\bar{R} = \bar{o} \otimes R$. Then it is well-known ([5], p. 174) that two R-lattices M,N are in the same genus if and only if $\bar{o} \otimes M$ and $\bar{o} \otimes N$ are isomorphic $\bar{R}$-lattices. Thus genera of R-lattices have unique decomposition if and only if $\bar{R}$-lattices have. Since $\bar{o}$ is semi-local, this can be decided by means of Theorem 1 and 2.

<u>Theorem 3.</u> Let o be a Dedekind ring whose quotient field is a global field and R an o-order in a separable algebra. Then R-lattices have unique decomposition if and only if

1) the genera of R-lattices have unique decomposition

2) all R-lattices in the same genus are isomorphic.

The proof follows from a result about genera ([3], p.16 and p.25). Let M be an R-lattice and put $N = M \oplus M$ and let M' be an arbitrary R-lattice in the same genus as M. Then there exists $M'' \sim M$, such that $N \cong M' \oplus M''$. From this follows at once the necessity of condition 2). It is at this point we use the assumption that k is a global field.

Group rings.

We now apply the general result to the case of the group ring oG of a finite group G. We always assume that kG is separable. Let q be a prime of o that does not divide the group order $|G|$. Then $o_q G$ is a maximal order. Let $\bar{o}$ be the ring of quotients $\frac{a}{b}, a,b \in o$ with $(b,|G|) = 1$. Then genera of oG-lattices have unique decomposition if and only if $\bar{o}G$-lattices have unique decomposition.

The following necessary condition for unique decomposition of oG-lattices was proved by Dress [1] by means of group theoretical methods.

Proposition 2. If oG-lattices have unique decomposition, then the group order generates a prime-power in o, i.e. $o|G| = p^a$.

Proof. If oG-lattices have unique decomposition, then $\bar{o}G$-lattices have so too. Now $e = \frac{1}{|G|} \sum g_i$ is a primitive central idempotent in kG. From the corollary to Theorem 2 we obtain that $p^a e \subset \bar{o}G$ and so $o|G| = p^a$.

Now assume o to be semi-local. By applying Theorems 2 and 3 we obtain necessary and sufficient conditions for unique decomposition of oG-lattices. Let B be a simple algebra that occurs in a decomposition of kG, e its unit element and l a simple B-module.

Theorem 4. Let o be a semi-local Dedekind ring and G a finite group. Then oG-lattices have unique decomposition if and only if

1) $o|G| = p^e$, p a prime ideal of o, and

2) for every simple algebra B one of the following conditions holds

(i) oG contains a maximal order of B, or

(ii) $1_p^\times$ is a simple $B_p^\times$-module, or

(iii) $1_p^\times = L \oplus L'$, with L, L' simple $B_p^\times$-modules, the unit element e lies in oG and $e(o_p^\times G)$ is a $\mathfrak{G}$-order in $B_p^\times$.

The three alternatives in condition 2) are equivalent to the conditions of Theorem 3. The simplification comes from a formula for the conductor of a maximal order with respect to the group ring (Jacobinski [4], Th.3). This formula implies that if $B_p^\times = B_1 \oplus \ldots \oplus B_t$ and if $o_p^\times G$ contains a maximal order of one of the B_i, then it contains a maximal order of all of the others.

References

[1] Dress, A.: On the Krull-Schmidt theorem for integral group representations of rank 1, Mich. Math. J. 17 (1970), 273-277.

[2] Heller, A.: On group representations over a valuation ring, Proc. Nat. Acad. Sci. USA 47 (1961), 1194-1197.

[3] Jacobinski, H.: Genera and decompositions of lattices over orders, Acta Math. 121 (1968), 1-29.

[4] Jacobinski, H.: On extensions of lattices, Mich. Math. J. 13 (1966), 471-475.

[5] Reiner, I.: Survey of integral representation theory, Bull. Amer. Math. Soc. 76 (1970), 159-227.

[6] Reiner, I.: Failure of the Krull-Schmidt Theorem for integral representations, Mich. Math. J. 9 (1962), 225 -231.

Chalmers University of Technology,

Gotenburg, Sweden

THE LOCAL INDEX OF ELEMENTS IN THE SCHUR GROUP

Gerald J. Janusz

For a field K, the Schur group, $S(K)$, is the subgroup of the Brauer group of K consisting of those classes $[A]$ which contain an algebra A that is K-isomorphic to a simple direct summand of the group algebra $K[G]$ of some finite group G. The elements in $S(K)$ are represented by crossed products of the following special kind. Let ε be a root of unity, $H = Gal(K(\varepsilon)/K)$, and α a factor set on H which has roots of unity for its values. Then

$$A = (K(\varepsilon)/K,\alpha) = \sum_{\sigma \text{ in } H} K(\varepsilon)u_\sigma$$

is called a cyclotomic algebra. The multiplication rules are

$$u_\sigma u_\tau = \alpha(\sigma,\tau)u_{\sigma\tau} \quad \text{for } \sigma,\tau \text{ in } H,$$
$$u_\sigma x = \sigma(x)u_\sigma \quad \text{for } x \text{ in } K(\varepsilon).$$

The elements u_σ, σ in H, along with a root of unity ε selected so that all values of α are powers of ε, generate a finite group, G, and $K[G]$ can be mapped onto A. It is clear then that A represents an element of $S(K)$. The converse is a theorem of Brauer and Witt.

We are concerned with the problem of computing the index of A. We will assume throughout that K is an algebraic number field so the index of A is equal to its exponent -- that is the order of the class $[A]$ in $S(K)$.

A beautiful theorem of Benard-Schacher gives a uniform bound on all the elements of $S(K)$.

<u>THEOREM</u>. If the group of roots of unity in K has order n, then the order of each element in $S(K)$ divides n.

The detailed study of elements in the Brauer group is made by passing to the completions of K. If $\mathfrak{q}$ is a prime, $K_{\mathfrak{q}}$ the completion, then an element $[A]$ in $S(K)$ is uniquely determined by the classes $[K_{\mathfrak{q}} \otimes A]$ in $S(K_{\mathfrak{q}})$ as $\mathfrak{q}$ ranges over all primes of K. Thus it is necessary to study the complete case. Here Yamada has done the computation of $S(K_{\mathfrak{q}})$.

<u>THEOREM</u>. Let $K_{\mathfrak{q}}$ be the completion of K at a prime $\mathfrak{q}$ which divides the rational prime q. Suppose q is odd. Then $S(K_{\mathfrak{q}})$ is the cyclic group of order $(q-1)/e_0$, where e_0 is the factor of $e(K_{\mathfrak{q}}/Q_q)$, the ramification index, which is prime to q. If $q = 2$, $S(K_{\mathfrak{q}})$ has order 1 or 2 and conditions on $K_{\mathfrak{q}}$ can be given to distinguish the two cases. In particular if ϵ_4 is in $K_{\mathfrak{q}}$, then $S(K_{\mathfrak{q}})$ has order 1.

This theorem is helpful because the index of an algebra A over the global field is the least common multiple of the $\mathfrak{q}$-local indices -- that is the indices of $K_{\mathfrak{q}} \otimes A$ -- as $\mathfrak{q}$ ranges over the primes. A theorem of Benard simplifies this situation somewhat. It says that whenever $\mathfrak{q}_1$ and $\mathfrak{q}_2$ are primes of K lying over the same rational prime q , then the $\mathfrak{q}_1$-local index of an element in S(K) is the same as its $\mathfrak{q}_2$-local index. Thus we may speak of the q-local index as the index of $K_{\mathfrak{q}} \otimes A$ for $\mathfrak{q}$ any of the primes over q .

The Schur group is the direct sum of its primary subgroups so it is sufficient to determine $S(K)_p$, the subgroup of elements of p-power order. If [A] is in $S(K)_p$, then the results above give the following bounds:

Let p^a be the order of the group of p-power roots of unity in K and for a prime q , let p^c be the highest power of p dividing q - 1 . Then

(B) q - local index of $A \leq \min\{p^a, p^c\}$.

For certain K , this bound is sharp. For example take $K = Q(\varepsilon_p a)$. If p is odd, then for every prime q , except q = p , we have an algebra A_q such that

$$q\text{ - local index } A_q = \min\{p^a, p^c\} .$$

In fact we may arrange r-local index $A_q = 1$ for all primes $r \neq q$.

For other fields this bound is not sharp. Take p an odd prime, r a prime such that p divides $r - 1$, and q a prime such that p divides $q - 1$ and p divides the order of q modulo r . For example take $(p,q,r) = (3,13,7)$. If $K = Q(\varepsilon_p,\varepsilon_r)$ then the q-local index of each element of $S(K)_p$ is 1 so the inequality in (B) is proper.

For the rest of the paper let us only consider $S(K)_p$ when p is odd. As usual, the case $p = 2$ requires exceptions be made in many theorems.

The problem of computing sharp bounds for the q-local index of elements in $S(K)_p$ is greatly reduced by the following result.

<u>THEOREM</u>. Let $L = Q(\varepsilon_m)$ and K a subfield of L . The group $S(K)_p$ is generated by classes of algebras which contain representatives of the type $A_q = (L(\varepsilon_q)/K,\alpha)$ where α is a factor set whose values are p-power roots of unity in L , and q is a prime not dividing m . The r-local index of A_q can be greater than 1 only if $r = q$ or r divides m .

Since L and K are fixed, this leaves only a finite amount of computation to determine the possible r-local indices. Once the factor set α is given, it is straight foreward, though tedious, to compute the r-local index of A_q . The main difficulty is the construction of all possible factor sets needed to maximize the r-local index. This can be done in a way that is sufficiently explicit to permit the computation of the local index of each element in $S(K)_p$. The complete statement of the results requires a great deal of notation. We shall limit ourselves to the determination of the q-local index for the primes q which do not divide m.

Fix the field K which is a subfield of $L = Q(\varepsilon_m)$ and the integer m is taken as small as possible for the given field K.

Notation: p^{a+b} = exact power of p dividing m,

ζ = root of unity of order p^{a+b} in L,

$K \cap \langle \zeta \rangle = \zeta^{p^b}$, root of order p^a in K,

$G = \mathrm{Gal}(L/K)$,

$C = \mathrm{Gal}(L/K(\zeta))$,

σ = element of G such that $\sigma(\zeta) = \zeta^{1+p^a}$,

q = prime not dividing m,

$\mathfrak{q}$ = prime of K over q,

f(q) = residue class degree of $\mathfrak{q}$ over q,

ϕ = Frobenius automorphism of $\mathfrak{q}$ acting in L,

$\phi = \sigma^k \xi$, $0 \leq k < p^b$ ξ in C,

X(q) = integer which satisfies the congruence

$$X(q)\ \frac{(1+p^a)^{p^b}-1}{p^a} \equiv \frac{q^{f(q)}-(1+p^a)^k}{p^a}$$

This congruence is taken modulo p^{a+b} and does always have a solution $X(q)$. Finally let $N(q)$ denote the order of the coset containing $\sigma^{-X(q)p^b}\zeta$ in the group C/C^{p^d} where $p^d = \gcd(p^a, q-1)$; let $n(q)$ be the non-negative integer defined as follows. If p^a divides $f(q)$, then $n(q) = 0$; if p^a does not divide $f(q)$, then $p^{a-n(q)}$ is the exact power of p which does divide $f(q)$.

<u>THEOREM</u> The maximum q-local index of an element in $S(K)_p$ is $\max\{N(q), p^{n(q)}\}$. This maximum is attained by an algebra class which is split at all primes other than q.

This result must be modified in several ways to discuss the r-local index of elements in $S(K)_p$ when r is a prime divisor of m. Another term which reflects the inertia of r between K and L must be included in addition to the two terms analogous to $N(q)$ and $p^{n(q)}$. In addition to this, there is a further difficulty because one cannot always attain the maximum r-local index by an algebra which is split at all other primes. Examples of this phenomenon can be given for each p by selecting a suitable K. Full proofs of these results will appear in [2].

References

1. G. J. Janusz, Generators for the Schur group of local and global number fields, (To appear in Pacific J. Math.)
2. ____________ The Schur group of an algebraic number field, (To appear)
3. T. Yamada, The Schur subgroup of the Brauer group, Lecture Notes in Math. 397, Springer 1974.

University of Illinois
Urbana, Illinois 61801

QUASI-FROBENIUS-ALGEBRAS OF FINITE REPRESENTATION TYPE

H. Kupisch

1.

In this paper we are concerned with Quasi-Frobenius-Algebras of finite representation type, more precisely, with properties of such algebras, which are closely related to the Cartan-invariants.

Let R be an algebra of finite dimension over an algebraically closed field K , N = rad R , n = number of simple R-left-moduls,

$$R = \sum_{i=1}^{n} \sum_{j=1}^{f_i} Re_{ij} \ , \ Re_{i1} \cong Re_{ij}, \ Re_{i1} \not\cong Re_{k1} \quad \text{for } i \neq k \ ,$$

where the e_{ij} are local idempotents,

$e_i = e_{i1}$, $e = \sum_i e_i$, $R^o = eRe$ the basic algebra of R ,

c_{ij} , $i,j = 1,\ldots,n$ the Cartan-invariants of R .

Theorem 1. If R is a Quasi-Frobenius-Algebra of finite representation type and $E = \sum_{c_{ii}>2} e_{ij}$, then ERE is a generalized uniserial Quasi-Frobenius-Algebra.

Theorem 2. If R is a symmetric algebra of finite representation type, which is indecomposable (as a two-sided ideal), then there is a number c such that

$$c_{ii} = c + 1 \quad \text{for all } i \text{ with } c_{ii} > 2 \ .$$

Theorem 2 is well known in case R = B is a block of finite representation type of a group-algebra KG of a finite group

G [2] . Moreover in this case the number $a(R)$ of non-isomorphic indecomposable R-moduls is given [5, 9] by

(1.1.) $$a(R) = n(cn + 1) .$$

The same is true for indecomposable symmetric generalized uniserial algebras, in particular for simple algebras.
It is, more generally, true for all indecomposable symmetric algebras of finite representation type whose Cartan-matrix comes from a Brauer-tree [5, 9] in the sense of the following definition.

<u>Definition</u>. An indecomposable symmetric algebra R is a <u>Ca-algebra</u> if

(1) R is of finite representation type .

(2) The Cartan-matrix (c_{ij}) of R is given by a connected graph with n+1 vertices $\{V_\alpha\}$, one vertex V_o singled out, and n edges, corresponding to the n indecomposable projective moduls Re_i , such that for $i \neq j$

$$c_{ij} = \begin{cases} c \text{ if } Re_i \text{ and } Re_j \text{ meet in } V_o \\ 1 \text{ if } Re_i \text{ and } Re_j \text{ meet in } V_\alpha \neq V_o \\ 0 \text{ else.} \end{cases}$$

<u>Theorem 3.</u> If R is an indecomposalbe symmetric algebra and $a(R)$ is the number of non isomorphic indecomposable R-left-moduls, then we have

(1) $a(R) \geq n(cn + 1)$.

(2) If $a(R) = n(cn + 1)$, then R is a Ca-algebra (or simple).
In particular $Ne_i = L_{i1} + L_{i2}$, $L_{i1} \cap L_{i2} = \text{Soc } Re_i$ with uniserial moduls L_{i1} and L_{i2} .

It is easy to construct all Ca-algebras by means of generalized uniserial algebras [1o].

We shall prove Theorem 2 in section 2, whereas Theorem 3 will be proved in sections 2 - 6 .
Theorem 1, which has been proved for symmetric algebras in [7], is contained in [1o] and will not be proved here. However we mention the following lemma which, in the non symmetric case, is essential.
For a Quasi-Frobenius-Algebra R there is a permutation $p \in S_n$ such that

(1.2) $$\mathrm{Soc}\, Re_i \cong Re_{p(i)}/Ne_{p(i)} \quad .$$

Let $X_1,\dots,X_s$ be the orbits of p and

$$E_t = \sum_{i \in X_t} e_{ij}$$

<u>Lemma.</u> If R is of finite representation type then E_tRE_t is a generalized uniserial Quasi-Frobenius-Algebra for every $t = 1,\dots,s$.

Observing that the basic algebra R^o has an automorphism h (Nakayama's automorphism) with the property

$$R^o e_{p(i)} \cong R^o h(e_i) \quad , \qquad i = 1,\dots,n$$

the proof follows immedeately from the fact, that for an ideal of R the corresponding graph (resp. quiver) has no cycles [4, 3]. This is aquivalent to the condition [7, Hilfssatz 1]:

For every sequence of pairwise independant elements

(1.3) $u^{m_1}_{i_1 i_2}$, $u^{m_2}_{i_3 i_2}$, $u^{m_3}_{i_3 i_4}$, $u^{m_4}_{i_5 i_4}$,...., $u^{m_{r-1}}_{i_r i_{r-1}}$
of A_R [6, (2.3)] we have $i_r \neq i_1$ $(r \geq 2)$,

2.

We now assume, that $R = R^o$ is a weakly symmetric K-algebra (i.e. p = id in 1.2) of finite representation type which is indecomposable as a two-sided ideal.

A basic concept in the proofs of theorem 2 and theorem 3 is that of a V-sequence which, in a slightly different form, hat been used in [9] .

2.1. Definition. A sequence $J = (j_1,\dots,j_s)$, where

$$j_i \in I = \{1,\dots,n\}$$

is called a V-sequence, provided

$$c_{j_i j_k} \neq 0 \quad \text{iff} \quad |i-k| \leq 1 .$$

$|J| = s$ is called the length of J .

As an easy consequence of the fact, that R is a block [1] and that

$$c_{ij} = c_{ji} \quad \text{for } i,j = 1,\dots,n$$

holds in a weakly symmetric algebra we obtain

(2.2.) For every pair $i,k \in I$ there is a V-sequence $J = (i = j_1,\dots$ $\dots, j_s = k)$ between i and k.

Proof of Theorem 2.

Let

$$c_i = c_{ii} - 1 \quad \text{for } i = 1,\dots,n .$$

Because of [7, Satz 6] we have to show that ERE is indecomposable or aquivalently

(2.3) $$c_{ik} \neq 0 \quad \text{if } c_i, c_k > 1 .$$

Assume, that (2.3) fails. Choose a minimal V-sequence $J=(i=j_1,\dots,j_s = k)$ between i and k (i.e. a V-sequence J between i and k for which $|J|$ is minimal). Then we have $s > 2$ and

(2.4) $$c_{j_q} = 1 \text{ iff } q \neq 1,s \quad .$$

In A_R [6, 2.3] we consider the following sequences of elements

(2.5a) $$u_{j_1j_2},\ u_{j_3j_2},\ u_{j_3j_4},\dots,u_{j_{s-1}j_s},\ u_{j_sj_s}\ ,\ u_{j_sj_{s-1}}\ , \\ u_{j_{s-2}j_{s-1}},\dots,u_{j_2j_3},\ u_{j_2j_1}\ ,\ u_{j_1j_1}$$

and

(2.5b) $$u_{j_1j_2},\ u_{j_3j_2}\ ,\dots,\ u_{j_sj_{s-1}},u_{j_sj_s}\ , \\ u_{j_{s-1}j_s}\ ,\ u_{j_{s-1}j_{s-2}}\ ,\dots,\ u_{j_2j_1}\ ,\ u_{j_1j_1}$$

according to whether s is even or not.

From (2.4) and [6, section 2] and from the definition 2.1. it follows that the elements in each of these sequences are pairwise independent. This is impossible because of (1.3). So (2.3) holds.

3.

This section contains some simple facts concerning V-sequences which will be used later.

By Theorem 2 we can number the idempotents e_i in such a way that

(3.1) $$\begin{aligned} c_i &= c \quad \text{for } i \in I_1 = \{1,\dots,m\} \\ c_i &= 1 \quad \text{for } i \in I_2 = \{m+1,\dots,n\} \quad . \end{aligned}$$

From Theorem 2 and (2.2) we get at once

(3.2) For every pair $i_1, k_1 \in I_2$ one of the following assertions holds

(a) There is some $p \in I_1$ and minimal V-sequences $J_1 = (i_1, \ldots, i_s, p)$ and $J_2 = (k_1, \ldots, k_r, p)$ such that

$$i_j \, , \, k_t \in I_2 \quad \text{for } j = 1, \ldots, s \, , \; t = 1, \ldots, r \, .$$

(b) There is no $p \in I_1$ such that (a) holds. Then there are p and q in I_1, $p < q$ and minimal V-sequences $J_1 = (i_1, \ldots, i_s, q, p)$ and $J_2 = (k_1, \ldots, k_r, p)$ such that

$$i_j \, , \, k_t \in I_2 \text{ for } j = 1, \ldots, s \, , \; t = 1, \ldots, r \, .$$

We now assign to each pair i_1, $k_1 \in I_2$ an element $p \in I_1$ which has the property (3.2a) or (3.2b) and denote it by

$$p = p(i_1, k_1) \, .$$

3.3. <u>Definition</u>. Let i_1, $k_1 \in I_2$. Two V-sequences J_1 and J_2 are called a <u>normal pair</u> with respect to i_1 and k_1 if either

(a) J_1 and J_2 are of the form $J_1 = (i_1, \ldots, i_s, p)$, $J_2 = (k_1 = i_r, i_{r+1}, \ldots, i_s, p)$, $r \geq 1$, where $p \in I_1$ and $i_j \in I_2$ for $j = 1, \ldots, s$.

or

(b) In case there is no $p \in I_1$ such that (a) holds, J_1 and J_2 are minimal V-sequences between i_1 and $p = p(i_1, k_1)$ resp. k_1 and $p(i_1, k_1)$.

(3.4) Let $J_1 = (i_1,\dots,i_s,p)$ and $J_2 = (k_1,\dots,k_r,p)$ be minimal V-sequences between i_1 and p resp. k_1 and p with the property

$$i_1 = k_t \text{ for some } t, \quad 1 \le t \le r .$$

a) If $J = (k_t, k_{t+1},\dots,k_r,p)$ then $|J| = |J_1|$.

b) $J_3 = (k_1,\dots,k_{t-1}, k_t = i_1,\dots,i_s,p)$ is a V-sequence.

Proof is immediate by the minimality of J_1 and J_2 .

(3.5) Let $J_1 = (i_1.,,,.i_s,p)$, $J_2 = (k_1,\dots,k_r,p)$ and $J_1' = (i_1',\dots,i_t',p)$, $J_2' = (k_1',\dots,k_f',p)$ resp. be normal pairs and let

$$\{i_1,\dots,i_s\} \cup \{k_1,\dots k_r\} = \{i_1',\dots,i_t'\} \cup \{k_1',\dots,k_f'\} \text{ and}$$

(1)

$$s+r = t+f$$

Then $J_1 = J_1'$ and $J_2 = J_2'$.

Proof. In case (3.3a) holds for the pair J_1, J_2 (1) implies that (3.3a) also holds for the pair J_1', J_2' . So by (1) the assertion follows.

In case (3.3b) holds for the pair J_1, J_2 , (1) implies

$$k_1' = k_d \text{ , } i_1' = i_m \text{ and } k_1 = k_q' \text{ or } i_1 = k_j'$$
for some d, m, q, j .

$J = (k_1,\dots,k_d = k_1',\dots,k_f',p)$ is by (3.4) a V-sequence. Consequently because of (3.3b) $i_1 = k_j'$ is not possible and $k_1 = k_q'$ is only possible for $d = 1$. Similar we have $m = 1$.

Remark. (3.5) is also valid, if the pair J_1', J_2' replaced by a minimal V-sequence J' between i_1' and p .

4.

In this section we discuss two types of modules which will be needed in the next section to construct enough indecomposable moduls.

1. Modules of type M(J).

Let $J = (j_1, \dots, j_s)$ be a V-sequence. In the direct sum

$$Re_{j_2} \oplus Re_{j_4} \oplus \dots \oplus Re_{j_{2k}} \quad , \quad 2k \leq s \leq 2k+1$$

we choose elements

$$v_{12}, v_{32}, v_{34}, \dots, v_{2k-1,2k}, v_{2k+1,2k} \quad , \quad \text{where}$$

(4.1) $$v_{iq} \in e_{j_i} Re_{j_q} \quad \text{and} \quad e_{j_i} Ne_{j_i} v_{iq} = 0 \quad ,$$

$$v_{2k+1,2k} = 0 \quad \text{for} \quad s = 2k$$

and define M(J) by

(4.2) $$M(J) = Rv_{12} + R(v_{32}+v_{34}) + \dots + R(v_{2k-1,2(k-1)} + v_{2k-1,2k}) + \\ + Rv_{2k+1,2k} \quad .$$

Because of (2.1) $M = M(J)$ has the following properties

(4.3) a) $\mathrm{Soc}\ R(v_{2i-1,2(i-1)} + v_{2i-1,2i}) \cong F_{2(i-1)} \oplus F_{2i}$,

where $F_j \cong Re_j/Ne_j$.

b) $\mathrm{Soc}\ M \cong F_{j_2} \oplus \dots \oplus F_{j_{2k}}$.

c) $M/NM \cong F_{j_1} \oplus \dots \oplus F_{j_{2k+1}}$ resp.

$\cong F_{j_1} \oplus \dots \oplus F_{j_{2k-1}}$ according to whether s is even or not .

d) $0 \leq g(M/NM) - g(\mathrm{Soc}\ M) \leq 1$ for the number $g(\ )$ of composition factors.

2. <u>Submodules of Re_p, $c_p > 1$.</u>

(4.4) Let $c_p > 1$, $c_{ip} \leq c_{kp} \leq c_{jp} < c_p$, $c_k < c_p$ and let

$v_i = e_i v_i$ and $v_j = e_j v_j$ be elements in Re_p such that

$e_i N e_j v_j = e_k N e_j v_j = 0$, $e_p N e_p N e_i v_i = 0$,

$(e_p N e_p)^{s-1} e_p R e_j v_j \neq 0$, $(e_p N e_p)^s e_p R e_j v_j = 0$.

The submodule $M = Rv_i + Rv_j$ of Re_p is called

a) <u>of type (i,p,k^s)</u> if $j = k$ $(s \geq 1)$

b) <u>of type (i,p,p^s)</u> if $j = p$ $(s \geq 2)$

c) <u>of type (i,p,q^s)</u> if $j = q$, where $c_{qp} = c_p$ and $c_{kp} = c_{iq} = 0$ $(s \geq 2)$.

(4.5) Let $c_p = c > 1$, $0 \leq c_{ip} \leq c_{kp} < c_p$, $c_{kp} \neq 0$.

a) There are at least $c_p - c_{kp}$ non-isomorphic modules of type (i,p,p^s).

b) There are at least $c_{kp} - 1$ non-isomorphic modules of type (i,p,k^s) and (k,p,i^s).

<u>Proof.</u> Because of $c_p > 1$ we know [7, Folgerung 2], that Re_p is right-regular. This means [6, Satz 1.1], that $e_j Re_p$, $j=1,\dots,n$, has a K-basis of the following form

$$\{ x_{jp} x_{pp}^m \,,\ m = 0,\dots,c_{jp}-1 \} ,$$

(4.6) where $x_{jp} \in e_j Re_p$, $x_{pp} \in e_p Ne_p$, $x_{pp} \notin (e_p Ne_p)^2$;

i.e. $\{ x_{pp}^m \,,\ m = 1,\dots,c_p \}$ is a K-basis of $e_p Ne_p$.

a) By [6, 2.4] we have

$$u_{kp} u_{pp}^{c_{kp}} = u_{ip} u_{pp}^{c_{kp}} = 0 .$$

This and [6, 1.14, 2.3, 2.4] imply that the modules

$$Rx_{ip}x_{pp}^{c_{ip}-1} + Rx_{pp}^{s}, \qquad s = c_{kp}, c_{kp}+1, \ldots, c_p-1$$

are of type (i,p,p^s).

b) Without restriction we can assume $c_{kp} \geq 2$.

In case $u_{kp} \nmid u_{ip}^j$ for every j we have

$$u_{ik}u_{kp} = 0 \quad .$$

Thus, using [6, F_5] and [7, 2.4]:

$$(4.7) \qquad u_{kk}u_{kp} = u_{ii}u_{ip} = 0$$

we find that the modules

$$Rx_{kp}x_{pp}^{s} + Rx_{ip}x_{pp}^{c_{ip}-1}, \qquad s=0,1,\ldots,c_{kp}-2$$

are of type (i,p,k^s). This case includes by (4.7) $i=k$.

In case $u_{kp} \mid u_{ip}^q$ for some q , let q be minimal. Then (4.7) implies

$$u_{ki}u_{ip}^j = 0 \quad \text{for every} \quad j \geq q \ .$$

Therefore in this case we find that the $c_{ip}-q$ modules

$$Rx_{ip}x_{pp}^{s} + Rx_{kp}x_{pp}^{c_{kp}-1}, \qquad s=q-1,\ldots,c_{ip}-2$$

are of type (k,p,i^s) and that the $c_{kp}-c_{ip}+ q-1$ modules

$$Rx_{kp}x_{pp}^{s} + Rx_{ip}x_{pp}^{c_{ip}-1}, \qquad s=c_{ip}-q+1,\ldots,c_{kp}-1$$

are of type (i,p,k^s).

(4.8) Let $c_p > 1$, $c_{qp} = c_p$, $c_{ip} = c_{kq} = 1$, $c_{kp} = c_{iq} = 0$. Then there are at least c_p-1 non-isomorphic modules of type (i,p,q^s).

<u>Proof</u>. Because of $u_{qp} \mid u_{pp}$ similar to proof of (4.5a).

5.

To prove part a) of theorem 3 we have to show that R has at least n^2c non-isomorphic non-projective indecomposable modules. To this aim we generalize in this section the construction [9] of indecomposable modules for Ca-algebras. Using the modules of section 4 we assign to each pair $(d,j) \in I^2$ at least c non-isomorphic non-projective indecomposable Modules.

We consider the following cases

(5.1) (a) $d,j \in I_1$

(b) $d \in I_1$, $j \in I_2$

(c) $d,j \in I_2$.

In case (a) we take the following submodules of Re_j

(5.2) $$Rx_{dj}x_{jj}^s \ , \quad s=0,\dots,c-1 \quad ,$$

In case (c) we choose a minimal V-sequence $J_o = (d=j_1,\dots,j_f=j)$ between d and j and a normal pair $J_1 = (d=i_t,\dots,i_1,p)$, $J_2 = (j=k_r,\dots,k_1,p)$ with respect to d,j .

Case (b) we obtain from case (c) by ommitting J_1,J_2 .

Therefore we can restrict to case (c).

According to (3.2),(3.3) and [7, Folgerung 2] we then have the following possibilities

(5.3) (a) $i_1, k_1 \in I_2$

(b) $i_1 \in I_2$, $k_1 = q \in I_1$. In this case

$$c_{i_1p} = c_{k_2q} = 0 \ , \quad c_{i_1q} = c_{k_2p} = 1 \ .$$

(5.4) Modules of type (V).

With the V-sequences J_o, J_1, J_2 we are given

$\bar{J}_1 = (i=i_1,\ldots,i_t)$,

$\bar{J}_2 = (k=k_1,\ldots,k_r)$, resp. $\bar{J}_2 = (k=k_2,\ldots,k_r)$ in case (5.3b) ,

$J_2^{-1} = (p,k_1,\ldots,k_r)$.

Let U_1, U_2, U_3 be submodules of Re_p of type (i,p,k^s) , (i,p,p^s), (i,p,q^s) resp. and let $L(J)$ be the sum of all the terms in (4.2) except the first one,

$$L(J) = R(v_{32}+v_{34}) + \ldots + Rv_{2k+1,2k} \cdot$$

Case $d \leq j$.

We define M_o by

$$(V_o) \qquad M_o = M(J_o)$$

and $M_1 \subset M(\bar{J}_1) \oplus U_1 \oplus M(\bar{J}_2)$, $M_2 \subset M(\bar{J}_1) \oplus U_2 \oplus M(J_2^{-1})$, and $M_3 \subset M(\bar{J}_1) \oplus U_3 \oplus M(\bar{J}_2)$ by

$$(V_1) \quad M_1 = R(v_i+v_{i_1i_2}) + R(v_k+v_{k_1k_2}) + L(\bar{J}_1) + L(\bar{J}_2)$$

$$(V_2) \quad M_2 = R(v_i+v_{i_1i_2}) + R(v_p+v_{pk_1}) + L(J_1) + L(J_2^{-1})$$

$$(V_3) \quad M_3 = R(v_i+v_{i_1i_2}) + R(v_q+v_{qk_2}) + L(\bar{J}_1) + L(\bar{J}_2) .$$

Case $d > j$.

Instead of the left-modules M_j we construct the corresponding right-modules M_j' and define M_j^* by

$$M_j^* = \mathrm{Hom}_K(M_j',K) \quad , \; j=0,\ldots,3 .$$

M_j, M_j^* and the modules (5.2) are called of type (V), (of type (V_j) resp.)

(5.5) Let $M = M_1$ be of type (V_1). Then we have

a) $\operatorname{Soc} M \cong \operatorname{Soc} M(\overline{J}_1) \oplus F_p \oplus \operatorname{Soc} M(\overline{J}_2)$.

b) $M/NM \cong M(\overline{J}_1)/NM(\overline{J}_1) \oplus M(\overline{J}_2)/NM(\overline{J}_2)$.

c) $|g(M/NM) - g(\operatorname{Soc} M)| \leq 1$.

Corresponding assertions hold for the other M_j .

<u>Proof</u> is immediate because of (4.3).

(5.6) a) To every pair $(d,j) \in I^2$ can be assigned at least c non-isomorphic non-projective modules of type (V).

b) Different pairs yield non-isomorphic modules.

<u>Proof</u>. a) follows from (4.4), (4.5), (4.8), (5.2) and from the fact that for modules M_j of type (V) we have

(5.7) $e_pM_j = e_pU_j$, $e_pNe_tM_j = e_pNe_tU_j$, $j=1,2,3$, $t=i,k,q$

b) follows from (3.3), (3.5), (4.3) and (5.5).

(5.8) Every module of type (V) is indecomposable.

<u>Proof</u>. Let M be of type (V_1). Then we have

$$M = \sum_j Rv'_{i_{2j-1}} + \sum_t Rv'_{k_{2t-1}} ,$$

where i_j and k_t are the elements of the V-sequences J_1 and J_2 in (5.4), and

$$v'_{i_1} = v'_i = v_i + v_{i_1 i_2} , \quad v'_{k_1} = v'_k = v_k + v_{k_1 k_2} ,$$

and the other v'_{i_j} and v'_{k_t} are as in (4.2).

By (5.7) we have $g_p(M) = g_p(U_1)$ for the number $g_p(\ )$ of composition factors which are isomorphic to F_p , in particular

(5.9) $$g_p(\mathrm{Soc}\ M) = 1 .$$

Let $h \in \mathrm{End}_R(M)$ with the property

$$h^2 = h \neq 0 .$$

Because of (5.9) we can assume without restriction

$$h(e_p \mathrm{Soc}\ M) = e_p \mathrm{Soc}\ M = e_p \mathrm{Soc}\ U_1 .$$

This implies

$$h(e_p M) = e_p M = e_p U_1 \quad \text{and}$$

(5.1o) $$h(v_p) = v_p$$

for a generating element $v_p = e_p v_p$ of $Re_p M$.

From (2.1), (5.7) and (5.1o) it follows

(5.11) $$h(v'_{k_{2j-1}}) = v'_{k_{2j-1}} + x_{2j-1} , \qquad x_{2j-1} \in (\sum_t Rv'_{i_{t-1}}) \cap \ker h$$

and a corresponding assertion for $v'_{i_{2t-1}}$.

Finally (4.3), (5.5) and (5.11) yield

$$h(\mathrm{Soc}\ M) = \mathrm{Soc}\ M$$

and thus $h = 1_M$.

For modules of type (V_2) and (V_3) the proof is similar.

6.

In this section we prove the second part of theorem 3.

Thus we now assume that

(6.1) $$a(R) = n(cn+1)$$

holds.

(6.2) For every $i \in I$ we have

a) Re_i is regular .

b) $g(\mathrm{Soc}\ Re_i/A_i) \leq 2$ for every submodule A_i of Re_i .

c) If Ne_i/N^2e_i is simple then Re_i is uniserial .

d) If u^m_{ji} and u^t_{ki} are independant then $c_{jk} = 0$.

Proof follows from (4.3), (5.5), (5.6) and (6.1).

(6.3) For every $i \in I$ there are uniserial submoduls L_{i1} and L_{i2} of Ne_i , such that

$Ne_i = L_{i1} + L_{i2}$, $L_{i1} \cap L_{i2} = \mathrm{Soc}\ Re_i$.

Proof follows from theorem 2, (6.2) and [6, F_7] and [7, Satz 1 and Satz 2] .

To show that R is a Ca-algebra it is sufficient because of [9,(5.5)] to prove that R has the properties Ca1-4 of [9, (1.6)] .

We define Q_{i1} and Q_{i2} by

$$Q_{it} = \{ j \in Q_i \mid g_j(L_{it}) \neq 0 \} , \quad t=1,2 ,$$

where Q_i is defined as in [9] .

(6.4) R has the properties Ca1-4.

Proof. Ca1 follows from (6.2d), Ca2 follows from (6.2b) and [6, F_7] . The first part of Ca3 followa from (6.2a), the second follows from the fact that in case $c_i=1$, c_{ki}, $c_{pi} \neq 0$, c_k, $c_p > 1$ there is a submodul in Re_p which is not of type (V).

Similar to the proof of theorem 2 it follows from (1.3) that those sequences which appear in Ca4 are V-sequences in the sense of the definition 2.1. Therefore Ca4 follows from the definition 2.1 and theorem 2.

(6.5) Remark. That R is a Ca-algebra follows already from (6.3) and the assumption that R is of finite representation type[1o]. That means, (6.1) is also aquivalent to: R has property (6.3) and is of finite representation type.

References.

1. Curtis, C.W. and Reiner, J: Representation theory of finite groups etc. New York 1962

2. Dade, E.C.: Blocks with cyclic defect groups, Ann. of Math. 84 (1966) 2o-48.

3. Gabriel, P.: Indecomposable representations II. Symposia Mathematica, INDAM Rome 11 (1973), 81-1o4.

4. Jans, J.P.: Indecomposable representations of algebras, Ann. of Math. 66 (1957).

5. Janusz, G.: Indecomposable moduls for finite groups, Ann. of Math. 89 (1969) 2o9-241.

6. Kupisch, H.: Symmetrische Algebren mit endlich vielen unzerlegbaren Darstellungen I, Reine Angew.Math.219 (1965) 1-25.

7. Kupisch, H.: Symmetrische Algebren II, Reine Angew.Math.245 (197o) 1-14.

8. Kupisch, H.: Projektive Moduln endlicher Gruppen mit zyklischer p-Sylow-Gruppe, J. Algebra 1o (1968) 1-7.

9. Kupisch, H.: Unzerlegbare Moduln endlicher Gruppen mit zyklischer p-Sylow-Gruppe, Math.Z. 1o8 (1969) 77-1o4.

1o. Kupisch, H.: Quasi-Frobenius-Algebren von endlichem Modul-Typ I. Erscheint.

11. Nakayama, T.: On Frobenius Algebras II, Ann. of Math. 42 (1941) 1-21.

12. Osima, M.: Notes on Basic-Rings I. Math.J. Okayama University 2 (1953) 1o3-11o.

INDECOMPOSABLE REPRESENTATIONS

OF FINITE ORDERED SETS

Michèle Loupias

In this paper I shall denote a finite (partially) ordered set, k a (commutative) field and $\mathcal{E}$ the category of the finite dimensional vector spaces over k. If we consider I as a category we note $_k\underline{I} = \mathrm{Hom}(I, \mathcal{E}) = $ the category of functors from I to $\mathcal{E}$. An object E of $_k\underline{I}$ is called a <u>representation</u> of I. The category $_k\underline{I}$ is a Krull-Remak-Schmidt category.

The set I is said of <u>finite representation type</u> (F.R.T) if $_k\underline{I}$ has only a finite number of indecomposable objects (up to isomorphism). The purpose of this work is to determine all the sets of F.R.T. We shall suppose from hereon that I is connected.

1. - <u>GENERALITIES - CRUCIAL SETS - CRITICAL SETS</u>

Some well known sets of F.R.T are the sets A_n, D_n, E_6, E_7 E_8 defined in [2] and the cycles defined in [1].

<u>1.1</u> - <u>PROPOSITION</u> - <u>The following sets are not of F.R.T</u>

$$\widetilde{E}_6 = c_2 - c_1 - \overset{\overset{\textstyle b_2}{|}}{\underset{|}{\overset{b_1}{a}}} - d_1 - d_2 \qquad \widetilde{E}_7 = c_3 - c_2 - c_1 - \overset{\overset{\textstyle b_1}{|}}{a} - d_1 - d_2 - d_3$$

$$\widetilde{E}_8 = c_2 - c_1 - \overset{\overset{\textstyle b_1}{|}}{a} - d_1 - d_2 - d_3 - d_4 - d_5 \qquad \widetilde{D}_1 = b_4 - \overset{\overset{\textstyle b_1}{|}}{\underset{\underset{\textstyle b_3}{|}}{a}} - b_2$$

$$\widetilde{A}_4 = a \overset{\nearrow \quad b}{\underset{\searrow \quad c}{}} \searrow d \qquad \qquad R_1 = a \overset{\nearrow \quad \overset{\textstyle d_1}{d}}{\underset{\searrow \quad c}{}} b - b_1 - b_2 - b_3 - b_4$$

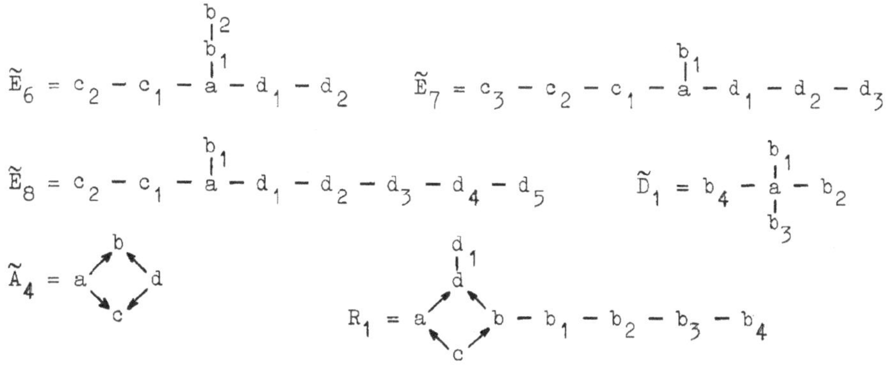

$$R_2 = \begin{array}{c} d_4 \\ | \\ d_3 \\ | \\ d_2 \\ | \\ d_1 \\ | \\ d \\ a \nearrow\searrow b - b_1 \\ \searrow\nearrow \\ c \end{array}$$

$$R_3 = a_2 - a_1 - a \begin{array}{c} d \\ \nearrow\searrow \\ \searrow\nearrow \\ c \end{array} b - b_1 - b_2$$

$$R_4 = a_1 - a \begin{array}{c} d \\ \nearrow\searrow \\ \searrow\nearrow \\ c \end{array} b - b_1 - b_2 - b_3 - b_4$$

$$R_5 = d \begin{array}{c} b \\ \nearrow\searrow \\ \searrow\nearrow \\ c \end{array} a - a_1 - a_2 - a_3 - a_4$$

$$R_6 = a_1 - a \begin{array}{c} d \\ \nearrow\searrow \\ x_1 \\ \nwarrow \\ c \end{array} b - b_1 - b_2 - b_3$$

$$R_7 = a_1 - a \begin{array}{c} d \\ z_1 \nearrow\searrow \\ x_1 \\ \nwarrow \\ c \end{array} b - b_1 - b_2$$

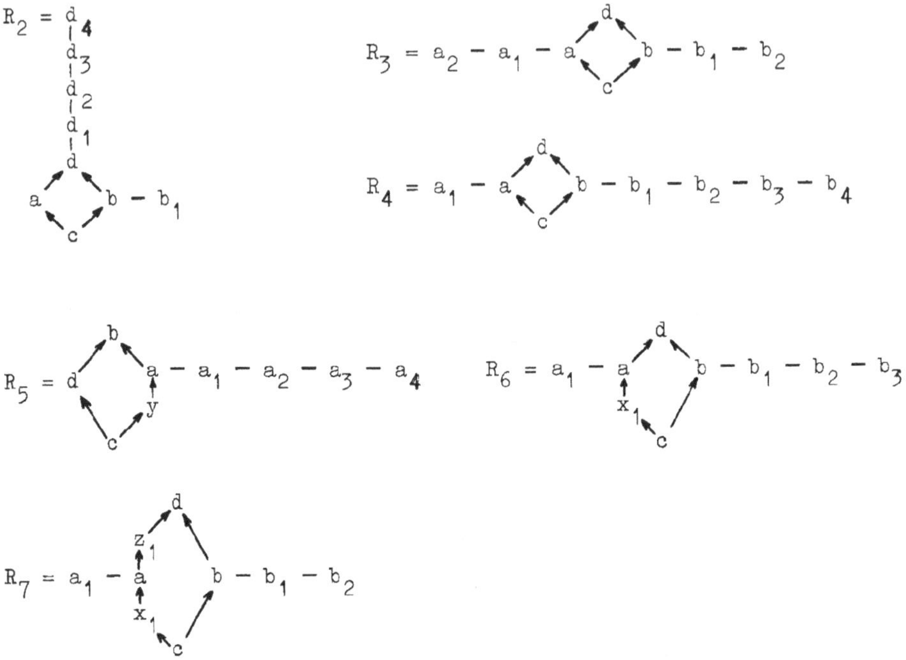

(when the orientation of the arrows is not indicated it is arbitrary).

The result is known for the sets $\widetilde{E}_6$, $\widetilde{E}_7$, $\widetilde{E}_8$, $\widetilde{D}_1$ and $\widetilde{A}_4$.

For R_1, R_2, R_3, R_4 it follows from an equivalence between the category of the representations E of the set

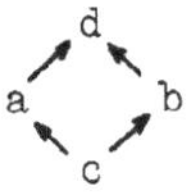

$a \begin{array}{c} d \\ \nearrow\searrow \\ \searrow\nearrow \\ c \end{array} b$ such that $\ker E(c \to a) \cap \ker E(c \to b) = 0$ and the category

of the representations F of the set $a' \to s' \overset{d'}{\underset{\uparrow}{\leftarrow}} b'$ such that
$\operatorname{Im} F(a' \to s') + \operatorname{Im} F(b' \to s') = F(s')$: to E we associate F, by
setting $F(s') =$ the fiber coproduct of $E(a)$ and $E(b)$ under $E(c)$,
and $F(a') = E(a)$, $F(b') = E(b)$, $F(d') = E(d)$.

Let $\mathbf{8}$ be the category of filtered vector spaces A , with
the filtration $B' \to X \to A \leftarrow A_1 \leftarrow A_2 \leftarrow A_3 \leftarrow A_4$.
$$\begin{array}{c} \uparrow \quad \uparrow \\ C \to Y \end{array}$$

Define a functor Φ from $\mathcal{S}$ to $R_{k=5}$ by setting

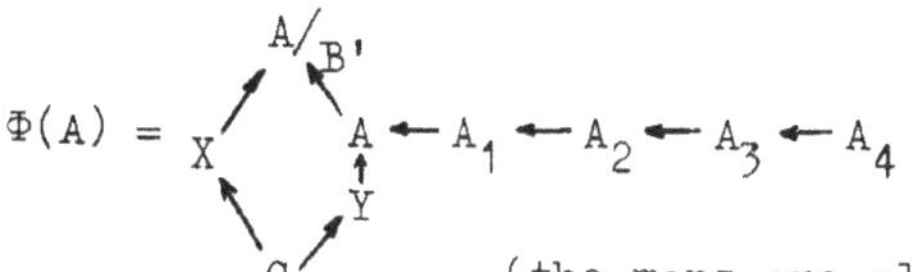

(the maps are all canonical)

It is obvious that Φ has a retraction, so the property is true for R_5 .

Similar proofs are used for R_6 and R_7 by taking $D_{k=1}$ instead of $\mathcal{S}$.

<u>DEFINITIONS</u> - We shall call the <u>opposite set of</u> I , the subjacent set of I , noted I^o , having the order opposite of that of I . Then I is of F.R.T if and only if I^o is.

We say that I is <u>crucial</u> if I or I^o is a set of the list 1.1.

<u>1.2</u> - <u>PROPOSITION</u> - <u>If I is of F.R.T, then every subset of I , provided with the induced order is of F.R.T.</u>

<u>1.3</u> - <u>PROPOSITION</u> - <u>Let I and J be two finite ordered sets, and f a surjective morphism of ordered sets from I onto J , such that $f^{-1}(j)$ is connected for every $j \in J$, then if I is of F.R.T so is J .</u>

<u>DEFINITION</u> - If I and J are as in 1.3, we say that J is a <u>contracted set of</u> I .

<u>Example of a contracted set</u> :

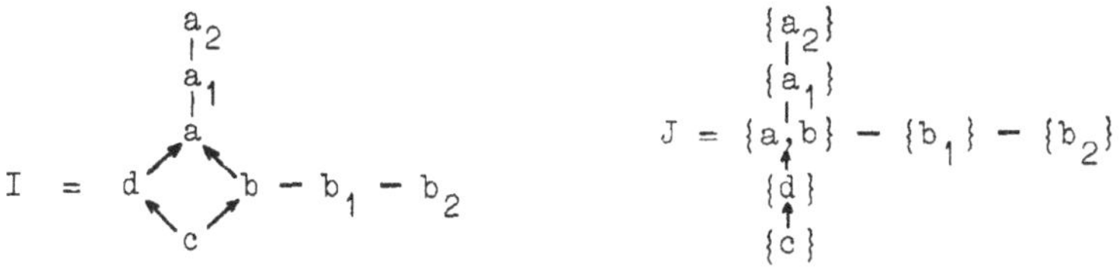

DEFINITIONS - I is critical if I is not of F.R.T, but every proper subset and every proper contracted set of I is of F.R.T. I is kind if none of its subsets is crucial or has a contracted set which is crucial. Every set of F.R.T is kind.

The purpose of this work is to show how we obtain the two following results.

1.4 - THEOREM - The critical sets are exactly the crucial sets.

1.5 - THEOREM - The sets of F.R.T are exactly the kind sets.

2. - KIND SETS

The notions of intervals and convex subsets are well known.

DEFINITION - Two elements a and b of I are said neighbours if they are different, comparable and if $]a,b[=]b,a[= \emptyset$.

2.1 - In a kind set, every element has at most three neighbours.

DEFINITION - Let J be a subset of I . We say that J is a thread if 1) J is linear and convex in I . Thus $J =$

$$a_0 - a_1 \dots - a_{n-1} - a_n \quad (n \geq 1)$$

2) a_i has only two neighbours in I , a_{i-1} and a_{i+1} , $1 \leq i \leq n-1$

a_n has only one neighbour in I , a_{n-1}

a_0 has at least three neighbours in I .

We say that a_0 is the bond point of J and a_n is the extremity of J .

2.2 - Let J be a thread of I . The changing of the orientation of the arrows in J does not alter the nature of I (concerning the properties of kindness or of F.R.T).

2.3 - A kind set has at most three threads.

2.4 - LEMMA - *Let* I *be kind,* H *and* K *two subsets of* I *such that*

1) card H *and* car $K \geq 2$;
2) $H \cup K$ *is connected* ;
3) $H \cap K = \{a\}$;
4) $\forall\, b \in K$ *with* $b \leq c$ (*resp* $b \geq c$) *one has* $b \leq a \leq c$ (*resp* $b \geq a \geq c$) .

Then, one of the two sets H *or* K *is linear and has* a *for extremity.*

2.5 - COROLLARY - *If* I *is kind, the bond points of two threads are the same if and only if* I *is equal to* D_n, E_6, E_7 *or* E_8 .

2.6 - PROPOSITION - *Let* I *be kind with* p *threads* $(p \leq 3)$. *The number* n_1 (*resp* n_2) *of maximal* (*resp minimal*) *points of* I *which do not belong to a thread is* $\leq 3 - p$.

Thus, we can class the kind sets by studying the diverse repartitions $p+n_1 \leq 3$ and $p+n_2 \leq 3$.

We give for instance the case of a set with three threads.

2.7 - *Let* I *be kind and* x *and* $y \in I$, x *and* y *not comparable* ; *then the set* $H = \{z \in I \mid z \leq x,\ z \leq y\}$ *is empty or is a linear interval.*

We note $x \wedge y$ the maximal element of H , when it exists. We have a similar definition for $x \vee y$.

2.8 - *Suppose* I *is kind with three threads. Then* I *is necessarily equal to* D_n, E_6, E_7, E_8 *or to one of the sets*

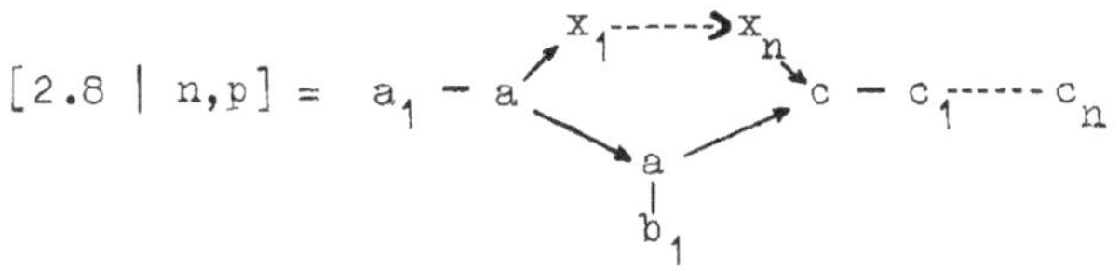

With $n,p \geq 1$, *or to its opposite set.*

Let J_1, J_2, J_3 the threads and a, b, c their respective bond points (supposed distincts) and α, β, γ their respective extremities. Set $I' = [I - (J_1 \cup J_2 \cup J_3)] \cup \{a,b,c\}$. From 2.6 $\{a, b, c\}$ contains the extremal elements of I' and one has $b \leftarrow a \rightarrow c$ or $a \rightarrow b \rightarrow c$ (up to duality).

In the first case one has $b \wedge c = a$ and $[a,b]$ and $[a,c]$ must be linear which is impossible.

In the second case, b is the neighbour of β, a and c have two neighbours in I', b is a neighbour of a and c and then $[a,c]$ is a cycle and $I = [2.8 \mid n,p]$ or $I = [2.8 \mid n,p]^0$.

$\underline{\underline{2.9}}$ - A $\underline{\text{necessary}}$ $\underline{\text{condition}}$ $\underline{\text{for}}$ $I = [2.8 \mid n,p]$ $\underline{\text{to be kind is either}}$ $n \leq 2$ $\underline{\text{and}}$ $p = 1$ $\underline{\text{or}}$ $n = 1$ $\underline{\text{and}}$ $p \leq 4$.

$I - \{a, a_1\}$ must be kind, as the contracted set of I equal to $a_1 - a - x_1 \cdots x_n - \{b,c\} - c_1 \cdots c_n$
$$\underset{b_1}{\overset{|}{}}$$

3. - SETS OF FINITE REPRESENTATION TYPE

In this paragraph we describe, on an example, technics which we shall use frequently.

Set $I = [2.8 \mid 2, 1] = a_1 \rightarrow a_2 \begin{smallmatrix} \nearrow x_1 \searrow \\ \searrow b \nearrow \end{smallmatrix} c \leftarrow c_1 \leftarrow c_2$
$$\underset{b_1}{\overset{\uparrow}{}}$$

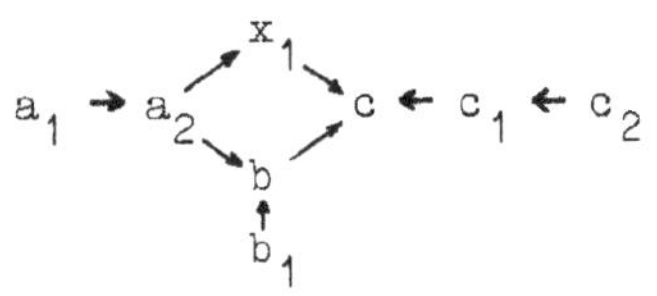

and set $J = I - \{c, c_1, c_2\}$.

Let E be an object of ${}_k\underline{J}$ (resp ${}_k\underline{I}$); we set $\Lambda(E) =$ the inductive limit of E (resp of the restriction of E to J). We say that E is Λ-faithful if we cannot have $E = E_1 \oplus E_2$ with $E_1 \neq 0$ and $\Lambda(E_1) = 0$. One sees that if E is an indecomposable object of ${}_k\underline{J}$ we have $\Lambda(E) = 0$ or k.

Let $\mathbf{C}_o$ be the full subcategory of ${}_k\underline{\underline{J}}$ formed with the Λ-faithful representations and let $\mathbf{C}_\Lambda$ be the category such that its objects are identical to the objects of $\mathbf{C}_o$ and its morphism set between two objects M and N is equal to $\{\Lambda u \in \mathrm{Hom}_{\mathcal{C}}(\Lambda M, \Lambda N) \mid u \in \mathrm{Hom}_{\mathbf{C}_o}(M, N)\}$. We define the functor $\overline{\Lambda}$ from $\mathbf{C}_o$ to $\mathbf{C}_\Lambda$ by $\overline{\Lambda} M = M$ and if $u \in \mathrm{Hom}_{\mathbf{C}_o}(M, N)$, $\overline{\Lambda} u = \Lambda u$.

3.1 - *$\mathbf{C}_\Lambda$ is a Krull-Remak-Schmidt category. The functor $\overline{\Lambda}$ reflects the representations* (i.e. *it is full, representative, and it reflects the isomorphisms*).

3.2 - *If E and F are indecomposable objects in $\mathbf{C}_\Lambda$, the relation $\mathcal{R}$ defined between their isomorphy classes by $E \,\mathcal{R}\, F \Leftrightarrow \mathrm{Hom}_{\mathbf{C}_\Lambda}(E,F) \neq 0$ is an ordering relation noted $\leq$.*

Let S be the set of isomorphy classes of indecomposable objects of $\mathbf{C}_\Lambda$, then S is equal to :

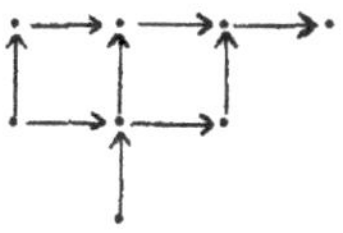

3.3 - *Let D be an object of $\mathbf{C}_\Lambda$ and $X \in S$. Let $D = \bigoplus_{\alpha \in A} D_\alpha$ be a decomposition of D in a direct sum of indecomposable objects of $\mathbf{C}_\Lambda$. Then the subobject of D equal to $\sum_{\mathrm{cl} D\alpha \geq X} D_\alpha$ depends only on X and not on the decomposition of D. We note it $\Phi_X(D)$ and Φ_X defines a functor from $\mathbf{C}_\Lambda$ to $\mathbf{C}_\Lambda$.*

3.4 - Let ψ be the functor from $\mathbf{C}_\Lambda$ to the category of S^o-spaces which to E associates $\psi(E) = V = \Lambda(E)$, the S^o-filtration being defined by $V(X) = \Lambda\, \Phi_X(E)$ for every $X \in S^o$. *The functor ψ is an equivalence between $\mathbf{C}_\Lambda$ and the category $\mathcal{P}$ of projective S^o-spaces V such that $\sum_{X \in S^o} V(X) = V$.*

3.5 - $I = [2.8 \mid 2,1]$ is of F.R.T

Let $\mathfrak{D}_0$ be the category of Λ-faithful representations of I. By following the previous method, one shows there exists a reflecting representations functor Ξ, from $\mathfrak{D}_0$ to the category $\mathfrak{L}$ of which the objects are the 5-uples (V, C, C_1, C_2, γ) where V is an object of $\mathfrak{P}$, C an object of $\mathcal{E}$, γ is a surjective k-space morphism from V onto C and C_1 and C_2 are two subspaces of C, such that $C_2 \subset C_1$. We provide C with the S^o-filtration defined by $C(X) = \gamma V(X)$ for every $X \in S^o$. Let P be the projective cover of C in the category of S^o-spaces. Then $V = P \oplus V_1$ and $(V, C, C_1, C_2, \gamma) = (P, C, C_1, C_2, \gamma) \oplus (V_1,0,0,0,0)$. The indecomposable objects of $\mathfrak{L}$ of the form $(V_1,0,0,0,0)$ correspond bijectively with the indecomposable faithful representations E of I such that $E(c) = 0$.

The indecomposable objects of $\mathfrak{L}$ of the form (P, C, C_1, C_2, γ) with $C \neq 0$, are entirely determined by the gift of the T-space C by setting $T = S^o \amalg \{c_2 \to c_1\}$.

This process allows us to count the number of indecomposable representations of I, there are 117 of them.

4. - CONCLUSION

Methods analogous to that of paragraph 2 and natural combinations enable us to establish necessary conditions for a set to be kind. By methods identical to those of paragraph 3 we show that the so founded sets are of F.R.T. So theorem 1.5 is proved.

We have obtained all the sets of F.R.T, and it is easy to see that the proper subsets and the proper contracted sets of the crucial sets are in this list. Thus every crucial set is critical. Conversely, if I is critical, it is not kind, thus exists a subset J of I which is contracted in a crucial set but this is impossible if $J \neq I$ or if the contraction is not injective. Thus theorem 1.4 is proved.

We obtain further the two following results.

4.1 - *If* I *is of* F.R.T *and if* E *is an indecomposable representation of* I *then we have* $\operatorname{Sup} \{\dim_k E(i) \mid i \in I\} \leq 6$.

4.2 - *If* I *is an interval* $[a,b]$ *of* F.R.T *the indecomposable representations* E *of* I *are such that : either* $E(a) = 0$, *either* $E(b) = 0$, *or* $E(i) = k \quad \forall\, i \in I$, *and* $E(i \to j) = id_k \quad \forall\, i,j \in I$, $i \leq j$.

BIBLIOGRAPHY

[1] N. CHAPTAL - Objets indécomposables dans certaines catégories de foncteurs, C.R. Acad. Sc., Paris, 268 (1969), 934-936.

[2] P. GABRIEL - Unzerlegbare Darstellungen I, Man. Math. 6 (1972), 71-103.

[3] P. GABRIEL - Représentations indécomposables des ensembles ordonnés d'après Nazarova-Roiter, Sém. Dubreil, Paris (1972/73).

[4] M.M. KLEINER - Partially ordered sets of finite type, Zapiski naučnykh Seminarov Leningr. Otd. Mat. Inst. Seklova, t. 28 (1972), 32-41.

[5] B. MITCHELL - Theory of categories, Academic Press, New-York, (1965).

[6] L.A. NAZAROVA and A.V. ROITER - Representations of partially ordered sets, Zapiski naučn. Sem. Leningr. Otd. Mat. Inst. Seklova, t. 28 (1972), 5-31.

Michèle Loupias
Département de Mathématiques
Faculté des Sciences, Parc de Grandmont
F-37200 TOURS

GREEN CORRESPONDENCE BETWEEN BLOCKS WITH CYCLIC DEFECT GROUPS II

Gerhard O. Michler

Introduction

Based on J. A. Green's new methods (9) in his theory of corresponding modules in this series of articles the module theory of blocks with cyclic defect groups is studied over arbitrary fields and complete discrete rank one valuation rings. The main results of the author's talk at the "Ottawa conference on representations of algebras" are collected in

THEOREM 0.1. Let F be an arbitrary field of characteristic $p > 0$ dividing the order $|G|$ of the finite group G. If $B \leftrightarrow e$ is a block of the group algebra FG with a cyclic defect group D and inertial index t, then the following assertions hold:

a) t divides $p-1$.

b) B contains t non-isomorphic simple FG-modules.

c) D is a vertex of every simple FG-module of B.

d) B contains $t|D|$ non-isomorphic indecomposable FG-modules.

The proof of this theorem appears in (14).

R. Brauer's definition of the inertial index of a block ((2), p.508) although meaningful even for non-splitting fields, gives in general a number which is too large for our purposes. Therefore in (14) another inertial index t of a block is introduced. Let $H = N_G(D)$ be the normaliser of D in G, and let $B_1 \leftrightarrow e_1 = \sigma_D(e)$ be the unique block of FH with defect group D corresponding to $B \leftrightarrow e$ under the Brauer correspondence σ_D. Then in (14) the inertial index t of $B \leftrightarrow e$ is defined as the number of non-isomorphic simple FH-modules of B_1, and it is shown ((14), Proposition 4.6) that t coincides with R. Brauer's inertial index, if D is cyclic, and if F is a splitting field for G and all its subgroups. Therefore Theorem 0.1 generalises the corresponding results of E. C. Dade (4), W. Feit and B. Rothschild (18), G. Janusz (11) and H. Kupisch (12). Whereas their work depends heavily on E. C. Dade's character theory of blocks with cyclic defect groups (4) the proofs of (14) are purely ring theoretical and avoid characters. On the other side several results of P. Gabriel's article (6) suggest that their theorems extend to blocks with cyclic defect groups over arbitrary fields and arbitrary complete discrete rank one valuation rings of characteristic zero.

In order to generalise E. C. Dade's theorem on the decomposition numbers of a block with cyclic defect groups (4), in this and the following part of this series of articles we are concerned with the question: Which indecomposable FG-modules M of a block $B \leftrightarrow e$ with cyclic defect group $D = (x)$ are liftable?

Throughout this article R denotes an arbitrary complete discrete rank one valuation ring of characteristic zero such that its residue class field $F \cong R/\pi R$ modulo its maximal ideal $\max(R) = \pi R$ has characteristic $p > 0$. The quotient field of R is denoted by K. If G is an arbitrary finite group, then the group algebra KG is a semi-simple artinian classical quotient ring of the group ring RG, and there is a natural epimorphism $\rho : RG \rightarrow FG$ with kernel πRG. If p divides the order $|G|$ of G, the triplet (K,R,F) is called a p-modular system for G. Since R is complete, for every block $B \leftrightarrow e$ of FG there is a unique block $\hat{B} \leftrightarrow \hat{e}$ of RG such that $\rho(\hat{B}) = B$, $\rho(\hat{e}) = e$, and B and $\hat{B}$ have the same defect groups

((5), p.300). Following J. A. Green (8), p.165, an indecomposable FG-module M of B is called <u>liftable</u>, if there exists an indecomposable lattice W of B such that $W/W\pi \cong M$. As is usual a finitely generated RG-module W is called a lattice, if W is torsion free as an R-module.

In order to find the liftable FG-modules of a block $B \leftrightarrow e$ with cyclic defect groups D in this article we study minimal projective resolutions of indecomposable FG-modules of B whose source is the trivial FD-module F_{FD}. Our main result is stated as

THEOREM 0.2. Let (K,R,F) be an arbitrary p-modular system for the finite group G. Let $B \leftrightarrow e$ be a block of the group algebra FG with a cyclic defect group $D = (x)$ and inertial index t. Then the following assertions hold:

a) B contains t non-isomorphic indecomposable FG-modules M_i whose source is the trivial FD-module F_{FD}.

b) If $0 \to \Omega M_i \to P_i \to M_i \to 0$ is a minimal projective resolution of M_i, then the FD-module $(1-x)FD$ is a source of ΩM_i for $i = 1,2,\dots,t$.

c) If $0 \to \Omega^2 M_i \to Q_i \to \Omega M_i \to 0$ is a minimal projective resolution of ΩM_i, then the trivial FD-module F_{FD} is a source of $\Omega^2 M_i$ for $i = 1,2,\dots,t$.

d) There exist indecomposable lattices W_i of $\hat{B} = \hat{e}RG$ whose source is the trivial RD-module R_{RD} such that $W_i/W_i\pi \cong M_i$ for $i = 1,2,\dots,t$.

e) Among the indecomposable lattices of $\hat{B} = \hat{e}RG$ with source R_{RD} each lattice W_i is uniquely determined by M_i up to RG-module isomorphism.

f) If $0 \to \Omega W_i \to \hat{P}_i \to W_i \to 0$ is a minimal projective resolution of W_i, then the RD-module $(1-x)RD$ is a source of ΩW_i, and $\Omega W_i/(\Omega W_i)\pi \cong \Omega M_i$ for $i = 1,2,\dots,t$.

g) Each ΩW_i is an indecomposable lattice of $\hat{B} = \hat{e}\, RG$.

h) ΩW_i is uniquely determined by W_i up to RG-module isomorphism.

i) If $0 \to \Omega^2 W_i \to \hat{Q}_i \to \Omega W_i \to 0$ is a minimal projective resolution of ΩW_i, then the trivial RD-module R_{RD} is a source of $\Omega^2 W_i$, and $\Omega^2 W_i/(\Omega^2 W_i)\pi \cong \Omega^2 M_i$ for $i = 1,2,\ldots,t$.

j) The numbering of the t non-isomorphic FG-modules M_i of B with trivial source F_{FD} can be so chosen that $M_i \cong \Omega^{2(i-1)} M_1$ for $i = 2,3,\ldots,t$, and $M_1 \cong \Omega^{2t} M_1$, $W_i \cong \Omega^{2(i-1)} W_1$ for $i = 2,3,\ldots,t$, and $W_1 \cong \Omega^{2t} W_1$.

k) The 2t FG-modules $\Omega^i M_1$, $i = 1,2,\ldots,2t$, are mutually non-isomorphic.

l) The 2t RG-modules $\Omega^i W_1$, $i = 1,2,\ldots,2t$, are mutually non-isomorphic.

If $B \leftrightarrow e$ is the principal block of FG, then Theorem 0.2 immediately implies the main theorem of J. Alperin and G. Janusz (1), p. 403, which asserts that each term of the minimal projective resolution of the trivial FG-module F_{FG} is indecomposable, and that such a projective resolution is periodic of period 2t. Another application of Theorem 0.2 is Corollary 5.1: The principal block B_o of a group algebra FG of a finite group G over an arbitrary field F of characteristic $p > 0$ is uniserial if and only if every simple FG-module M of B_o has trivial source F_{FD}, and the Sylow p-subgroups of G are cyclic.

By means of the results of (14) Theorem 0.2 is easily proved for blocks with cyclic normal defect groups D (Proposition 2.5). Therefore in general it is true for the block $B_1 \leftrightarrow e_1 = \sigma_D(e)$ of FH, $H = N_G(D)$, where σ_D denotes the Brauer correspondence with respect to D. Let $Y \neq 1$ be the socle of the cyclic defect group D of the block $B \leftrightarrow e$ of FG, and let N be the normaliser of Y in G, $\overline{N} = N/Y$, $\overline{D} = D/Y$ and $\overline{H} = H/Y$. By R. Brauer's first main theorem on blocks there is a unique block $B_2 \leftrightarrow e_2 = \sigma_Y(e)$ of FN with defect group D corresponding to $B \leftrightarrow e$ under the Brauer correspondence σ_Y with respect to Y. By Lemma 6.2 and Corollary 7.4 of (14) the following diagram of blocks is commutative,

$$\begin{array}{ccccc} B & \xrightarrow{\sigma_Y} & B_2 & \xrightarrow{\sigma_D} & B_1 \\ & & \downarrow{\scriptstyle \tau_N} & & \downarrow{\scriptstyle \tau_H} \\ & & \overline{B}_2 & \xrightarrow[\sigma_{\overline{D}}]{} & \overline{B}_1 \end{array}$$

where τ_N and τ_H are the natural epimorphisms $FN \twoheadrightarrow F\overline{N}$ and $FH \twoheadrightarrow F\overline{H}$ respectively. Since $\overline{D}$ is a defect group of $\overline{B}_2$ and $\overline{B}_1$, Theorem 0.2 is proved by induction using the Green correspondence g between the non-projective indecomposable FN-modules of B_2 and the non-projective indecomposable FG-modules of B and by "lifting" Theorem 0.2 from $\overline{B}_2$ to B_2.

Concerning our terminology we refer to (15) and the books of L. Dornhoff (5), D. Gorenstein (7) and J. Lambek (13). All results on A. Heller's operator Ω used in this article can be found in J. A. Green's set of notes (8). His paper (9) contains all the required knowledge on the Green correspondence g.

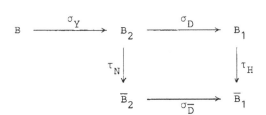

1. Blocks of defect zero and liftable modules

For later use we restate in this section a characterisation of the blocks of defect zero which is due to R. Brauer and C. Nesbitt (see (5), Theorem 62.5, p.382). We include a (different) proof for the sake of completeness.

PROPOSITION 1.1. Let $B \leftrightarrow e$ be a block of FG with defect group $\delta(B) =_G D$. Then the following statements are equivalent:

(1) $\delta(B) = 1$

(2) B is a simple artinian ring.

(3) $\hat{B} = \hat{e}\,RG$ is isomorphic to a full ring of $n \times n$ matrices over a (not necessarily commutative) complete discrete rank one valuation ring E with maximal ideal $J(E) = \pi E$.

(4) $\hat{B}_K = \hat{e}\,KG$ is a simple artinian ring.

Proof. By Corollary 3.2 of (14) the assertions (1) and (2) are equivalent. As $\hat{B}$ is semiperfect and $\pi\hat{B} \leq J(\hat{B})$, also the conditions (2) and (3) are equivalent. Since $K \otimes_R E_n \cong (K \otimes_R E)_n$ for every full ring E_n of $n \times n$ matrices over E , assertion (4) follows immediately from (3). The converse is well known ((3), Theorem 77.14, p.548).

DEFINITION 1.2. Let G be a finite group. A finitely generated FG-module V is <u>liftable</u>, if there is a finitely generated torsion free RG-module M such that $V \cong M/M\pi$. Furthermore, V is <u>uniquely liftable</u>, if M is uniquely determined by V up to RG-module isomorphism.

COROLLARY 1.3. Let $B \leftrightarrow e$ be a block of FG with defect zero. Then every simple FG-module of B is uniquely liftable.

Proof. Since B has defect zero, Proposition 1.1 asserts that every simple FG-module $V \cong \hat{P}/\hat{P}\pi$, where P is an indecomposable projective RG-module of $\hat{B}$. Hence V is liftable. Since V is also a simple RG-module of $\hat{B}$, it follows that P is a projective cover of V. Hence P is uniquely determined by V up to RG-module isomorphism.

2. Blocks with normal cyclic defect groups

In this section Theorem 0.2 is proved for the special case of normal defect groups.

The following notation is standard. The augmentation ideal of the group G is denoted by $\omega G = \sum_{1 \neq g \in G}(1-g)FG$. If D is a normal subgroup of G, then $(\omega D)FG$ is a two-sided ideal of FG which is equal to the kernel of the natural epimorphism $\tau : FG \to F\bar{G}$, $\bar{G} = G/D$.

LEMMA 2.1. Let $B \leftrightarrow e$ be a block of FG with normal defect group D. Then for every simple FG-module V of B there is an indecomposable lattice W of $\hat{B} = \hat{e}RG$ on which D acts trivially such that $W/W\pi \cong V$.

Furthermore, among the RG-modules on which D acts trivially W is uniquely determined by V up to RG-module isomorphism.

Proof. Let $\bar{G} = G/D$, and let τ be the natural algebra epimorphism $FG \to F\bar{G}$. Then by Lemma 3.3 of (14) the Jacobson radical $J(B)$ of B is $J(B) = e(\omega D)FG$ and $e(\omega D) = e \ker\tau$, where $\ker\tau$ denotes the kernel of τ. Hence $\tau(B)$ is a direct sum of blocks $\bar{B}_i$ of $F\bar{G}$

having defect zero, $i = 1,2,\dots,s < \infty$. By Proposition 1.1 each $\overline{B}_i$ has (up to isomorphism) a unique simple $F\overline{G}$-module V_i . Since $\ker \tau$ is nilpotent every simple FG-module of B is isomorphic to precisely one of the modules V_i . By Corollary 1.3 V_i is liftable to a finitely generated torsion free $R\overline{G}$-module M_i . Hence M_i is a torsion free finitely generated RG-module belonging to $\hat{B} = \hat{e}\,RG$ and satisfying $M_i/M_i\pi \cong V_i$. Therefore all simple FG-modules of B are liftable.

The uniqueness part follows now immediately from Corollary 1.3.

DEFINITION. Let A be a ring, and let K be a normal subgroup of the finite group G . If $b \leftrightarrow f$ is a block of the group ring AK , then its <u>inertial subgroup</u> is

$$T_G(b) = \{g \in G \mid g^{-1} f g = f\} .$$

If U is an indecomposable AK-module, then its inertial subgroup is

$$T_G(U) = \{g \in G \mid U \otimes_{AK} g \cong U \text{ as AK-modules}\} .$$

In this article the above definition will only be used for the rings $A = F$ or $A = R$.

Combining the results of Lemma 2.1 and of Lemma 4.2 of (14) we easily obtain

LEMMA 2.2. Let $B \leftrightarrow e$ be a block of FG with cyclic defect group D . Let $H = N_G(D)$, $K = D\,C_G(D)$ and σ the Brauer homomorphism from FG to FH . Let $B_1 \leftrightarrow e_1 = \sigma(e)$. Then:

a) The block $B_1 \leftrightarrow e_1$ of FH has defect group D and is uniquely determined by B .

b) There exists (up to conjugacy in H) a uniquely determined block $b \leftrightarrow f$ of FK with defect group D having only one simple FK-module M such that α) $T = T_H(M) = T_H(b)$

β) $e_1 = \sum_{j=1}^{h} g^{g_j}$, where

$\{g_j \in H \mid j = 1,2,\ldots,h\}$ is a

transversal of T in H .

c) $M^T = M \otimes_{FK} FT$ is a completely reducible FT-module.

d) If $M_1, M_2, \ldots, M_t$ are the non-isomorphic composition factors of the FT-module M^T , then there is an integer v such that

$$M^T = M_1^{\ v} \oplus M_2^{\ v} \oplus \ldots \oplus M_t^{\ v} ,$$

and every simple FT-module M_i, $i = 1,2,\ldots,t$, belongs to the block $b' = fFT$ of FT.

e) $\{V_i = M_i \otimes_{FT} FH \mid i = 1,2,\ldots,t\}$ is the set of all non-isomorphic simple FH-modules of B_1 .

f) There is an indecomposable torsion free RK-module W of $\hat{b} = \hat{f}RK$ on which D acts trivially such that $W/W\pi \cong M$. Furthermore, W is uniquely determined by M up to RK-module isomorphism among the RK-lattices on which D acts trivially.

g) $T = T_H(b) = T_H(W)$.

h) There are t non-isomorphic indecomposable torsion free RT-modules W_i of $b' = fRT$ on which D acts trivially such that $W_i/W_i\pi \cong M_i$. Furthermore, each W_i is uniquely determined by M_i up to RT-module isomorphism among the RT-lattices on which D acts trivially.

i) $L_i = W_i \otimes_{FT} FH$ is an indecomposable torsion free RH-module of $\hat{B}_1 = \hat{e}_1 RH$ on which D acts trivially such that $L_i/L_i\pi \cong V_i$ for $i = 1,2,\ldots,t$.

j) Among the indecomposable RH-lattices on which D acts trivially L_i is uniquely determined by V_i up to RH-module isomorphism for $i = 1,2,\ldots,t$.

Proof. Assertions a) through e) are a restatement of Lemmas 4.2 and 4.4 of (14). f) follows at once from Lemma 2.1, because b is a block of FK with normal defect group D .

g) Let $h \in T_H(M)$. Then W and $W \otimes h$ are indecomposable torsion free RK-modules of b on which D acts trivially such that $W/W\pi = \overline{W} \cong \overline{W \otimes h}$. From the uniqueness part of f) we deduce $W \cong W \otimes h$ as RH-modules. Thus $T = T_H(M) = T_H(W)$, because the reverse inclusion is trivial.

h) As D acts trivially on W , also $W^T = W \otimes_{RH} RT$ may be considered as an R(T/D)-module, because D is normal in T . Let τ the R-algebra epimorphism $RT \rightarrow R\overline{T}$, $\overline{T} = T/D$. Then W^T is a module of the block $\tau(f)R\overline{T}$ which has defect zero by Lemma 3.3 of (14). From g) follows that $W^T / W^T\pi \cong M^T$. Hence W^T is a projective cover of the right $R\overline{T}$-module M^T . Therefore assertion d) implies that there are t indecomposable torsion free $R\overline{T}$-modules W_i of $\tau(f)R\overline{T}$ such that

$$W^T \cong \sum_{i=1}^{t} \oplus W_i^{v} \quad , \text{ and } \quad W_i/W_i\pi \cong M_i \; , \; i = 1,2,\dots,t \; ,$$ because projective covers are uniquely determined. Hence each W_i is an indecomposable torsion free RT-module of $\hat{b}' = \hat{f}\,RT$ on which D acts trivially such that $W_i/W_i\pi \cong M_i$ as FT-modules. Furthermore, W_i is uniquely determined by M_i for $i = 1,2,\dots,t$.

i) As D is normal in H , it follows from h) that D acts trivially on $L_i = W_i \otimes_{RT} RH$ for $i = 1,2,\dots,t$. Since D is cyclic, T/K may be considered as a subgroup of the abelian automorphism group A(D) of D. A(D) also contains H/K . Therefore T is normal in H . Hence Lemma 52.2 of (5), p.327, asserts that each RH-module L_i is indecomposable and torsion free, because $T_H(W_i) = T$ by g) and h) . Furthermore, $L_i/L_i\pi \cong V_i$ for $i = 1,2,\dots,t$ by e) and (13) , Proposition 1, p.132.

The final assertion j) is immediate by Lemma 2.1.

REMARK. The integer t occurring in Lemma 2.2 is the inertial index of the block $B \leftrightarrow e$ of FG by Definition 4.3 of (14).

The following subsidiary result appears to be well known.

LEMMA 2.3. Let $A \in \{R,F\}$. Let M be an indecomposable right AG-module with vertex D. Let $Y \neq 1$ be a normal subgroup of G contained in D, and let $\bar{D} = D/Y$ and $\bar{G} = G/Y$.

Then the trivial right AD-module A_{AD} is a source of the right AG-module M if and only if Y acts trivially on M, and the trivial right $A\bar{D}$-module $A_{A\bar{D}}$ is a source of the right $A\bar{G}$-module M with vertex $\bar{D}$.

Proof. If the right AG-module M is a component of $A^G = A_{AD} \otimes_{AD} AG$ then Y acts trivially on M, because Y is normal in G. Hence A^G and M are $A\bar{G}$-modules, and M is a component of $A^{\bar{G}} = A_{A\bar{D}} \otimes_{A\bar{D}} A\bar{G}$ As D is a vertex of M, it follows that $\bar{D}$ is a vertex of $M_{A\bar{G}}$.

Since a normal p-subgroup lying in the kernel of an indecomposable AG-module M is always contained in a vertex of M by (5), Theorem 53.9, the converse of Lemma 2.3 also follows.

Another well known fact (e.g. K.Morita (16), Lemma 1) is

LEMMA 2.4. Let B be an indecomposable, uniserial, symmetric algebr with indecomposable projective B-modules $U_1, U_2, \ldots, U_t$, Jacobson radical J and simple B-modules $V_i = U_i/U_iJ$, $i = 1,2,\ldots,t$. Then the projective modules U_i can be ordered such that they have the following composition series:

$U_1 : V_1, V_2, \ldots, V_t, \; V_1, V_2, \ldots, V_t, \; V_1, V_2, \ldots, V_t, \; V_1$

$U_2 : V_2, V_3, \ldots, V_1, \; V_2, V_3, \ldots, V_1, \; V_2, V_3, \ldots, V_1, \; V_2$

$U_3 : V_3, V_4, \ldots, V_2, \; V_3, V_4, \ldots, V_2, \; V_3, V_4, \ldots, V_2, \; V_3$

. . . .

$U_t : V_t, V_1, \ldots, V_{t-1}, V_t, V_1, \ldots, V_{t-1}, V_t, V_1, \ldots, V_{t-1}, V_t$

PROPOSITION 2.5. If $B \leftrightarrow e$ is a block of FG with inertial index t and normal cyclic defect group $D = (x)$, then all assertions a) through b) of Theorem 0.2 hold for B.

Proof. By Lemma 2.2 the block $B \leftrightarrow e$ has t non-isomorphic simple FG-modules M_i, $i = 1,2,\ldots,t$. Each M_i has trivial source F_{FD} by Propositions 5.1 and 8.1 of (14), which proves a).

If $0 \to \Omega M_i \to P_i \to M_i \to 0$ is a minimal projective resolution of M_i, then ΩM_i is a maximal submodule of the indecomposable projective FG-module P_i. Hence the FD-module $(1-x)FD$ is a source of each ΩM_i by Propositions 5.1 and 8.1 of (14). Thus b) holds.

Since B is uniserial by Proposition 5.1 of (14), Lemma 2.4 asserts that $\Omega^2 M_i$ is a simple FG-module of B for $i = 1,2,\ldots,t$. Therefore a) implies c).

Assertions d) and e) now follow easily from Lemma 2.2 i) and j). In particular, there are indecomposable lattices W_i of $\hat{B} = \hat{e}RG$ on which D acts trivially such that $W_i/W_i\pi \cong M_i$. Hence W_i may be considered as an indecomposable $R\overline{G}$-module of the block $\tau(B)$ with defect zero, where τ is the natural epimorphism $RG \to R\overline{G}$, $\overline{G} = G/D$. Therefore by Lemma 2.3 the trivial RD-module R_{RD} is a source of each lattice W_i.

By Proposition 5.1 of (14) there are t orthogonal primitive idempotents $a_i \in B$ such that $M_i \cong a_i(1-x)^{q-1}FG$ for $i = 1,2,\ldots,t$, where $q = |D|$. Again by Proposition 5.1 of (14) and Lemma 2.4 $\Omega M_i \cong a_i(1-x)FG$ for each i, and $0 \to a_i(1-x)FG \to a_iFG \to a_i(1-x)^{q-1}FG \to 0$ is a minimal projective resolution of M_i. As R is a complete discrete rank one valuation ring, there are t orthogonal primitive idempotents $\hat{a}_i$ of $\hat{B}$ such that $\rho(\hat{a}_i) = a_i$, $i = 1,2,\ldots,t$, where ρ denotes the natural epimorphism $RG \to FG$ with kernel πRG. Since D is normal in G, and since $\hat{a}_i RG$ is a free RD-module, the exactness of the sequence

$$0 \to (1-x)RD \to RD \to (1-x)^{q-1}RD \to 0$$

implies that

$$0 \to \hat{a}_i(1-x)RG \to \hat{a}_iRG \to \hat{a}_i(1-x)^{q-1}RG \to 0$$

yields a minimal projective resolution of the right RG-module $Z_i = \hat{a}_i(1-x)^{q-1}RG$. As Z_i is torsion free, $\Omega Z_i = \hat{a}_i(1-x)RG$ is a pure submodule of $\hat{a}_iRG$ by Theorem 16.15 of (3). Hence $(\Omega Z_i)\pi = a_i\pi RG \cap \Omega Z_i$, and

(*) $\quad \Omega Z_i \,/\, (\Omega Z_i)\pi \cong a_i(1-x)FG \cong \Omega M_i$ for $i = 1,2,\ldots,t$.

Since $\hat{a}_i$ is a primitive idempotent of RG, and since D acts trivially on $Z_i = \hat{a}_i(1-x)^{q-1}RG$, the torsion free RG-module Z_i is indecomposable. As D is a defect group of $\hat{B} = \hat{e}\,RG$, Corollary 53.8 of (5), p.341, asserts that D is a vertex of each Z_i. Furthermore,

$$(1-x)^{q-1}RG = \hat{a}_i(1-x)^{q-1}RG \oplus (1-\hat{a}_i)(1-x)RG,$$

because D is normal in G. Hence the trivial RD-module $R_{RD} \cong (1-x)^{q-1}RD$ is a source of Z_i for $i = 1,2,\ldots,t$.

As D is a vertex of Z_i, Theorem 1.1 of (14), which is due to J.A.Green, asserts that D is also a vertex of ΩZ_i. As D is normal in G,

$$(1-x)RG = \hat{a}_i(1-x)RG \oplus (1-\hat{a}_i)(1-x)RG.$$

Hence the RD-module $(1-x)RD$ is a source of $\Omega Z_i = \hat{a}_i(1-x)RG$ for $i = 1,2,\ldots,t$. This together with (*) proves assertion f), if we can show $Z_i \cong W_i$, because then $\Omega W_i/(\Omega W_i)\pi \cong \Omega M_i$ by Lemma 17.6 of J.A.Green (8).

Clearly $Z_i \cong W_i$ by e), if $Z_i = \hat{a}_i(1-x)^{q-1}RG$ is a pure submodule of a_iRG for $i = 1,2,\ldots,t$. Now $(1-x)^{q-1}RG$ is the right annihilator of $(1-x)$ by Lemma 2 of (13), p.154. Since $(1-x)^{q-1}KG$ is the right annihilator of $(1-x)$ in KG, it follows that $(1-x)^{q-1}RG = (1-x)^{q-1}KG \cap RG$. Hence $(1-x)^{q-1}RG$ is a pure submodule of RG by Theorem 16.18 of (3). Therefore (**) implies that $Z_i = \hat{a}_i(1-x)^{q-1}RG$ is a pure submodule of $\hat{a}_iRG$ for $i = 1,2,\ldots,t$, which completes the proof of f).

The same argument proves i), and assertions g) and h) are well known, e.g. (8), p.70.

Finally the statements j), k) and therefore l) follow immediately from Lemma 2.4, because B is uniserial. This completes the proof of Proposition 2.5.

3. Projective resolutions of indecomposable FG-modules with trivial source

In this section the characteristic $p > 0$ part of Theorem 0.2 is proved.

LEMMA 3.1. Let $B \leftrightarrow e$ be a block of FG with cyclic defect group $D = (x)$. Let $1 \neq Y \lneqq D$ be a normal subgroup of G having order $|Y| = p$. Let $\bar{G} = G/Y$, $\bar{D} = D/Y = (\bar{x})$, and let τ be the canonical epimorphism from FG onto $F\bar{G}$.

Suppose that $0 \to \Omega M \to P \to M \to 0$ is a minimal projective resolution of the non-projective indecomposable FG-module M of B. Then the following assertions hold:

a) If Y acts trivially on M, then

$$0 \to \tau(\Omega M) \to \tau(P) \to M_{F\bar{G}} \to 0$$

is a minimal projective resolution of the right $F\bar{G}$-module $M_{F\bar{G}}$.

b) If Y acts trivially on ΩM, then

$$0 \to \Omega M_{F\bar{G}} \to \tau(P) \to \tau(M) \to 0$$

is a minimal projective resolution of the right $F\bar{G}$-module $\tau(M)$.

Proof. By Corollary 7.4 of (14) B is uniserial. Let $q = |D|$ and $\bar{q} = |\bar{D}|$. Then $q = \bar{q} \cdot p$, and $e J^{\bar{q}} = e(\omega Y)FG$, where J denotes the Jacobson radical of FG. Hence $P J^{\bar{q}} = P \cap e(\omega Y)FG$ for every indecomposable projective right FG-module P of B. By Corrollary 7.4 of (14) also $\bar{B} = \tau(B)$ is a uniserial block of $F\bar{G}$, and $\bar{D}$ is its defect group. Furthermore, $\tau(P)$ is an indecomposable projective $F\bar{G}$-module of $\bar{B}$ with composition length $l(\tau(P)) = \bar{q}$, and every indecomposable $F\bar{G}$-module U of $\bar{B}$ has length $l(U) \leq \bar{q}$.

If Y acts trivially on M, then M may be considered as an indecomposable right $F\bar{G}$-module. Hence $l(M) \leq \bar{q}$, and $l(\Omega M) = q - l(M) \geq \bar{q}p - \bar{q} = \bar{q}(p-1)$. Since $l(P J^{\bar{q}}) = \bar{q}(p-1)$, $P J^{\bar{q}} \leq \Omega M$, and

$$0 \to \tau(\Omega M) \to \tau(P) \to M_{F\bar{G}} \to 0$$

is a minimal projective resolution of the right $F\bar{G}$-module $M_{F\bar{G}}$, where $\tau(\Omega M) = \Omega M / P J^{\bar{q}}$ and $\tau(P) = P/P J^{\bar{q}}$.

If Y acts trivially on ΩM, then a similar length arguement as above shows that $l(M) \geq \bar{q}(p-1)$. If Q denotes an injective hull of M, then $Q J^{\bar{q}} = \ker \tau \cap Q$ has length $l(Q J^{\bar{q}}) = \bar{q}(p-1)$. Hence $M \geq Q J^{\bar{q}}$. By Theorem 10.1 of (14) the number t of simple FG-modules of B divides $p-1$. As $l(P J^{\bar{q}}) = \bar{q}(p-1)$, it follows from Lemma 2.4 that ΩM, $\Omega M_{F\bar{G}}$, P and $\tau(P)$ all have the same simple FG-module as their socle. Hence $\tau(P)$ is an injective hull of $\Omega M_{F\bar{G}}$, and $l(\tau(P)/\Omega M_{F\bar{G}}) = l(\tau(M))$, where $\tau(M) = M/Q J^{\bar{q}}$. Since $\bar{B} = \tau(B)$ is uniserial and $\tau(P)$ and $\tau(M)$ have the same top simple factor, it follows that $0 \to \Omega M_{F\bar{G}} \to \tau(P) \to \tau(M) \to 0$ is a minimal projective resolution, completing the proof.

PROPOSITION 3.2. Let $B \leftrightarrow e$ be a block of the group algebra FG with a cyclic defect group $D = (x)$ and inertial index t. Then the following assertions hold:

a) B contains t non-isomorphic indecomposable FG-modules M_i whose source is the trivial FD-module F_{FD}.

b) If $0 \to \Omega M_i \to P_i \to M_i \to 0$ is a minimal projective resolution of M_i, then the right FD-module $(1-x)FD$ is a source of ΩM_i for $i = 1,2,\dots,t$.

c) If $0 \to \Omega^2 M_i \to Q_i \to \Omega M_i \to 0$ is a minimal projective resolution of ΩM_i, then the trivial FD-module F_{FD} is a source of $\Omega^2 M_i$ for $i = 1,2,\dots,t$.

d) The numbering of the t non-isomorphic indecomposable FG-modules M_i of B with trivial source F_{FD} can be so chosen that $M_i \cong \Omega^{2(i-1)} M_1$ for $i = 2,3,\dots,t$, and $M_1 \cong \Omega^{2t} M_1$.

e) The 2t FG-modules $\Omega^i M_1$, $i = 1,2,\dots,2t$ are mutually non-isomorphic.

Proof. This proposition is proved by induction on $|G|$. It is trivial for $G = 1$. Suppose it has been proved for all groups of order less than $|G|$. Let $1 \neq Y \leq D$ be the socle of D, and $N = N_G(Y)$. Let σ_Y be the Brauer correspondence between the blocks of FG and of FN having defect groups conjugate to D. Then $B_1 \leftrightarrow e_1 = \sigma_Y(e_1)$ is the unique block of FN with defect groups $\delta(B_1) =_N D$ corresponding to $B \leftrightarrow e$. By Corollary 7.4 of (14) B_1 is a uniserial block of FN with inertial index t. Therefore B_1 contains t non-isomorphic indecomposable FN-modules A_i whose source is the trivial FD-module F_{FD} by Proposition 9.4 of (14).

Let g be the Green correspondence between the non-projective indecomposable FN-modules of B_1 and the non-projective indecomposable FG-modules of B . Since $|Y| = p$,

$$\mathfrak{X} = \{D \cap D^g \mid g \in G,\ g \notin N\} = 1 .$$

Hence g is one-to-one and preserves vertices and sources by Theorem 4. of J.A.Green (9). Let $M_i = g(A_i)$, $i = 1,2,\dots,t$. Then a) holds. Furthermore, $\Omega M_i = \Omega g(A_i) \cong g(\Omega A_i)$, and $\Omega^2 M_i \cong g(\Omega^2 A_i)$ for $i = 1,2,\dots,t$ by Theorem 4.5 of J.A.Green (9). Therefore we may assume that $G = N$ and $B = B_1$.

If $D = Y$, then all assertions of Proposition 3.2 hold by Proposition 2.5. Therefore we can assume that $D \neq Y$. Let $\bar{G} = G/Y$, $\bar{D} = D/Y$, and let τ be the natural epimorphism $FG \twoheadrightarrow F\bar{G}$. Then by Corollary 7.4 of (14) $\bar{B} = \tau(B) = \tau(e)F\bar{G}$ is a uniserial block of $F\bar{G}$ with inertial index t and defect group $\bar{D} = (\bar{x})$. Since $|\bar{G}| < |G|$ the block $\bar{B}$ satisfies all assertions b) through e) of Proposition 3.2 by induction.

Suppose that M denotes one of the t non-isomorphic indecomposable FG-modules M_i of B . As F_{FD} is a source of M , the normal subgroup Y of G acts trivially on M . Thus M can be considered as a right $F\bar{G}$-module with source $F_{F\bar{D}}$ by Lemma 2.3. Let

$$0 \to \Omega M \to P \to M \to 0$$

be a minimal projective resolution of the right FG-module M . Since $Y \neq D$ and Y is normal in G , Lemma 3.1 asserts,

$$0 \to \tau(\Omega M) \to \tau(P) \to M_{F\bar{G}} \to 0$$

is a minimal projective resolution of the right $F\bar{G}$-module $M_{F\bar{G}}$. Therefore by induction hypothesis the $F\bar{D}$-module $(1-\bar{x})F\bar{D}$ is a source of $\tau(\Omega M)$. Now Lemma 9.1 of (14) asserts that $(1-x)FD$ is a source of ΩM , which proves b) .

Let $K_i = \Omega M_i$ for $i=1,2,\dots,t$. Then $K_i \cong K_j$ if and only if $i = j$ by Corollary 17.5 c) of J.A.Green (8). Again it follows from

Proposition 9.4 of (14) that $\{K_i \mid i = 1,2,\dots,t\}$ is a full set of non-isomorphic, indecomposable FG-modules of B having source $(1-x)FD$. Furthermore, the proof of b) shows that $\{\tau(K_i) \mid i = 1,2,\dots,t\}$ is a full set of non-isomorphic, indecomposable $F\overline{G}$-modules of $\overline{B}$ having source $(1-\overline{x})F\overline{D}$. Let K be one of the FG-modules K_i.

Suppose that $0 \to \Omega\tau(K) \to \tau(P) \to \tau(K) \to 0$ is a minimal projective resolution of the right $F\overline{G}$-module $\tau(K)$ of $\overline{B}$. Then by induction hypothesis $\Omega\tau(K)$ has source $F_{F\overline{D}}$. Since Y is normal in G, and B and $\overline{B}$ both only have t non-isomorphic indecomposable modules with trivial source, there is an FG-module M of B with trivial source F_{FD} such that $M \cong \Omega\tau(K)$. By Theorem 10.1 of (14) t divides $p-1$. As $P \cap \ker\tau = P\,J^{\overline{q}}$, $\overline{q} = |\overline{D}|$, has composition length $l(P\,J^{\overline{q}}) = \overline{q}(p-1)$, it follows from Lemma 2.4 that M, $M_{F\overline{G}}$, P and $\tau(P)$ all have the same simple FG-module as their socle. Therefore P is an injective hull of the right FG-module M. Let $V = P/M$. Then $\tau(V) \cong \tau(K)$ by Lemma 3.1 b). Since B is uniserial, it follows that $V \cong K$. Therefore $M \cong \Omega K$ has source F_{FD}, which completes the proof of Proposition 3.2.

Since $Y \neq 1$ is normal in G Proposition 9.4 of (14) asserts that all indecomposable FG-modules U of the uniserial block $B \leftrightarrow e$ of FG of equal composition length $l(U)$ have the same source. Therefore d) follows immediately from b) and c) by Lemma 2.4.

The final assertion e) holds by Lemma 3.1 and our induction hypothesis. This completes the proof.

4. Indecomposable modules with trivial source are liftable

In this section the proof of Theorem 0.2 is completed by showing that indecomposable FG-modules with trivial source F_{FD} belonging to a block $B \leftrightarrow e$ with cyclic defect group D are liftable.

The first assertion of the following auxiliary result is an obvious generalization of Lemma 3 of J.G.Thompson (19); as for the second assert it is contained in J.A.Green's paper (9). All quoted results can be proved for group algebras over arbitrary complete discrete valuation rings R by an easy modification of the known proofs.

LEMMA 4.1. Let $B \leftrightarrow e$ be a block of FG with cyclic defect group $D = (x)$ and inertial index t. Let $1 \neq Y \leq D$ be the socle of D, $N = N_G(Y)$, and denote by σ_Y the Brauer correspondence between the blocks of FG and FN having defect groups conjugate to D. Let g_1 be the Green correspondence between the indecomposable non-projective FN-modules of $B_1 \leftrightarrow e_1 = \sigma_Y(e)$ and the indecomposable non-projective FG-modules of $B \leftrightarrow e$. Denote by g_2 the Green correspondence between the indecomposable torsion free non-projective RN-modules of $\hat{B}_1 = \hat{e}_1 RN$ and the indecomposable non-projective torsion free RG-modules of $\hat{B} = \hat{e}\, RG$.

If M is an indecomposable non-projective FN-module of B_1 such that there is an indecomposable non-projective torsion free RN-module W of $\hat{B}_1$ satisfying $W/W\pi \cong M$, then

a) $g_2(W)/g_2(W)\pi \cong g_1(M)$,

b) $\Omega g_1(M) \cong g_1(\Omega M) \cong \Omega(g_2(W)/g_2(W)\pi) \cong g_2(\Omega W)/g_2(\Omega W)\pi$.

Proof. a) follows from J.G.Thompson's lemma (see (5), Lemma 68.11, p.445) and Theorem 4.2 of J.A.Green (9).

Assertion b) follows from a) and Theorem 4.5 and Theorem 3.6 of J.A.Green (9).

REMARK. In the following we identify the notation of the functors g_1 and g_2, whenever we have the situation of Lemma 4.1. Both Green correspondences are then denoted by g.

LEMMA 4.2. Let $B \leftrightarrow e$ be a block of FG with cyclic defect group $D = (x)$, and inertial index t. Suppose that M denotes an indecomposable FG-module of B having the trivial FD-module F_{FD} as a source, and that $0 \to \Omega M \to P \to M \to 0$ is a minimal projective resolution of M. Then the following assertions hold:

a) There exists an indecomposable lattice W of $\hat{B} = \hat{e}\,RG$ such that $W/W\pi \cong M$, and the trivial RD-module R_{RD} is a source of W.

b) Among the indecomposable lattices of $\hat{B} = \hat{e}\,RG$ with source R_{RD} W is uniquely determined by M up to RG-module isomorphism.

c) If $0 \to \Omega W \to \hat{P} \to W \to 0$ is a minimal projective resolution of W, then the RD-module $(1-x)RD$ is a source of ΩW.

d) The $2t$ indecomposable lattices $\Omega^i W$ of $\hat{B} = \hat{e}\,RG$, $i = 1,2,\dots,2t$, are mutually non-isomorphic, and $W \cong \Omega^{2t} W$.

Proof. This lemma is proved by induction on $|G|$. It certainly holds for $G = 1$. Suppose it is proved for all groups of order less than $|G|$. Let $N = N_G(Y)$, where $Y \neq 1$ denotes the socle of D. As the Green correspondence g preserves sources and commutes with the Heller operator Ω by Lemma 4.1, we may assume $G = N$. If $D = Y$, then Lemma 4.2 holds by Proposition 2.5. Thus we are left with the case $D \neq Y$. By induction all assertions of Lemma 4.2 hold for $\overline{G} = G/Y$.

Let $\overline{D} = D/Y$, and let τ be the natural epimorphism $FG \to F\overline{G}$. Then Corollary 7.4 of (14) asserts that $\overline{B} = \tau(e)F\overline{G}$ is a uniserial block of $F\bar{G}$ with defect group $\bar{D}$.

Since F_{FD} is a source of M , the normal subgroup Y of G acts trivially on M by Lemma 2.3, and the trivial $F\bar{D}$-module $F_{F\bar{D}}$ is a source of the indecomposable right $F\bar{G}$-module M . Hence by induction hypothesis there is an indecomposable torsion free right $R\bar{G}$-module W such that $W/W\pi \cong M$ as $F\bar{G}$-modules, and $R_{R\bar{D}}$ is a source of $W_{R\bar{G}}$. Therefore, a) holds by Lemma 2.3.

Suppose Z is another torsion free indecomposable right RG-module of $\hat{B} = \hat{e}\, RG$ such that R_{RD} is a source of Z and $Z/Z\pi \cong M$ as FG-modules. Then Y acts trivially on Z by Lemma 2.3, and $R_{R\bar{D}}$ is a source of the torsion free indecomposable $R\bar{G}$-module Z . Furthermore, $Z/Z\pi \cong M$ as $F\bar{G}$-modules. By induction hypothesis we conclude that $Z \cong W$ as $R\bar{G}$-modules. Hence $Z \cong W$ as RG-modules, because Y acts trivially on both. This proves b) .

If $0 \to \Omega W \to \hat{P} \to W \to 0$ is a minimal projective resolution of W , then $\Omega W/(\Omega W)\pi \cong \Omega M$ by Lemma 17.6 of (8). By Proposition 3.2 ΩM has vertex $(1-x)FD$. Hence Proposition 8.1 of (14) asserts that there are non-zero primitive idempotents $a, b \in B$ such that:

(α) $(1-x)FG = a(1-x)FG \oplus (1-a)(1-x)FG$,

(β) $(1-x)^{q-1}FG = b(1-x)^{q-1}FG \oplus (1-b)(1-x)^{q-1}FG$,

(γ) $\Omega M \cong a(1-x)FG$,

(δ) $M \cong b(1-x)^{q-1}FG$,

(ε) $0 \to a(1-x)FG \to aFG \to b(1-x)^{q-1}FG \to 0$

is a minimal projective resolution.

Since R is a complete discrete rank one valuation ring there are primitive idempotents $\hat{a}, \hat{b} \in \hat{B} = \hat{e}\, RG$ such that $\hat{a}\, RG \cong \hat{P}$, $\rho(\hat{a}) = a$ and $\rho(\hat{b}) = b$, where ρ denotes the canonical epimorphism $RG \to FG$ with kernel $\pi\, RG$.

By J. Lambek ((13), Lemma 2, p.154) the right ideal $(1-x)^{q-1}RG$ of RG is the right annihilator of $(1-x)$. Hence $(1-x)^{q-1}RG = (1-x)^{q-1}KG \cap RG$, and therefore $(1-x)^{q-1}RG$ is a pure submodule of RG. As KG is a quasi-Frobenius ring $(1-x)KG$ is the right annihilator of $(1-x)^{q-1}$. Hence $(1-x)RG$ is the right annihilator of $(1-x)^{q-1}$, because RG is torsion free. Thus $(1-x)RG$ is a pure submodule of RG, and so $(1-x)RG \cap \pi RG = \pi(1-x)RG$. Therefore we obtain from (α)

$$(1-x)RG = \hat{a}(1-x)RG + (1-\hat{a})(1-x)RG + \pi(1-x)RG, \text{ and so}$$

$$(*) \qquad (1-x)RG = \hat{a}(1-x)RG \oplus (1-\hat{a})(1-x)RG$$

by Nakayama's lemma.

Since $(1-x)^{q-1}RG$ is a pure submodule of RG we obtain from (β) by the same argument that

$$(**) \qquad (1-x)^{q-1}RG = \hat{b}(1-x)^{q-1}RG \oplus (1-\hat{b})(1-x)^{q-1}RG .$$

Thus $\hat{b}(1-x)^{q-1}RG$ is a pure submodule of $\hat{b}RG$, and (δ) implies $M \cong b(1-x)^{q-1}FG \cong \hat{b}(1-x)^{q-1}RG / \hat{b}(1-x)^{q-1}RG\pi$. From the uniqueness part of Lemma 4.2 b) it follows that $W \cong \hat{b}(1-x)^{q-1}RG$, because the trivial RD-module $R_{RD} \cong (1-x)^{q-1}RD$ is a source of $\hat{b}(1-x)^{q-1}RG$ by $(**)$. Because of $(*)$ $\hat{a}(1-x)RG$ is a pure submodule of $\hat{a}RG \cong \hat{P}$. Therefore (ε) implies that the sequence

$$0 \to \hat{a}(1-x)RG \to \hat{a}RG \to \hat{b}(1-x)^{q-1}RG \to 0$$

is exact. Hence $\Omega W \cong \hat{a}(1-x)RG$. By Theorem 1.1 of (14) the p-subgroup D of G is a vertex of ΩW, because D is a vertex of W. Thus $(1-x)RD$ is a source of ΩW by $(*)$, proving assertion c).

The final assertion d) follows at once from Lemma 4.2 a), Proposition 3.2 and Lemma 17.6 of (8).

<u>4.3 Proof of Theorem 0.2:</u> Because of Proposition 3.2 the assertions a), b), c), k) and the first part of j) hold. The second part of statement j) follows from Lemma 4.2 a) and b) and from Lemma 17.6 of (8). Now assertion b), d) and e) are restatements of Lemma 4.2 d), a) and b) respectively. Again Lemma 17.6 of (8) and Lemma 4.2 imply f). Both assertions g) and h) are well known, e.g. see (8). Finally

statement i) is now an easy consequence of Proposition 3.2 and Lemma 17.6 of (8). This completes the proof of Theorem 0.2.

5. Uniseriality of the principal block

As an application of Theorem 0.2 we now prove

COROLLARY 5.1. The principal block $B \leftrightarrow e$ of the group algebra FG of a finite group G over an arbitrary field F of characteristic $p > 0$ is uniserial if and only if the Sylow p-subgroups D of G are cyclic, and every simple FG-module M of B has trivial source F_{FD}.

Proof. If B is uniserial, then B is of finite module type. Hence each p-Sylow subgroup D of G is cyclic by Theorem 64.1 of (3).
As B is the principal block of FG D is a defect group of B, and the trivial FG-module $M_1 = F_{FG}$ belongs to B. Clearly F_{FD} is a source of M_1. Let $M_i = \Omega^{2(i-1)} M_1$, for $i = 2,3,\dots,t$, where t is the inertial index of B. Then by Theorem 0.2 j) $\{M_i \mid i = 1,2,\dots,t\}$ is a full set of non-isomorphic indecomposable FG-modules of B with trivial source F_{FD}. Since M_1 is simple it follows from Lemma 2.4 that each M_i is simple. Hence all simple FG-modules of B have trivial source by Theorem 10.1 of (14).

Suppose that each Sylow p-subgroup D of G is cyclic, and that all the non-isomorphic simple FG-modules M_i, $i = 1,2,\dots,t$, of the principal block $B \leftrightarrow e$ of FG have trivial source F_{FD}. Let $1 \neq Y$ be the socle of D, $N = N_G(Y)$. Then by R. Brauer's third main theorem on blocks (see (10), p.21, for a proof in the general case of arbitrary fields) the principal block $B_1 \leftrightarrow e_1 = \sigma_Y(e)$ of FN

corresponds uniquely to $B \leftrightarrow e$ under the Brauer correspondence σ_Y. By Corollary 7.4 of (14) B_1 is uniserial and it has inertial index t. Hence the t non-isomorphic simple FG-modules V_i, $i = 1,2,\ldots,t$, of B_1 all have trivial source F_{FD} by the first part of this proof. Let g be the Green correspondence between the indecomposable non-projective FN-modules of B_1 and the indecomposable non-projective FG-modules of B. Then by Theorem 0.2 a) we may assume that $M_i = g(V_i)$ for $i = 1,2,\ldots,t$. Since B_1 is uniserial, Lemma 2.4 and Theorem 3.9 of R. Peacock (17) now assert that g preserves the composition length of each non-projective indecomposable FG-module of B_1. Hence B is uniserial by Theorem 10.1 of (14).

R E F E R E N C E S

(1) J. Alperin and G. Janusz, Resolutions and periodicity. Proc. Amer. Math. Soc. 37 (1973), 403 - 406.

(2) R. Brauer, Some applications of the theory of blocks of characters IV. J. Algebra 17 (1971), 489 - 521.

(3) C. W. Curtis and I. Reiner, Representation theory of finite groups and associative algebras. Interscience, New York, 1962.

(4) E. C. Dade, Blocks with cyclic defect groups, Annals of Math. 84 (1966), 20 - 48.

(5) L. Dornhoff, Group representation theory, Marcel Dekker, New York, 1972.

(6) P. Gabriel, Indecomposable representations II. Symposia Mathematica INDAM Rome 11 (1973), 81 - 104.

(7) D. Gorenstein, Finite groups. Harper and Row, New York, 1968.

(8) J. A. Green, Vorlesungen über modulare Darstellungstheorie endlicher Gruppen. Vorlesungen Math. Inst. Giessen, Heft 2, 1974.

(9) J. A. Green, Walking around the Brauer tree. J. Austral. Math. Soc. 17 (1974), 197 - 213.

(10) W. Hamernik und G. Michler, Hauptblöcke von Gruppenalgebren. Archiv d. Math. 24 (1973), 21 - 24.

(11) G. Janusz, Indecomposable modules for finite groups. Annals of Math. 89 (1969), 209 - 241.

(12) H. Kupisch, Unzerlegbare Moduln endlicher Gruppen mit zyklischer p-Sylow Gruppe. Math. Z. 108 (1969), 77 - 104.

(13) J. Lambek, Lectures on rings and modules. Blaisdell, Waltham, Mass. 1966.

(14) G. Michler, Green correspondence between blocks with cyclic defect groups I. J. Algebra (to appear).

(15) G. Michler, Blocks and centers of group algebras. Lecture Notes in Mathematics 246 (1973), Springer, Berlin, 430 - 565.

(16) K. Morita, On group rings over a commutative field which possess radicals expressible as principal ideals. Science reports Tokyo Bunrika Daigaku 4 (1951), 177 - 194.

(17) R. M. Peacock, Blocks with a cyclic defect group. J. Algebra (to appear).

(18) B. Rothschild, Degrees of irreducible modular characters of blocks with cyclic defect groups. Bull. Amer. Math. Soc. 73 (1967), 102 - 104.

(19) J. Thompson, Vertices and sources. J. Algebra 6 (1967), 1 - 6.

Mathematisches Institut
Justus Liebig - Universität
63 G i e ß e n
Arndtstraße 2

ON ARTIN RINGS OF FINITE REPRESENTATION TYPE

Wolfgang Müller

The aim of this note is to give a short approach to the theory of indecomposable modules over artin rings (an artin ring is a right- and left-artinian ring) with radical square zero. We characterize the self-dual rings of finite type and construct all indecomposable modules (they are finitely generated [2]) over these rings. We need neither the theory of k-species in [4] nor the construction of the SQF-ring in [5], but our theorem 2 is in relation with theorem 3.6 in [5].

Throughout this paper we use the notations of [5] p.199. All modules will be finitely generated.

Let A be an artin ring with radical N. By [1 , 3] there is a correspondence T between $\mathfrak{M}_A$ and ${}_A\mathfrak{M}$: A non-projective indecomposable module $M \in \mathfrak{M}_A$ resp. $M \in {}_A\mathfrak{M}$ is transformed into a non-projective indecomposable module $TM \in {}_A\mathfrak{M}$ resp. $TM \in \mathfrak{M}_A$ in the following way. Apply the functor $^* = \mathrm{Hom}_A(- , A)$ to f, where

$$P_1 \xrightarrow{f} P_0 \longrightarrow M \longrightarrow 0$$

is a minimal projective presentation of M, and setting TM:= cok f^* obtain

$$P_0^* \xrightarrow{f^*} P_1^* \longrightarrow TM \longrightarrow 0$$

which is a minimal projective presentation, too.

If A is a self-dual (that means there exists a Morita duality $D: \mathcal{M}_A \longleftrightarrow {}_A\mathcal{M}$) artin ring, then DT induces an injective map $\omega: \underline{\mathcal{M}}_A \longrightarrow \underline{\underline{\mathcal{M}}}_A$ resp. $\omega: {}_A\underline{\mathcal{M}} \longrightarrow {}_A\underline{\underline{\mathcal{M}}}$ where $\underline{\underline{\mathcal{M}}}_A$ resp. ${}_A\underline{\underline{\mathcal{M}}}$ is the set of all isomorphism classes of indecomposable right- resp. left-A-modules.

Definition 1. a) A ring A is called weakly-symmetric-self-dual (ws-self-dual) if there exists a duality $D: \mathcal{M}_A \longleftrightarrow {}_A\mathcal{M}$ such that $D(eA/eN) \cong Ae/Ne$ for all primitive idempotents $e \in A$. (A finite dimensional algebra over a field is ws-self-dual.)

b) If M is an indecomposable right- or left-A-module, then the sequence

$$[M],\ \omega[M],\ \omega^2[M],\ \ldots$$

is called ω-sequence of M. If it stops at a projective module, more precisely at an isomorphism class of projective modules, then the sequence is called finite.

Theorem 2. Let A be a ws-self-dual artin ring with radical square zero. Then the following two statements are equivalent.

a) A is of finite type.

b) The ω-sequence of each simple right-A-module and each simple left-A-module is finite or periodic.

If a) and b) hold, then every indecomposable module lies in a ω-sequence of either a simple or an indecomposable injective module.

Proof. It is trivial that a) implies b). The converse will be a consequence of lemma 3 - 6.

We begin by introducing additional notations. For convenience we shall not make a difference between a module and the isomorphism class it belongs to.

Denote by

F_* resp. $_*F$ the free abelian group over $\mathfrak{C}_A$ resp. $_A\mathfrak{C}$ (every semisimple module can be considered as an element of F_* resp. $_*F$),

$\varrho : F_* \longrightarrow F_*$ resp. $\varrho : {_*F} \longrightarrow {_*F}$ the homomorphism defined by

$\varrho(eA/eN) = eN$ resp. $\varrho(Ae/Ne) = Ne$ for all primitive idempotents $e \in A$,

$H_* = F_* \times F_*$ resp. $_*H = {_*F} \times {_*F}$ the product of F_* resp. $_*F$ with itself,

$\ker : H_* \longrightarrow {_*H}$ resp. $\ker : {_*H} \longrightarrow H_*$ the isomorphism defined by

$\ker(M,N) = (D\varrho M - DN,\ DM)$ for all semisimple modules M,N,

$\mathrm{cok} : H_* \longrightarrow {_*H}$ resp. $\mathrm{cok} : {_*H} \longrightarrow H_*$ the inverse isomorphism $\mathrm{cok} = \ker^{-1}$ (we have

$\mathrm{cok}(M,N) = (DN,\ \varrho DN - DM)$ for all semisimple modules M,N),

and

$\varphi : \underline{\underline{\mathfrak{K}}}_A \longrightarrow H_*$ resp. $\varphi : {_A\underline{\underline{\mathfrak{K}}}} \longrightarrow {_*H}$ the map defined by

$\varphi(M) = (M/\mathrm{rad}\,M,\ \mathrm{rad}\,M)$ for all $M \in \underline{\underline{\mathfrak{K}}}_A$ resp. $M \in {_A\underline{\underline{\mathfrak{K}}}}$.

In the following $\overline{M}$ denotes the radical factor module $M/\mathrm{rad}\,M$.

<u>Lemma 3.</u> Let $M \in \underline{\mathfrak{K}}_A$ resp. $M \in {_A\underline{\mathfrak{K}}}$. Then

$$\ker^2(\varphi(M)) = \begin{cases} (0\,,\omega(M)) & \text{if } \omega(M) \text{ is simple,} \\ \varphi(\omega(M)) & \text{otherwise.} \end{cases}$$

Proof. If $P_1 \longrightarrow P_0 \longrightarrow M \longrightarrow 0$ is a minimal projective presentation of M, then

$$\varphi(M) = (\overline{P}_0 , \varrho\overline{P}_0 - \overline{P}_1) , \quad \varphi(\tau M) = (D\overline{P}_1 , \varrho D\overline{P}_1 - D\overline{P}_0),$$

and

$$\varphi(\omega(M)) = \begin{cases} (\overline{P}_1 , 0) & \text{if } \omega(M) \text{ is simple,} \\ (D\varrho D\overline{P}_1 - \overline{P}_0 , \overline{P}_1) & \text{otherwise.} \end{cases}$$

Thus

$$\ker^2(\varphi(M)) = \ker^2(\overline{P}_0, \varrho\overline{P}_0 - \overline{P}_1) = \ker(D\overline{P}_1 , D\overline{P}_0) =$$

$$= (D\varrho D\overline{P}_1 - \overline{P}_0, \overline{P}_1) = \begin{cases} (0 , \overline{P}_1) & \text{if } \omega(M) \text{ is simple,} \\ \varphi(\omega(M)) & \text{otherwise.} \end{cases}$$

Now, for each simple module $E = \overline{eA}$ resp. $E = \overline{Ae}$ where e is a primitive idempotent in A, we consider the "kernel-sequence"

$$E = E^0, E^1, E^2, \ldots$$

where

$$\left.\begin{aligned} E^{2k} &= \omega^k(\overline{eA}) \\ E^{2k+1} &= \omega^k(D(eA)) \end{aligned}\right\} \text{ for } k = 0,1,2, \ldots$$

resp.

$$\left.\begin{aligned} E^{2k} &= \omega^k(\overline{Ae}) \\ E^{2k+1} &= \omega^k(D(Ae)) \end{aligned}\right\} \text{ for } k = 0,1,2, \ldots .$$

Let l_E resp. m_E denote the least positive upper index (when it exists - otherwise $l_E = \infty$ resp. $m_E = \infty$) such that E^{l_E-1} is projective resp. E^{m_E} is simple.

The name "kernel-sequence" is motivated by the next lemma.

<u>Lemma 4.</u> If each module of the sequence $E^1, E^2, \ldots , E^k$, $k < l_E$, is not simple, then

$$\ker(\varphi(E^{i-1})) = \varphi(E^i) \quad \text{for } i = 1,2, \ldots ,k.$$

Proof. We proceed by induction on i. If $E = \overline{eA}$ (if $E = \overline{Ae}$, the proof is similar) and $i = 1$, then

$$\ker(\varphi(E^0)) = \ker(\varphi(\overline{eA})) = \ker(\overline{eA}, 0) = (D(eN), D(\overline{eA})) =$$
$$= \varphi(D(eA)) = \varphi(E^1).$$

Assume $\ker(\varphi(E^{i-2})) = \varphi(E^{i-1})$, then

$$\ker(\varphi(E^{i-1})) = \ker^2(\varphi(E^{i-2})) = \varphi(\omega(E^{i-2})) = \varphi(E^i) .$$

Lemma 5. $l_E = m_E$.

Proof. E^{l_E-1} is projective. Thus

$$\varphi(E^{l_E-1}) = (\overline{E^{l_E-1}}, \varrho \overline{E^{l_E-1}})$$

and

$$\ker(\varphi(E^{l_E-1})) = (0, D\overline{E^{l_E-1}}).$$

By the previous lemma we can say E^{l_E} is simple. Hence $m_E \leq l_E$.

Since $E^1, E^2, \ldots, E^{m_E-1}$ are not simple, we have

$$(0, E^{m_E}) = \ker^2(\varphi(E^{m_E-2})) = \ker(\varphi(E^{m_E-1})) .$$

This implies

$$\mathrm{cok}(0, E^{m_E}) = (D(E^{m_E}), \varrho D(E^{m_E})) = \varphi(E^{m_E-1}) .$$

Therefore, E^{m_E-1} is a projective cover of $D(E^{m_E})$, whence the assertion follows.

We call l_E resp. m_E the kernel-length of E.

From now we assume the condition b) of theorem 2 holds. Consequently the kernel-lengths l_E are finite for all simple modules E.

<u>Lemma 6.</u> Let M be an indecomposable non-simple right- or left-A-module and $M/\text{rad}\,M = \bigoplus_{i\in I} E_i$, $\text{rad}\,M = \bigoplus_{j\in J} F_j$ where E_i, F_j are simple modules for all $i \in I$, $j \in J$. Then

a) there exists a natural number l_M such that
$l_{E_i} = l_M$ for all $i \in I$ and $l_{DF_j} = l_M$ for all $j \in J$.

b) there exists a simple or an indecomposable injective module M' and a number $n \in \mathbb{N}\cup\{0\}$ such that $M = \omega^n(M')$.

Proof. a) If we consider the indecomposable module M as submodule of its injective envelope and simultaneously as factor-module of its projective cover, we can reduce the problem to the following two cases:

1) M is injective, $J = \{1\}$ and $l_{DF_1} \leq l_{E_i}$ for all $i \in I$.

2) M is projective, $I = \{1\}$, $J \neq \emptyset$ and $l_{E_1} \leq l_{DF_j}$ for all $j \in J$.

In the first case let $l = l_{DF_1}$, $I' = \{i \mid i \in I,\ l_{E_i} = l\}$ and $I'' = I \setminus I'$. Then applying $\ker^l$ to the equation

$$\ker(\varphi(DF_1)) = (0,F_1) + \sum_{i\in I'}(E_i,0) + \sum_{i\in I''}(E_i,0)$$

yields

$$-(D(DF_1)^l,0) = -\varphi((DF_1)^{l-1}) + \sum_{i\in I'}(0,E_i^l) + \sum_{i\in I''}\varphi(E_i^l)\,.$$

Since $(DF_1)^{l-1}/\text{rad}(DF_1)^{l-1} = D(DF_1)^l$ we conclude $I'' = 0$.

The other case can be reduced to 1) if we consider DM instead of M.

b) Suppose the ω-sequence of M is an infinite sequence of non-simple modules. If l_M is even resp. odd, apply $\ker^{\ell_M}$ resp. $\ker^{\ell_M+1}$ to the equation

$$\varphi(M) = \sum_{i\in I}(E_i, 0) + \sum_{j\in J}(0, F_j)$$

and obtain

$$\varphi(\omega^{\ell_M/2}(M)) = \sum_{i\in I}(0, E_i^{\ell_M}) - \sum_{j\in J}\varphi((DF_j)^{\ell_M - 1})$$

resp.

$$\varphi(\omega^{(\ell_M+1)/2}(M)) = -\sum_{i\in I}(D(E_i^{\ell_M}), 0) - \sum_{j\in J}(0, (DF_j)^{\ell_M}) .$$

These equations are obviously wrong. Hence there exists either a simple or a projective module in the ω-sequence of M. So M lies in the kernel-sequence of a simple module. This implies b).

With this result theorem 2 is completely proved.

Since the method of constructing indecomposable modules by means of ω is not restricted to the case of radical square zero we may ask whether every indecomposable module over an artin ring A of finite type is connected by a ω-sequence with an A-module which belongs to a special class of modules, for example, the class of all non-faithful A-modules?

Questions of this kind will be discussed in another paper.

References

[1] M. Auslander, M. Bridger, Stable module theory, Memoirs Amer. Math. Soc. 94, Providence 1969.

[2] M. Auslander, Representation theory of artin algebras II, Comm. in Algebra 1 (4), 269 - 310 (1974).

[3] V. Dlab, C.M.Ringel, Decomposition of modules over right uniserial rings, Carleton Math. Series No. 75, 1972.

[4] V. Dlab, C.M.Ringel, On algebras of finite representation type, Carleton Math. Lect. Notes No. 2 , 1973.

[5] W. Müller, Unzerlegbare Moduln über artinschen Ringen, Math. Z. 137, 197 - 226 (1974).

Wolfgang Müller
Mathematisches Institut der Universität
8000 München 2
Theresienstr. 39

PARTIALLY ORDERED SETS WITH AN INFINITE NUMBER OF INDECOMPOSABLE REPRESENTATIONS

L. A. Nazarova

In accordance with [1], [2], we shall define a representation of a partially ordered set $\mathcal{N}$ over a field k by prescribing, to each element $i \in \mathcal{N}$, a subspace V_i of a finite dimensional vector space V over a field k in such a way that $i \leq j$ implies $V_i \subseteq V_j$. The decomposition and the equivalence of representations is defined in the usual way.

We say that a partially ordered set is a set of finite type if there is only a finite number of indecomposable representations and that it is a set of infinite type otherwise. A partially ordered set is said to be a set of a solvable type if the problem of a classification of its representations does not embody the problem of a classification of pairs of matrices under simultaneous similarity transformations.

A criterion for a partially ordered set to be of finite type has been given by M. M. Kleiner [3]. We shall formulate this criterion using a graphical illustration of the partially ordered sets as follows: the elements of sets will be represented by points and the order relation by edges so that $\alpha < \beta$ whenever α is connected with β through a sequence of edges directed from below upwards.

A partially ordered set $\mathcal{N}$ is a set of finite type if and only if it does not contain any subsets of the form

1) $(1, 1, 1, 1)$ = ;

2) $(2, 2, 2)$ =  ;

3) (1, 3, 3) =  ;

4) (1, 2, 5) =  ;

5) $\mathcal{R}$ =  .

After having determined the structures of finite type, it was only natural to ask for classification of structures of solvable type. Here we have

Theorem. A partially ordered set $\mathcal{N}$ is a set of solvable type if and only if it does not contain any subsets of the form

1) (1, 1, 1, 1, 1) =  ;

2) (1, 1, 1, 2) =  ;

3) (2, 2, 3) = 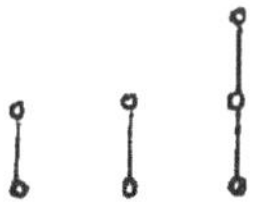 ;

4) (1, 3, 4) =  ;

5) (1, 2, 6) =  ;

6) $\bar{\mathcal{R}}$ = 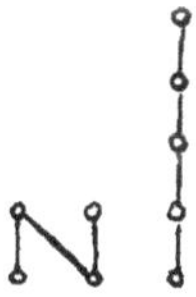 .

The partially ordered sets which appear in the formulation of the theorem will be called critical.

We shall present the main idea and the course of the proof without giving any detailed calculations. The proof of the necessity of the theorem is standard and sufficiently trivial. For the proof of the sufficiency one needs to recall the definition of the derivative of a set given in [1].

If $\mathcal{H}$ is a partially ordered set and a a maximal element of $\mathcal{H}$ which is not contained in a subset of the form $(1, 1, 1, 1)$, then $\mathcal{H}'_a$ will stand for the partially ordered set defined as follows: The elements of $\mathcal{H}'_a$ are

1) all elements of $\mathcal{H}$ with the exception of a ;

2) all pairs (u, v) of elements u, v of $\mathcal{H}$ such that a, u, v are mutually incomparable.

The order relation in $\mathcal{H}'_a$ is given by

1) the induced order on $\mathcal{H} \setminus a$;

2) $x \leqslant (c, b)$ if $x \leqslant c$ and $x \leqslant b$;

3) $x \geqslant (c, b)$ if $x \geqslant c$ and $x \geqslant h$;

4) $(c, h) \leqslant (d, f)$ if each of c, h is smaller than one of d, f .

It has been proved in [1] that, up to a finite number of representations whose form can be described, there is a one-to-one correspondence between all indecomposable representations of $\mathcal{H}'_a$ and all indecomposable representations of $\mathcal{H}$.

In a similar way, one can construct (see [2]) $\mathcal{H}'_b$, where b is a minimal element of $\mathcal{H}$ which is not contained in a quadruple of mutually incomparable elements and establish a one-to-one correspondence between the representations of $\mathcal{H}'_b$ and $\mathcal{H}$

The relation between the representations of $\mathcal{H}'_a$ and $\mathcal{H}$ turned out to be decisive in the proof of the finite type criterion of partially ordered sets [3] . However, it is not difficult to observe that, among the solvable partially ordered set there is a sufficient number of examples in which all minimal and all maximal elements belong to a subset of the form $(1, 1, 1, 1)$ and therefore the above differentiation cannot be applied.

The problem relates to the fact that, by differentiating with respect to a maximal (or a minimal) element which belongs to a quadruple (1, 1, 1, 1), we proceed as follows: If N is an arbitrary representation of $\mathcal{N}$ and a a maximal element of $\mathcal{N}$, we bring N to the form

$$N = \left(\begin{array}{c|c|c|c|c|c|c} \multicolumn{1}{c}{\overbrace{\quad}^{a}} & \multicolumn{3}{c}{\overbrace{\qquad\qquad}^{B}} & \multicolumn{3}{c}{\overbrace{\qquad\qquad}^{A}} \\ E & & & & & & \\ \hline 0 & 0 & \cdots & 0 & & \cdots & \end{array}\right)$$

where A is the family of the matrices corresponding to the elements incomparable to a and B is the family of the matrices corresponding to the elements comparable to a. When a does not belong to a quadruple (1, 1, 1, 1), the set A has width 2, i.e. for no $x \in A$ are there two incomparable elements which are both incomparable with x, as well. Therefore, decomposing A in the lower horizontal strip and applying the transformation defined in [1] to the upper horizontal strip, we get in the upper horizontal strip a representation of the partially ordered set $\mathcal{N}'_a$. If a belongs to (1, 1, 1, 1), then A has width 3. From the description of the exact finite partially ordered sets [4], it follows that all exact sets, with the exception of the sets

$$(1,\ 1)\ ,\quad (1,\ 1,\ 1)$$

contain as a subset

$$(1,\ 1,\ 2) = \circ \quad \bullet \quad$$  ;

this means that if A is one of the exact finite sets different to (1, 1) or (1, 1, 1), then $\mathcal{N}$ contains a subset (1, 1, 1, 2), and thus does not satisfy the assumptions of our theorem. Since we consider the sets which do not contain any critical subsets, N has to decompose in the lower horizontal strip into a direct sum of indecomposable representations of the sets (1, 1) and (1, 1, 1).

If no representations of the form

1) $a_1 = \binom{1}{0}, \quad a_2 = \binom{0}{1}, \quad a_3 = \binom{1}{1}, \qquad$ where $\{a_1, a_2, a_3\} = \circ \quad \circ \quad \circ$;
2) $a_1 = (1), \quad a_2 = (1), \quad a_3 = (1)$

appear among the direct summands of A, then N can be differentiated in the usual way. If there is a summand of the type 2) among the componenets of A, then, performing a decomposition of A into indecomposable summands and transformations of the upper horizontal strip given in [1] , we obtain in the upper horizontal strip a pencil of matrices; this is a problem which does not correspond to any partially ordered set, i.e. the differentiation is in this case not possible.

We assume that there are no components of the type 2) in the decomposition of A , but that components of type 1) appear. In this case, we construct a new differentiation as follows:

$\mathcal{K}'_a$ consists of

1) singletons: all elements of $\mathcal{K} \setminus a$;

2) pairs (u, v) such that a, u, v are mutually incomparable;

3) triples (u, v, w) such that a, u, v, w are mutually incomparable.

The order relation is defined in the following way:
Let $\alpha, \beta \in \mathcal{K}'_a$, where α and β can be singletons, pairs or triples of elements of $\mathcal{K}$. Then $\alpha \leq \beta$ if, for every element α_i from α there is an element β_j from β such that $\alpha_i \leq \beta_j$.

In analogy, we can define a differentiation of $\mathcal{K}$ with respect to a minimal element belonging to a quadruple $(1, 1, 1, 1)$ if the decomposition of the representation of the set of the elements which are incomparable to it, into indecomposable summands does not contain any summands of the type 1) (but contains summands of the type 2)).

If, however, by attempting to differentiate a representation with respect to a maximal element a , one obtains components of the type 2, then such a representation of the set $\mathcal{K}$ cannot be obtained from a representation of $\mathcal{K}'_a$. Similarly, there are representations of $\mathcal{K}$ which cannot be obtained from representations of $\mathcal{K}'_b$, where b is a minimal element of $\mathcal{K}$. Nevertheless, we have the following

<u>Proposition 1.</u> If all minimal and all maximal elements of $\mathcal{K}$ belong to a subset of the form $(1, 1, 1, 1)$ and $\mathcal{K}$ does not contain any critical subset and if it differs from the set

$\mathcal{K} \neq$

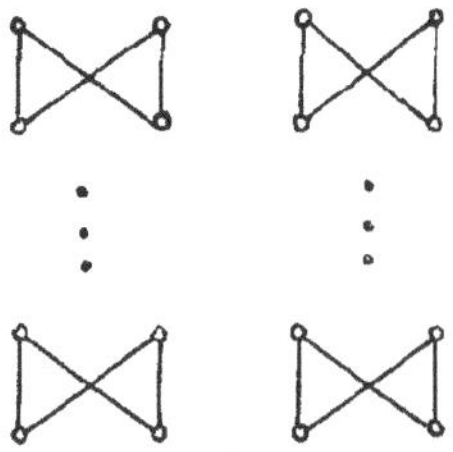

,

then, with the exception of a finite number of representations whose form can easily be described, every exact representation of $\mathfrak{N}$ can be obtained either from a representation of $\mathfrak{N}'_a$ or from a representation of $\mathfrak{N}'_b$ (where a is a fixed maximal point and b a minimal point of $\mathfrak{N}$).

We shall present the proof of this proposition in an important special case well illustrating the idea and the lines of the general proof.

Let

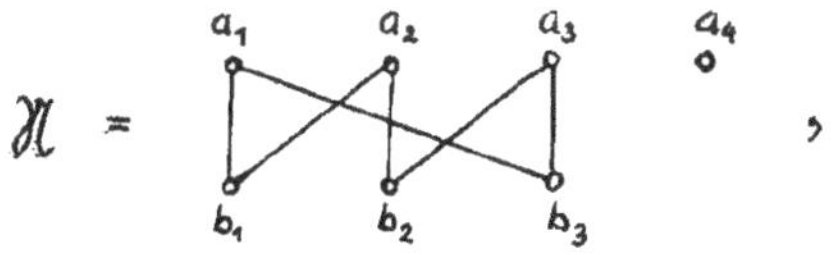

We bring an arbitrary representation N to the form

$N =$

$b_1 b_2 b_3$	b_1	b_2	b_3	a_1		a_2		a_3		a_4	
B	0	0	0	0	0	0	0	0	0	X	0
0	E	0	0	0	0	0	0	C_1	C_2	T_3	
	0	E	0	F_1	F_2	0	0	0	0	T_2	
	0	0	E	0	0	D_1	D_2	0	0	T_1	
	0			0	A_1	0	A_2	0	A_3	0	A_4

where B consist of representations of the triad $\{b_1, b_2, b_3\}$ whose dimensions are greater than one and which will be, in what follows, called non-singular in distinction from the singular, one-dimensional, ones. In the matrices A_i, there are no more zero columns; the remaining zeros of N are easily obtained by means of operations on the rows of the matrices A_i and the columns of the matrices b_i .

The matrix X can be non-zero only if the triad (T_1, T_2, T_3) is zero or if it consists of onedimensional direct summands, and such a representation $\{T_i\}$ cannot be linked with a non-zero representation $\{A_i\}$.

So, if N has non-trivial matrices A_i , then, the matrices corresponding to the minimal elements b_i, consist only of singular representations, and then it is easy to verify that such a representation can be differentiated with respect to a minimal point.

If all A_1, A_2, A_3, A_4 are zero, then the representation can be differentiated with respect to an arbitrary maximal point.

The following proposition is important for the proof of the theorem.

Proposition 2. If a partially ordered set $\mathcal{H}$ does not contain any critical subset, then also $\mathcal{H}'_a$ ($\mathcal{H}'_b$) does not contain any critical set.

For the proof of this proposition, the following geometrical interpretation of an arbitrary partially ordered set appeared to be helpful.

Let $\mathcal{H}$ be a partially ordered set of width 3, i.e. such that no point $x \in$ belongs to a set (1, 1, 1, 1) . Let $\mathcal{H} = C \cup B \vee a$, where a is an isolated point, i.e. a point which is incomparable to each of the points of B and C .

A partially ordered set will be called primitive if it is a union of several disjoint linearly ordered sets such that the elements of distinct subsets are mutually incomparable.

If $C = \{ c_1 \leq \cdots \leq c_n \}$, $B = \{ b_1 \leq \cdots \leq b_m \}$ and $C \cup B$ is a primitive set, then $\mathcal{H}'_a$ is a lattice:

$\mathcal{H}'_a =$ 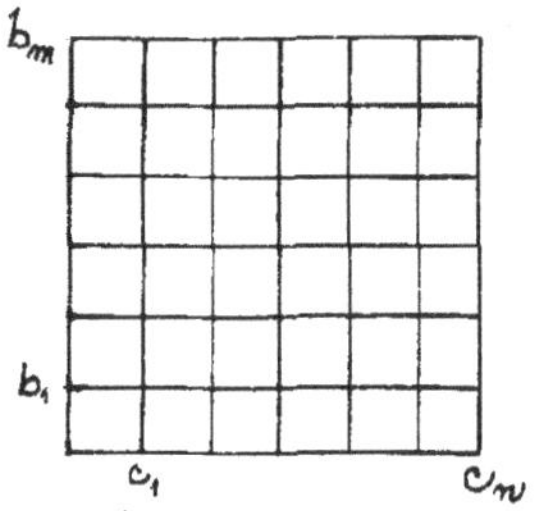

If there is a relation $c_i \leq b_j$, then

$\mathcal{H}'_a =$ 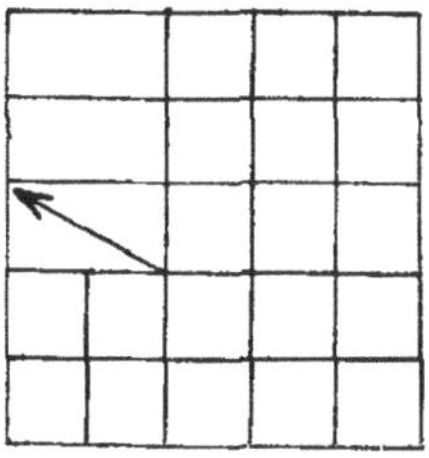

If $b_i \leqslant c_j$, then

$\mathcal{K}'_a =$ 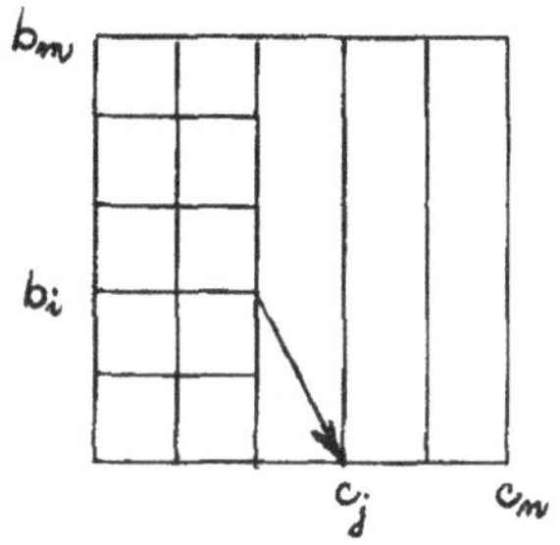

Thus in the case that $\mathcal{K}'$ is of width 3 and a is an isolated point, $\mathcal{K}'_a$ is a lattice with some points discarded, and endowed with arrows directed to the coordinate axes. The elements are in the relation $(c_i, b_j) \leqslant (c_r, b_s)$ if (c_i, b_j) can be connected with (c_r, b_s) by means of a sequence of segments directed upwards or from the left to the right and by means of arrows.

The following lemma is essential for the proof of Proposition 2.

<u>Lemma.</u> Let $\mathcal{K} = C \cup B \cup a$, where a is an isolated point, $C = \{c_1 \leqslant \cdots \leqslant c_n\}$, $B = \{b_1 \leqslant \cdots \leqslant b_m\}$. Let $T = T_1 \cup \cdots \cup T_k$, with $k > 2$ be a primitive subset of $\mathcal{K}'_a$ such that

1) in writing down its elements, all points of $\mathcal{K} \setminus a$ appear as entries;

2) T is maximal, i.e. any subset of $\mathcal{K}'_a$ containing properly T is no longer primitive.

Then $n + m = t_1 + t_2 + \ldots + t_k + k - 2$, where t_i is the order of T_i.

The above mentioned arguments do not apply (cf. Proposition 1) to the set

$\mathcal{K} =$ 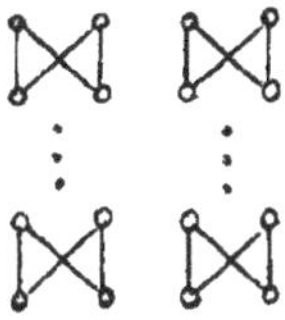 .

However, a complete classification of the representations of such a set can be deduced from the results in 5 if one decomposes the representations of its subset

into indecomposable ones.

References

[1] Nazarova, L.A. and Roiter, A.V.: Representations of partially ordered sets, Zap. Naučn. Sem. Leningrad. Otdel. Mat. Inst. Steklov 28 (1972), 5 - 32.

[2] Nazarova, L.A.: Representations of quivers of infinite type, Izv. Akad. Nauk SSSR, ser. Mat. 37 (1973), 752 - 791.

[3] Kleiner, M.M.: Partially ordered sets of finite type, Zap. Naučn. Sem. Leningrad. Otdel. Mat. Inst. Steklov 28 (1972), 32 - 42.

[4] Kleiner, M.M.: On exact representations of partially ordered sets of finite type, Zap. Naučn. Sem. Leningrad. Otdel. Mat. Inst. Steklov 28 (1972), 42 - 60.

[5] Nazarova, L.A. and Roiter, A.V.: On a problem of I.M. Gel'fand, Funkcional. anal. i Priložen. 7 (1973), 54 - 69.

Mathematical Institute AN USSR
Kiev 252004, Repina 3
U.S.S.R.

LOCALLY FREE CLASS GROUPS OF ORDERS

Irving Reiner*

§1. Introduction.

Throughout this article, let R be a Dedekind ring with quotient field K, and let Λ be an R-order in a semi-simple K-algebra A, where $(A:K)$ is finite. We shall denote by $\mathrm{Cl}\,\Lambda$ the (locally free) class group of Λ, to be defined below.

In particular, $\mathrm{Cl}\,R$ is the usual ideal class group of R, consisting of R-isomorphism classes of fractional R-ideals in K, and where the group operation is determined by multiplication of fractional ideals. Another special case, of great interest from the standpoint of applications, is that in which $\Lambda = RG$, the integral group ring of a finite group G over a Dedekind ring R of characteristic 0. Under some mild hypotheses (see (1.5)), $\mathrm{Cl}\,RG$ is precisely the "reduced projective class group" of RG defined in [25] and [29].

The purpose of this article is to survey the present state of knowledge about class groups of orders. No proofs will be given. We begin by recalling a number of definitions. A Λ-lattice M is a left Λ-module which is finitely generated and torsionfree as R-module. For P a prime ideal of R, let M_P denote the localization of M at P. If M and N are Λ-lattices such that $M_P \cong N_P$ as Λ_P-modules for each P, we say that M and N are in the same genus, and write $M \vee N$.

*This work was partially supported by a research grant from the National Science Foundation.

If $M \vee \Lambda^{(n)}$, a free Λ-module on n generators, call M <u>locally free of rank n.</u> The locally free rank one Λ-lattices play the role of fractional ideals, and are called <u>locally free ideals</u>.

Let us introduce an equivalence relation on the set of locally free Λ-lattices, writing $M \sim N$ if there exist non-negative integers r, s such that

$$M \dotplus \Lambda^{(s)} \cong N \dotplus \Lambda^{(s)} . \tag{1.1}$$

Let $[M]$ denote the equivalence class of M. Lattices in the class $[\Lambda]$ are called <u>stably free</u>.

Given two locally free Λ-lattices M and M', it is easily shown that

$$M \dotplus M' \cong \Lambda^{(t)} \dotplus M'' \tag{1.2}$$

for some locally free ideal M''. This result is essentially due to Swan [29]; for other proofs, see Reiner [19] or Roggenkamp [26]. This permits us to define "addition" of classes, by setting

$$[M] + [M'] = [M''] \tag{1.3}$$

whenever (1.2) holds true. It also shows that every class is represented by a locally free ideal.

We now define the (locally free) <u>class group</u> $\mathrm{Cl}\,\Lambda$ as the abelian additive group generated by classes $[M]$ of locally free Λ-lattices M, with addition defined by (1.3).

This class group Cl Λ is finite whenever K is a global field, by virtue of the Jordan-Zassenhaus Theorem (see [19] or [26]).

The restriction of this discussion to locally free lattices is justified to some extent by two basic results:

(1.4) Theorem. Let Λ be a maximal R-order in A, and let M be any Λ-lattice such that $K \otimes_R M$ is a free A-module. Then M is locally free.

(1.5) Theorem (Swan [29]). Let $\Lambda = RG$, where G is a finite group of order g, and R has characteristic 0. Suppose that every rational prime which divides g is a non-unit in R. Then every projective Λ-lattice is locally free.

For a proof of (1.4), see [19], [27] or [32]. For (1.5), see [1] or [29]. It follows from (1.5) that Cl RG is precisely the "reduced projective class group" of RG (see [29]).

§2. Maximal orders.

In this section we show how to compute class groups of maximal orders in terms of ray class groups of Dedekind rings. We begin with

(2.1) Definition. Let B be a central simple K-algebra. For each prime spot P of the field K, let $B_{\hat{P}}$ denote the P-adic completion of B. Call B ramified at P if $B_{\hat{P}}$ is not a full matrix algebra over its center $K_{\hat{P}}$. (Note that B cannot ramify at any complex prime of K.)

(2.2) Definition. Let B be a central simple K-algebra, where K is an algebraic number field. Call B a totally definite quaternion algebra if B ramifies at every infinite prime of K, and if also $(B:K) = 4$.

Let $I(R)$ denote the group of R-ideals in K, and let B be any central simple K-algebra. Denote by $P_B(R)$ the subgroup of $I(R)$ consisting of all principal ideals $R\alpha$, where α ranges over all elements of K whose image in $K_{\hat{P}}$ is positive, for every infinite prime P of K at which B ramifies. We now define

(2.3) $\quad \mathrm{Cl}_B\, R = I(R)/P_B(R) =$ ray class group of R relative to B.

This ray class group maps onto the usual ideal class group Cl R, and coincides with it in many cases.

Returning to the general case of a semisimple K-algebra A, we shall write

(2.4) $$A = \sum_{i=1}^{m}{}^{\oplus} A_i \quad \text{(simple components)}, \quad A_i = M_{n_i}(\Omega_i),$$

where $M_n(\Omega)$ denotes the ring of all $n \times n$ matrices over a skewfield Ω. We set (for $1 \leq i \leq m$)

(2.5) $$\begin{cases} K_i = \text{center of } \Omega_i, \quad R_i = \text{integral closure of } R \text{ in } K_i, \\ F = \sum_{i=1}^{m}{}^{\oplus} K_i = \text{center of } A, \quad C = \sum_{i=1}^{m}{}^{\oplus} R_i. \end{cases}$$

Now let Λ' be any maximal R-order in A, and assume* that K is an algebraic number field. Then (see [19] or [27]) we may write

(2.6) $$\Lambda' = \sum_{i=1}^{m}{}^{\oplus} \Lambda_i, \quad \text{where } \Lambda_i = \text{maximal } R_i\text{-order in } A_i, \quad 1 \leq i \leq m.$$

It follows at once that

(2.7) $$\text{Cl } \Lambda' \cong \sum_{i=1}^{m}{}^{\oplus} \text{Cl } \Lambda_i .$$

Thus, in order to compute the class group of a maximal order, it suffices to treat the central simple case.

(2.8) Theorem. Keep the above notation and hypotheses. Then for each i, $1 \leq i \leq m$, there is an isomorphism

$$\text{Cl } \Lambda_i \cong \text{Cl}_{A_i} R_i ,$$

induced by the reduced norm map nr_{A_i/K_i} from A_i into its center K_i.

---

*This assumption is needed for (2.8). Formulas (2.6) and (2.7) hold whenever char K = 0, and even more generally, whenever A is a separable K-algebra.

As shown by Swan [30], this result is an easy consequence of a theorem of Eichler; see also [19], [32]. An analogous result holds when K is a function field (see [19], Th. 35.14).

We conclude with the following definition, which will be used in later sections:

(2.9) Definition. We shall say that the semisimple K-algebra A satisfies the Eichler condition if no simple component of A is a totally definite quaternion algebra (see (2.2)).

§3. Explicit formulas for the case of arbitrary orders.

Throughout this section, let K be an algebraic number field. Given any R-order Λ in the semisimple K-algebra A, we may choose a maximal R-order Λ' in A containing Λ (see [19] or [27]). Since the class group $\mathrm{Cl}\ \Lambda'$ can be computed as in §2, we may attempt to determine $\mathrm{Cl}\ \Lambda$ by comparing it with $\mathrm{Cl}\ \Lambda'$.

(3.1) Theorem. There is a surjection $\mathrm{Cl}\ \Lambda \longrightarrow \mathrm{Cl}\ \Lambda'$, given by

$$[M] \longrightarrow [\Lambda' \otimes_\Lambda M], \quad [M] \in \mathrm{Cl}\ \Lambda.$$

The result is due to Swan [31]; other proofs are given in [4], [10] and [20]. For the remainder of this article, we use the following terminology:

(3.2) Definition. Let Λ' be a maximal order containing Λ, and let $D(\Lambda)$ denote the kernel of the surjection given in (3.1). Thus there is an exact sequence of groups

$$(3.3) \qquad 0 \longrightarrow D(\Lambda) \longrightarrow \mathrm{Cl}\ \Lambda \longrightarrow \mathrm{CL}\ \Lambda' \longrightarrow 0.$$

It follows readily from [4] that $D(\Lambda)$ is independent of the choice of Λ'.

As shown by Jacobinski [9,10], we can give explicit formulas for the groups occurring in (3.3). Keeping the notation of (2.4) and (2.5), let $nr_{A/F}$ be the reduced norm map (computed componentwise). Let $u(\Lambda)$ denote the group of units of the ring Λ. For the remainder of this section, let $\underline{f}$ be a nonzero ideal in R such that $\underline{f} \cdot \Lambda' \subset \Lambda$. We set

$$(3.4) \qquad \Lambda_{\underline{f}} = \{\lambda/\beta : \lambda \in \Lambda,\ \beta \in R,\ \underline{f} + R\beta = R\}.$$

(Of course, $\Lambda_{\underline{f}}$ is the semi-localization of Λ at the primes of R dividing $\underline{f}$.) Now put

$$(3.5) \qquad I^*(\Lambda_{\underline{f}}) = \{C \cdot nr_{A/F}\, x : x \in u(\Lambda_{\underline{f}})\},$$

where C is as in (2.5). Finally, let $I(R_i,\underline{f})$ be the subgroup of the ideal group $I(R_i)$ generated by all prime ideals of R_i which do not divide $\underline{f}$, and set

$$(3.6) \qquad I(C,\underline{f}) = \prod_{i=1}^{m} I(R_i,\underline{f}).$$

We are now ready to state Jacobinski's Theorem (see [9], [10]).

(3.7) Theorem. If A satisfies the Eichler condition, then in the following diagram each vertical arrow is an isomorphism induced by the reduced norm map:

$$\begin{array}{ccccccccc}
0 & \longrightarrow & D(\Lambda) & \longrightarrow & Cl\,\Lambda & \longrightarrow & Cl\,\Lambda' & \longrightarrow & 0 \\
 & & \downarrow & & \downarrow & & \downarrow & & \\
0 & \longrightarrow & I^*(\Lambda'_{\underline{f}})/I^*(\Lambda_{\underline{f}}) & \longrightarrow & I(C,\underline{f})/I^*(\Lambda_{\underline{f}}) & \longrightarrow & I(C,\underline{f})/I^*(\Lambda'_{\underline{f}}) & \longrightarrow & 0.
\end{array}$$

Further, each row is exact and each square commutes.

As shown in [10], there is an analogue of (3.7) which holds whether or not A satisfies the Eichler condition. Further, it may well happen that the vertical arrows in (3.7) are isomorphisms, even when A fails to satisfy the Eichler condition.

Fröhlich [4] uses the idèle-theoretic approach to the problem of finding explicit formulas for $Cl\ \Lambda$ and $D(\Lambda)$. Let $J(K)$ denote the idèle group of the algebraic number field K, and set

$$J(F) = \prod_{i=1}^{m} J(K_i) ,$$

using the notation of (2.4) and (2.5). For each prime ideal P of R, let $nr_{\hat{P}} : A_{\hat{P}} \longrightarrow F_{\hat{P}}$ be the reduced norm map from the semisimple $K_{\hat{P}}$-algebra $A_{\hat{P}}$ into its center $F_{\hat{P}}$. (The subscript $\hat{P}$ denotes P-adic completion, and the reduced norm is to be computed componentwise.)

(3.7) <u>Theorem</u> (Fröhlich [4]). Let R be the ring of all algebraic integers in an algebraic number field K and let Λ be an R-order in the semisimple K-algebra A. Then

$$Cl\ \Lambda \cong J(F)/\{u(F) \cdot \prod_{P} nr_{\hat{P}}\ u(\Lambda_{\hat{P}})\},$$

where $u(\ )$ denotes group of units, and where P ranges over all prime ideals of R.

Analogous formulas can be given for $Cl\ \Lambda'$ and $D(\Lambda)$. Wilson [35] has generalized the preceding theorem to the case where K is an arbitrary ground field.

§4. Group rings.

Throughout this section we shall take $\Lambda = RG$, the integral group ring of a finite group G over the ring R of all algebraic integers in an algebraic number field K. Let Λ' be a maximal R-order in KG containing Λ. As in (3.3), there is an exact sequence of finite abelian groups

$$0 \longrightarrow D(RG) \longrightarrow Cl\ RG \longrightarrow Cl\ \Lambda' \longrightarrow 0.$$

The calculation of $Cl\ \Lambda'$ is essentially an arithmetic question, as shown in (2.7) and (2.8). The difficulties in determining $Cl\ RG$ arise mainly in trying to describe $D(RG)$. In this section we shall list most of the results so far obtained; they deal primarily with the calculation of $D(ZG)$.

In the discussion below, the symbol p always denotes a prime.

(4.1) Theorem. $D(ZG) = 0$ if G is cyclic of order p.

This theorem was first proved by Rim [25], using results of Reiner; a simpler proof is given in Galovich-Reiner-Ullom [8]. In contradistinction to the above theorem, the case of cyclic groups of order p^n, $n > 1$, is considerably more difficult. The next theorem is due to Galovich [8]; for related results, see Kervaire-Murthy [13].

(4.2) Theorem. Let G_n be a cyclic group of order p^{n+1}, where p is an odd prime, and $n \geq 1$. Denote by $r \cdot C(k)$ the direct sum of r copies of the cyclic group $C(k)$ or order k.

i) If p is a regular* prime, then

$$(4.2a)\quad D(ZG_n) \cong a \cdot C(p^n) \oplus b \cdot C(p^{n-1}) \oplus \sum_{i=2}^{n-1} c_i \cdot C(p^{n-i}) ,$$

where

$$a = (p-3)/2, \quad b = (n-1)(p^2-3p+2)/2 \ + \ 1 ,$$

$$c_i = (n-i)p^{i-2}(p-1)^3/2 + (p^{i-1} - p^{i-2})/2 \ + \ 1, \quad 2 \leq i \leq n-1 .$$

ii) Suppose that p does not divide the class number h_0 of the maximal real subfield of $Q(\sqrt[p]{1})$. Let $p^* = (p-3)/2$, and denote by δ_p the number of Bernoulli numbers $B_1, B_2, \ldots, B_{p^*}$ which are multiples of p. Then

$$D(ZG_1) \cong (p^* + \delta_p) \cdot C(p).$$

For the case where p is irregular, Kervaire-Murthy [13] showed that the expression on the right hand side of formula (4.2a) is a direct summand of $D(ZG_n)$. Thus, in general, if G_n is cyclic of order p^{n+1} with $n \geq 1$, then $|D(ZG_n)|$ is large as compared to the order of G_n. An analogous remark holds for abelian groups, namely, $|D(ZG)|$ is much larger than $|G|$ if G is an abelian group not of prime order. This follows readily from the proof of the next theorem, due to Reiner-Ullom [22]:

(4.3) Theorem. Let $\{H_i\}$ be any sequence of abelian groups of composite order, such that $|H_i| \longrightarrow \infty$. Then also $|D(ZH_i)| \to \infty$.

---

*This means that p does not divide the class number of the cyclotomic field $Q(\sqrt[p]{1})$.

A more precise version of this result, for the case of abelian p-groups, has been given by Fröhlich [2, II]. In [36], Cassou-Noguès determined all abelian groups G for which D(ZG) = 0.

Turning next to a qualitative result, we have

(4.4) Theorem. If G is an arbitrary p-group (not necessarily abelian), then the order of D(ZG) is a power of p.

This result was established by Fröhlich [2, I] for the case of abelian p-groups. The general result is due to Reiner-Ullom [21]. An alternative proof by McCulloh is given in [21], and another proof may be found in [23]. As shown by Ullom [33], the conclusion of the theorem need not hold when Z is replaced by a larger ring of algebraic integers. Ullom also gave an example showing that |D(ZG)| may be divisible by primes not occurring as divisors of |G|; this may happen even for G cyclic of order pq, where p,q are distinct primes.

The next result gives some estimates on the exponent of the finite abelian group D(ZG).

(4.5) Theorem (Ullom [34]). Let G be a group of order p^n, where p is an odd prime. Then the exponent of D(ZG) divides p^{n-1}. If H has order 2^n, then the exponent of D(ZH) divides 2^{n-2}.

To conclude this section, we list results so far obtained for specific groups G. For the remainder of this section, let p denote an odd prime. Let ω be a primitive p-th root of 1, and let

(4.6) $$R = Z[\omega], \quad S_2 = Z[\omega + \omega^{-1}].$$

(4.7) Theorem. Let G be cyclic of order $2p$. Keeping the above notation, let $\bar{R} = R/2R$, and let $\varphi : R \longrightarrow \bar{R}$ be the canonical map. Then

$$D(ZG) \cong u(\bar{R})/\varphi\{u(R)\},$$

where $u(\ )$ denotes "group of units". Furthermore,

$$|Cl\ ZG| = |Cl\ R|^2 \cdot |D(ZG)|.$$

This result is due to Ullom [33]; other proofs are given in [21] and [22]. On the other hand, we have

(4.8) Theorem. Let G be a dihedral group of order $2p$. Then

$$D(ZG) = 0, \quad Cl\ ZG \cong Cl\ S_2,$$

with S_2 given by (4.6).

This result follows from Lee's classification [14] of all indecomposable ZG-lattices. Simpler proofs are given in [8], [21] and [22]. As a matter of fact, the techniques developed in these latter references yield the following theorem of Galovich-Reiner-Ullom [8]:

(4.9) Theorem. Let q be any divisor of $p-1$, and let

$$G = \langle x, y : x^p = 1,\ y^q = 1,\ y x y^{-1} = x^r \rangle,$$

where r is a primitive q-th root of 1 modulo p. Let $R = Z[\omega]$ as in (4.6), and let S_q be the subring of R fixed by that automorphism of R which maps ω onto ω^r. Let H be a cyclic group of order q. Then there are exact sequences of abelian groups

$$0 \longrightarrow D_0(ZG) \longrightarrow Cl\ ZG \longrightarrow Cl\ S_q \dotplus Cl\ ZH \longrightarrow 0\ ,$$

$$0 \longrightarrow D_0(ZG) \longrightarrow D(ZG) \longrightarrow D(ZH) \longrightarrow 0\ ,$$

where $D_0(ZG)$ is cyclic of order q, if q is odd, and of order q/2 if q is even.

The preceding theorem also follows from the results of Pu [17], who determined all indecomposable ZG-lattices for such groups G. In the special case where $q = 2$, we have $Cl\ ZH = D(ZH) = 0$ by (4.1), and so we recover Theorem 4.8. The following extension of (4.9) is due to Keating [12], who also calculated $K_1(ZG)$ for the groups $\{G_n\}$ below:

(4.10) <u>Theorem</u>. Let p be a regular* odd prime, let q be any divisor of p - 1, and let r be a primitive q-th root of 1 modulo p. Let G_n be the metacyclic group defined by

$$G_n = \langle x,y : x^{p^n} = 1,\ y^q = 1,\ y x y^{-1} = x^r \rangle.$$

Let ω_n denote a primitive p^n-th root of 1 over Q, and let S(n,q) be the subring of $Z[\omega_n]$ fixed under the automorphism which maps ω_n onto $(\omega_n)^r$. Then for each $n \geq 1$ there are exact sequences of abelian groups

$$0 \longrightarrow D_n \longrightarrow Cl\ ZG_n \longrightarrow CL\ ZG_{n-1} \oplus Cl\ S(n,q) \longrightarrow 0\ ,$$

$$0 \longrightarrow D_n \longrightarrow D(ZG_n) \longrightarrow D(ZG_{n-1}) \longrightarrow 0,$$

*See footnote to (4.2).

where D_n is a cyclic group of order $q/(q,2)$.

Our next result concerns class groups for the alternating and symmetric groups.

(4.11) Theorem. For $n \geq 1$, let A_n denote the alternating group on n symbols, and S_n the symmetric group. Then

$$\mathrm{Cl}\ ZA_n = 0, \quad \mathrm{Cl}\ ZS_n = 0, \quad 1 \leq n \leq 4.$$

Further, $\mathrm{Cl}\ ZS_5$ has order 2, and $\mathrm{Cl}\ ZA_5 = 0$.

The theorem is due to Reiner-Ullom [24]. Ullom [34] observed that $\mathrm{Cl}\ ZS_n = D(ZS_n)$ for all n, and proved that no prime greater than $n/2$ can divide $|\mathrm{Cl}\ ZS_n|$. Next we state

(4.12) Theorem. Let

$$G = \langle x, y : x^{2^{n-1}} = 1,\ y^2 = 1,\ yxy^{-1} = x^{-1} \rangle,$$

the dihedral group of order 2^n. Let

$$H = \langle x, y : x^{2^{n-1}} = y^2,\ y^4 = 1,\ yxy^{-1} = x^{-1} \rangle,$$

the generalized quaternion group of order 2^{n+1}. Then for $n \geq 2$,

$$D(ZG) = 0, \qquad |D(ZH)| = 2.$$

The preceding theorem was proved by Fröhlich-Keating-Wilson [5]. Special cases had been established previously: for G dihedral of order 8, Reiner-Ullom [22] showed that in fact $\mathrm{Cl}\ ZG = 0$. For H the quaternion group of order 8, it was already known that

$$|D(ZH)| = |Cl\ ZH| = 2,$$

a result proved in [15], [21] and [22].

Along these lines, we may quote a result due to Wilson [35]:

(4.13) Theorem. Let p be an odd prime, and let

$$H = \langle x, y : x^{p^n} = 1,\ y^4 = 1,\ y x y^{-1} = x^{-1} \rangle,$$

the quaternion group of order $4p^n$. Then for $n \geq 1$, the 2-primary component of the finite abelian group $D(ZH)$ is an elementary abelian 2-group of rank n.

The proof is based on related results obtained by Fröhlich [4a].

§5. Induction theorems.

Let R be any Dedekind ring of characteristic 0. For H a subgroup of the finite group G, every RG-module may be viewed (by restriction) as an RH-module. This restriction induces a homomorphism of additive groups:

$$\text{res}: \ \text{Cl } RG \longrightarrow \text{Cl } RH.$$

On the other hand, to each RH-module M there corresponds an induced RG-module M^G, defined by $M^G = RG \otimes_{RH} M$. This yields an induction map

$$\text{ind}: \text{Cl } RH \longrightarrow \text{Cl } RG,$$

which is also an additive homomorphism. While restriction and induction maps play an important role in the representation theory of groups, they have been used only marginally in the study of class groups (see Reiner-Ullom [24]).

Recall that a hyper-elementary group is one which is a semidirect product NP of a cyclic normal subgroup N, and a subgroup P of prime power order, where $(|N|,|P|) = 1$. The following theorem is due to Swan [29] (see also [32]):

(5.1) Theorem. Let H range over a full set of nonconjugate hyper-elementary subgroups of G. Then the map

$$\text{Cl } ZG \longrightarrow \prod_H \text{Cl } ZH,$$

defined by restriction at each H, is a monomorphism.

The preceding enables us to get upper bounds on the size of Cl RG, in terms of the sizes of the groups {Cl RH}. No corresponding result is known for D(RG), however.

To obtain lower bounds on the size of Cl RG, we may use

(5.2) Theorem. Let H be a self-normalizing subgroup of G such that for each $a \in G - H$, Cl RT = 0 for every hyper-elementary subgroup T of $H \cap aHa^{-1}$. Then the induction map of Cl RH into Cl RG is a monomorphism.

The above result is due to Reiner-Ullom [24]. Generalizations and improvements of (5.1) and (5.2) may be obtained by using the induction and restriction theorems of A. Dress.

§6. Mayer-Vietoris sequences.

For the calculation of the class group $\mathrm{Cl}\,\Lambda$ and its subgroup $D(\Lambda)$, when Λ is an R-order in a semisimple K-algebra, the most useful and powerful technique has been that of Mayer-Vietoris sequences. To begin with, consider a pullback diagram (or fibre product) of rings

$$(6.1)\qquad \begin{array}{ccc} \Lambda & \longrightarrow & \Lambda_1 \\ \downarrow & & \downarrow \varphi_1 \\ \Lambda_2 & \underset{\varphi_2}{\longrightarrow} & \bar{\Lambda} \end{array} ,$$

where each arrow represents a ring homomorphism. This means that there is an identification

$$\Lambda \cong \{(x_1,x_2) : x_i \in \Lambda_i,\quad \varphi_1 x_1 = \varphi_2 x_2\} .$$

The following basic theorem was proved by Milnor (see [16] for details and definitions of K_0, K_1 and K_2)

(6.2) Theorem. Given a fibre product of rings as in (6.1), in which either φ_1 or φ_2 is surjective, there is an exact sequence of additive groups:

$$(6.3)\qquad K_1(\Lambda) \longrightarrow K_1(\Lambda_1) \dotplus K_1(\Lambda_2) \longrightarrow K_1(\bar{\Lambda})$$
$$\longrightarrow K_0(\Lambda) \longrightarrow K_0(\Lambda_1) \dotplus K_0(\Lambda_2) \longrightarrow K_0(\bar{\Lambda}).$$

Furthermore, if both φ_1 and φ_2 are surjective, we may insert additional terms on the left:

$$(6.4)\qquad K_2(\Lambda) \rightarrow K_2(\Lambda_1) \dotplus K_2(\Lambda_2) \rightarrow K_2(\bar{\Lambda}) \rightarrow K_1(\Lambda)$$
$$\rightarrow \quad \cdots .$$

The sequence in (6.3) is usually called a Mayer-Vietoris sequence. Roughly speaking, $K_0(\Lambda)$ is analogous to the class group $\mathrm{Cl}\ \Lambda$, and $K_1(\Lambda)$ to the group of units $u(\Lambda)$ of the ring Λ. By adapting Milnor's techniques, Reiner-Ullom [22] proved

(6.5) Theorem. Let Λ be an R-order in a semisimple K-algebra A satisfying the Eichler condition, where K is an algebraic number field, and let (6.1) be a fibre product diagram in which both Λ_1 and Λ_2 are R-orders in semisimple K-algebras. Assume that $\bar{\Lambda}$ is a finite ring, and that either φ_1 or φ_2 is surjective. Let us set

$$u^*(\Lambda_i) = \varphi_i\{u(\Lambda_i)\}, \quad i = 1,2.$$

Then there are exact sequences of groups

$$(6.6) \quad 1 \longrightarrow u^*(\Lambda_1)\cdot u^*(\Lambda_2) \longrightarrow u(\bar{\Lambda}) \xrightarrow{\delta} \mathrm{Cl}\ \Lambda \longrightarrow \mathrm{Cl}\ \Lambda_1 \dotplus \mathrm{Cl}\ \Lambda_2 \longrightarrow 0 ,$$

$$(6.7) \quad 1 \longrightarrow u^*(\Lambda_1)\cdot u^*(\Lambda_2) \longrightarrow u(\bar{\Lambda}) \xrightarrow{\delta} D(\Lambda) \longrightarrow D(\Lambda_1) \dotplus D(\Lambda_2) \longrightarrow 0.$$

The maps $\mathrm{Cl}\ \Lambda \longrightarrow \mathrm{Cl}\ \Lambda_i$, $i = 1,2$, are defined by

$$[M] \longrightarrow [\Lambda_i \otimes_\Lambda M], \quad [M] \in \mathrm{Cl}\ \Lambda.$$

The "connecting homomorphism" δ is defined by

$$\delta(u) = \{(\lambda_1,\lambda_2) : \lambda_i \in \Lambda_i,\ (\varphi_1\lambda_1)u = \varphi_2\lambda_2\},$$

for each $u \in u(\bar{\Lambda})$. A slight modification of the theorem (see [22]) enables one to handle the case where A need not satisfy the Eichler condition.

As an illustration of how to use the above theorem for explicit calculations, we outline the proof given in [22] for Theorem 4.7 in §4 of this article. Let p be an odd prime, and let

$$G = \langle x,y : x^p = 1,\ y^2 = 1,\ xy = yx\rangle,$$

a cyclic group of order $2p$. Let H be the cyclic group $\langle y\rangle$ of order 2, and define R as in (4.6). Let $\Lambda = ZG$ and let I,J be the ideals of Λ with generators $(x^{p-1} + x^{p-2} + \ldots + x+1)$ and $(x-1)$, respectively. The fibre product diagram

$$\begin{array}{ccc} \Lambda/(I \cap J) & \longrightarrow & \Lambda/I \\ \downarrow & & \downarrow \\ I/J & \longrightarrow & \Lambda/(I+J) \end{array}$$

becomes

$$\begin{array}{ccc} \Lambda & \longrightarrow & RH \\ \downarrow & & \downarrow \\ ZH & \longrightarrow & \overline{Z}H \end{array},$$

where $\overline{Z} = Z/pZ$. Since QG is commutative, it automatically satisfies the Eichler condition, and the exact sequence (6.7) becomes

$$1 \longrightarrow u^*(ZH)\cdot u^*(RH) \longrightarrow u(\overline{Z}H) \longrightarrow D(\Lambda) \longrightarrow D(ZH) \dotplus D(RH) \longrightarrow 0.$$

But $D(ZH) = 0$ by (4.1), while a simple calculation shows that $u^*(RH) = u(\overline{Z}H)$. Thus we obtain an isomorphism $D(\Lambda) \cong D(RH)$.

In a similar manner, there is a fibre product diagram

$$\begin{array}{ccc} RH & \longrightarrow & R \\ \downarrow & & \downarrow \\ R & \longrightarrow & R/2R \end{array} ,$$

so again using (6.7) we obtain an exact sequence

$$1 \longrightarrow u^*(R)\cdot u^*(R) \longrightarrow u(R/2R) \longrightarrow D(RH) \longrightarrow D(R) \dotplus D(R) \longrightarrow 0 .$$

Since R is a maximal Z-order in $Q(\omega)$, we have $D(R) = 0$, and consequently there is an isomorphism

$$D(RH) \cong u(R/2R)/u^*(R) .$$

This proves the first assertion in Theorem 4.7; the second, dealing with $|\mathrm{Cl}\ ZG|$, follows in a similar manner.

For another approach to (6.5), see Fröhlich [4] and Wilson [35]. To conclude this section, we state the following "splitting theorem" proved by Reiner-Ullom [24]:

(6.8) <u>Theorem</u>: Let K be an algebraic number field, and let Λ, Λ_1 and Λ_2 be R-orders in semisimple K-algebras. Suppose that there is a fibre product diagram (6.1) in which $\bar{\Lambda}$ is a finite ring, and where either φ_1 or φ_2 is surjective. Suppose finally that

$$\varphi_1\{u(\Lambda_1)\} = u\{\varphi_1(\Lambda_1)\}.$$

Then both of the surjections

$$\mathrm{Cl}\ \Lambda \longrightarrow \mathrm{Cl}\ \Lambda_1, \quad D(\Lambda) \longrightarrow D(\Lambda_1),$$

are split surjections.

§7. Picard groups and class groups.

We have previously remarked that for a Dedekind ring R, the class group $\mathrm{Cl}\, R$ is the usual ideal class group of R. However, this usual class group is in fact a multiplicative group, with multiplication given by $(M)(M') = (MM')$ for each pair of fractional R-ideals M, M' in K; here, (M) denotes the R-isomorphism class of M.

If Λ is an R-order as in §1, then there is a theory of one-sided Λ-ideals, which gives rise to the class group $\mathrm{Cl}\, \Lambda$. But there is also a theory of two-sided Λ-ideals, which leads to the Picard group of Λ. In particular, let us consider all two-sided Λ-lattices X in A, such that $X \cong \Lambda$ as left Λ-modules. In this case, it is easily shown (see [3] or [6]) that also $X \cong \Lambda$ as right Λ-modules. Let (X) denote the isomorphism class of X as two-sided Λ-module. Each such X is an invertible ideal (see [3]), and the set of such classes (X) forms a multiplicative group, with multiplication defined by $(X)(X') = (XX')$. This group is denoted by $\mathrm{LFP}(\Lambda)$, and is called the _locally free Picard group of_ Λ. For a detailed discussion of such Picard groups, see [3] and [19].

The connection between $\mathrm{LFP}(\Lambda)$ and $\mathrm{Cl}\, \Lambda$ is given by a theorem of Fröhlich-Reiner-Ullom [6], as follows:

(7.1) _Theorem._ There is a homomorphism

$$\theta : \mathrm{LFP}(\Lambda) \longrightarrow \mathrm{Cl}\, \Lambda ,$$

given by $\theta(X) = [X]$ for each $(X) \in \mathrm{LFP}(\Lambda)$. If Λ is commutative, then θ is an isomorphism.

We shall say the the cancellation law holds for locally free (left) Λ-lattices if for every pair of locally free Λ-lattices M and N,

$$M \dot{+} \Lambda \cong N \dot{+} \Lambda \text{ implies that } M \cong N.$$

By virtue of [10] (see also [4]), this cancellation law holds whenever K is an algebraic number field and A satisfies the Eichler condition.

The next theorem gives a connection between the outer automorphism group of Λ, and the kernel of the map θ in (7.1). Let Autcent (Λ) denote the group of all automorphisms φ of Λ such that $\varphi(c) = c$ for each c in the center of Λ. Set

$$\text{Outcent } \Lambda = \text{Autcent } (\Lambda)/\text{In } (\Lambda) ,$$

where In (Λ) is the group of inner automorphisms of Λ. The following result is proved in [6]:

(7.2) Theorem. If the cancellation law holds for locally free (left) Λ-lattices, then

$$\ker \theta \cong \text{Outcent } \Lambda ,$$

where $\theta : \text{LFP}(\Lambda) \longrightarrow \text{Cl } \Lambda$ is the homomorphism defined in (7.1).

In those cases where LFP(Λ) and Cl Λ can be calculated explicitly, the preceding theorem yields information about Outcent Λ (see [6] for the case where $\Lambda = ZG$, with G a dihedral group of order $2p$.)

It is possible to give an explicit formula for the cokernel of θ, under the assumption that K is an algebraic number field. Using the notation in (2.4) and (2.5), let S be some finite set of primes of K containing all of the following:

i) all infinite primes of K,

ii) all prime ideals of R which ramify in at least one of the fields K_i,

iii) all prime ideals P of R for which the completion $\Lambda_{\hat{P}}$ is not a direct sum of full matrix rings over complete discrete valuation rings in finite extensions of $K_{\hat{P}}$.

Now let $\underline{f}$ be the ideal in R which is the product of the prime ideals listed in ii) and iii), and define $I(C,\underline{f})$ as in (3.6). Let

$$\psi : I(C,1) \longrightarrow I(C,\underline{f})$$

be the obvious surjection, obtained by neglecting prime factors which divide $\underline{f}$. Furthermore, for each prime P of K, let

$$N^*(\Lambda_{\hat{P}}) = \{nr_{\hat{P}}\, x : x \in u(A_{\hat{P}}),\ x\Lambda_{\hat{P}} = \Lambda_{\hat{P}}\, x\},$$

where $nr_{\hat{P}}$ is defined as in (3.7). We may now state the following result, which is proved in [6]:

(7.2) Theorem. If K is an algebraic number field, then the cokernel of the map $\theta : LFP(\Lambda) \longrightarrow Cl\,\Lambda$ satisfies

$$\operatorname{cok}\theta \cong I(C,\underline{f})/W_1 W_2 ,$$

where

$$W_1 = \{\psi(Ca) : a \in F, \quad a_{\hat{P}} \in N^*(\Lambda_{\hat{P}}) \quad \text{for each} \quad P \in S\} ,$$

and where

$$W_2 = \prod_{i=1}^{m} I(R_i,\underline{f})^{m_i}, \quad m_i^2 = (A_i : K_i).$$

Here we have used the notation of (2.4) and (2.5), and by definition,

$$I(R_i,\underline{f})^{m_i} = \{\underline{a}^{m_i} : \underline{a} \in I(R_i,\underline{f})\}, \quad 1 \leq i \leq m .$$

It follows as a corollary that the order of every element of the finite group $\operatorname{cok} \theta$ is a divisor of the least common multiple of the integers $\{m_i\}$.

§8. Hereditary orders.

The R-order Λ is *hereditary* if every (left) Λ-lattice is a projective Λ-module. By a theorem of Auslander-Goldman, Λ is left hereditary if and only if Λ is right hereditary (see also [19], Theorem 40.1).

(8.1) Theorem. Let Λ be a hereditary R-order in a semisimple K-algebra A, and let Λ' be any R-order in A containing Λ. Then there is an isomorphism

$$\mathrm{Cl}\,\Lambda \cong \mathrm{Cl}\,\Lambda' ,$$

given by the map defined in (3.1).

This theorem was proved by Jacobinski [11] for the case where K is an algebraic number field. A simpler proof, valid for the general case, may be found in Reiner [19] or [20]. As a corollary, we obtain

(8.2) Theorem. Keeping the above notation, let M be a Λ-lattice, and denote $\Lambda' \otimes_{\Lambda} M$ by $\Lambda'M$, for brevity. Set

$$\Gamma = \mathrm{Hom}_{\Lambda'}(\Lambda'M, \Lambda'M),$$

and assume that the cancellation law holds for locally free right Γ-lattices. Then for any Λ-lattice N in the genus of M, we have $\Lambda'M \cong \Lambda'N$ if and only if $M \cong N$.

For proofs, see [11] or [19].

References

[1] C. W. Curtis, I. Reiner, Representation theory of finite groups and associative algebras, Pure and Appl. Math., vol. XI, Interscience, New York, 1962, 2nd ed., 1966.

[2] A. Fröhlich, On the classgroup of integral group rings of finite abelian groups I, II, Mathematika 16(1969), 143-152; 19(1972), 51-56.

[3] ________, The Picard group of non-commutative rings, in particular of orders. Trans. Amer. Math. Soc. 180(1973) 1-46.

[4] ________, Locally free modules over arithmetic orders, J. für Math. (to appear).

[4a] ________, Module invariants and root numbers for quaternion fields of order $4\ell^r$, Proc. Camb. Phil. Soc. 76(1974), 393-399.

[5] A. Fröhlich, M. E. Keating, S. M. J. Wilson, The class groups of quaternion and dihedral 2-groups, Mathematika 21(1974), 64-71.

[6] A. Fröhlich, I. Reiner, S. Ullom, Picard groups and class groups of orders. Proc. London Math. Soc. (3)29(1974), 64-71.

[7] S. Galovich, The class group of a cyclic p-group. J. of Algebra 30(1974), 368-387.

[8] S. Galovich, I. Reiner, S. Ullom, Class groups for integral representations of metacyclic groups. Mathematika 19(1972), 105-111.

[9] H. Jacobinski, Über die Geschlechter von Gittern über Ordnungen, J. Reine Angew. Math. 230(1968), 29-39.

[10] ________, Genera and decompositions of lattices over orders, Acta Math. 121(1968), 1-29.

[11] ________, Two remarks about hereditary orders. Proc. Amer. Math. Soc. 28(1971), 1-8.

[12] M. E. Keating, Class groups of metacyclic groups of order $p^r q$, p a regular prime.

[13] M. A. Kervaire, M. P. Murthy, On the projective class group of cyclic groups of prime power order.

[14] M. P. Lee, Integral representations of dihedral groups of order 2p, Trans. Amer. Math. Soc. 110(1964), 213-231.

[15] J. Martinet, Modules sur l'algèbre du groupe quaternion. Ann. Sci. Ecole Norm. Sup. 4(1971), 399-408.

[16] J. Milnor, Introduction to algebraic K-theory, Ann. of Math. Studies, Princeton Univ. Press, 1971.

[17] L. C. Pu, Integral representations of non-abelian groups of order pq, Michigan Math. J. 12(1965), 231-246.

[18] I. Reiner, A survey of integral representations, Bull. Amer. Math. Soc. 76(1970), 159-227.

[19] ________ , Maximal orders. Academic Press, London, 1975.

[20] ________ , Hereditary orders. Rend. Sem. Mat. Univ. Padova 52(1974).

[21] I. Reiner, S. Ullom, Class groups of integral group rings. Trans. Amer. Math. Soc. 170(1972), 1-30.

[22] ________ , A. Mayer-Vietoris sequence for class groups, J. of Algebra 31(1974), 305-342.

[23] ________ , Class groups of orders, and a Mayer-Vietoris sequence. Springer Notes 353(1973), 139-151.

[24] ________ , Remarks on class groups of integral group rings. Symp. Math. Inst. Nazionale Alta Mat. (Rome) 13(1974), 501-516.

[25] D. S. Rim, Modules over finite groups, Ann. of Math. (2) 69(1959), 700-712.

[26] K. W. Roggenkamp, Lattices over orders II. Springer Lecture Notes 142, Berlin, 1970.

[27] W. Roggenkamp, V. Huber-Dyson, Lattices over orders I. Springer Lecture Notes 115, Berlin, 1970.

[28] M. Rosen, Representations of twisted group rings. Ph.D. thesis, Princeton Univ., 1963.

[29] R. G. Swan, Induced representations and projective modules, Ann. of Math. (2) 71(1960), 552-578.

[30] ________ , Projective modules over group rings and maximal orders, Ann. of Math. (2) 76(1962), 55-61.

[31] ________ , The Grothendieck ring of a finite group, Topology 2 (1963), 85-110.

[32] R. G. Swan, E. G. Evans, K-theory of finite groups and orders. Springer Lecture Notes 149, Berlin, 1970.

[33] S. Ullom, A note on the class group of integral group rings of some cyclic groups. Mathematika 17(1970), 79-81.

[34] _______ , The exponent of class groups, J. of Algebra 29(1974), 124-132.

[35] S. M. J. Wilson, Reduced norms in the K-theory of orders.

[36] P. Cassou-Noguès, Classes d'idéaux de l'algèbre d'un groupe abélien, C. R. Acad. Sc. Paris 276 (1973), 973-975.

THE REPRESENTATION TYPE OF LOCAL ALGEBRAS

Claus Michael Ringel

Let k be a (commutative) field, and $k\langle X_1,\ldots,X_n\rangle$ the free associative algebra in n (non-commuting) variables. Denote by M_i the ideal of $k\langle X_1,\ldots,X_n\rangle$ generated by all monomials of degree i. For any k-algebra A, let ${}_A\underline{\underline{m}}$ be the category of all A-modules which are finite dimensional as k-vector spaces. If I is a twosided ideal of $k\langle X_1,\ldots,X_n\rangle$, then for $A = k\langle X_1,\ldots,X_n\rangle/I$, the category ${}_A\underline{\underline{m}}$ is just the category of all (finite dimensional) vector spaces endowed with n endomorphisms which satisfy the relations expressed by the elements of I.

The k-algebra A is called <u>local</u>, provided $A = k\cdot 1 + \operatorname{rad} A$, where $\operatorname{rad} A$ is the Jacobson radical of A. If A is a local k-algebra, we will consider also its completion $\bar{A} = \varprojlim A/(\operatorname{rad} A)^n$. There is a canonical ring homomorphism $A \longrightarrow \bar{A}$, and A is said to be <u>complete</u> in case this homomorphism is an isomorphism. Since obviously every object in ${}_A\underline{\underline{m}}$ is annihilated by some power $(\operatorname{rad} A)^n$, the canonical homomorphism $A \longrightarrow \bar{A}$ induces an isomorphism of the categories ${}_A\underline{\underline{m}}$ and ${}_{\bar{A}}\underline{\underline{m}}$. Thus, in order to consider the behaviour of ${}_A\underline{\underline{m}}$ for a local algebras A, we may restrict to the case where A is complete.

The k-algebra A is said to be <u>wild</u> (or to be of wild representation type) provided there is a full and exact subcategory of ${}_{A}\underline{\underline{m}}$ which is representation equivalent to the category ${}_{k\langle X,Y\rangle}\underline{\underline{m}}$. The reason for calling it wild, is that there seems to be no hope to expect a complete classification of the indecomposable objects in ${}_{k\langle X,Y\rangle}\underline{\underline{m}}$, since for any finitely generated k-algebra B, there is a full and exact embedding of ${}_{B}\underline{\underline{m}}$ into ${}_{k\langle X,Y\rangle}\underline{\underline{m}}$. On the other hand, the algebra A is said to be tame (or to be of tame representation type), if there exists a complete classification of the indecomposable objects in ${}_{A}\underline{\underline{m}}$, and if there are not only finitely many indecomposables.

In order to distinguish the complete local algebras according to there representation type, we have to find the smallest possible wild algebras (that is, wild algebras for which all proper residue algebras are tame or of finite representation type), and the largest possible tame algebras (that is, tame algebras which do not occur as proper residue algebras of other tame algebras).

(1.1) We will have to consider several algebras which we want to introduce now. First, we mention

(a) $k\langle X,Y,Z\rangle/M_2$,

the local algebra of dimension 4 with radical square zero. Next, we single out certain residue algebras $k\langle X,Y\rangle/I$ of $k\langle X,Y\rangle$ of dimension 5, namely those with I the twosided ideal generated by the elements

(b) X^2, XY, Y^2X, Y^3;

(b^o) X^2, YX, XY^2, Y^3;

(c) X^2, $XY - \alpha YX$, Y^2X, Y^3 with $\alpha \neq 0$; and

(d) $X^2 - Y^2$, YX.

Also, we are interested in another set of local algebras $k\langle X,Y\rangle/I$, where the ideal I is generated by just two elements:

(1) YX , XY ;

(2) $YX - X^n$, XY , with $n \geq 2$;

(3) $YX - X^n$, $XY - Y^m$ $n \geq 2$, $m \geq 3$;

(4) $YX - X^2$, $XY - \alpha Y^2$ $0 \neq \alpha \neq 1$ in k ;

(5) $X^2 - (YX)^n Y$, $Y^2 - (XY)^n X$ $n \geq 1$;

(6) $X^2 - (YX)^n Y$, Y^2 $n \geq 1$;

(7) $X^2 - (YX)^n$, $Y^2 - (XY)^n$ $n \geq 2$;

(8) $X^2 - (YX)^n$, Y^2 $n \geq 2$;

(9) X^2 , Y^2 ;

Let us mention first which algebras are known to be tame or wild.

(1.2) The algebras (a), (b), (b^o), (c) and (d) are wild.

For (a), (b) and (b^o), this was proved by Heller and Reiner [7], for (c) this was proved by Drozd [4] and Brenner [2]. In section 3, we will deal with these algebras.

(1.3) The algebras (1) - (4) and (7) - (9) are tame. Namely, we have the following theorem:

Let A be a local algebra, and assume there are elements x_1, x_2, y_1, y_2 in rad A such that rad $A = Ax_1+Ay_1 = Ax_2+Ay_2$ and $x_1x_2 = y_1y_2 = 0$, then A is tame.

The case of the algebra (1) was proved by Gelfand and Ponomarev [6] and by Szekeres (unpublished, but see [12]). The case (9), which includes the decomposition of the modular representations of the dihedral 2-groups, was proved in [11]. An indication of the method of the proof of (1.3) will be given in the last section, we follow quite closely the ideas devellopped by Gelfand and Ponomarev in the case of algebra (1).*)

(1.4) Let k be an algebraically closed field. Let A be a complete local algebra. Then either (i) A has a residue ring of type (a) – (d), or (ii) A is a residue ring of the completion of one of the algebras (1) – (9), or (iii) char k = 2, and A is isomorphic to $k\langle X,Y\rangle/I$ with I the twosided ideal generated by

(5') $X^2 - (YX)^nY + \gamma(YX)^{n+1}, \quad Y^2 - (XY)^nX + \delta(YX)^{n+1}$, or

(6') $X^2 - (YX)^nY + \gamma(YX)^{n+1}, \quad Y^2 + \delta(YX)^{n+1}$,

with $(\gamma,\delta) \neq (o,o)$.

In section 2 we will prove this theorem. The first step in its proof is the classification of the local algebras $k\langle X,Y\rangle/I$ of dimension 5 given by Gabriel (unpublished). Certain partial results were obtained by Dade [3], Janusz [8] and Müller [1o], when they considered the problem to bring certain algebras (group algebras of 2-groups of maximal rank) into a normal form. Drozd [4] proved the result for commutative A .

With respect to representation theory, the case (iii) in the theorem is of no real importance. Namely, the algebras (5') and (6') – as well as (5) and (6) – are Frobenius algebras, and modulo the socle, (5') and (5), as well as (6') and (6), are isomorphic (for fixed n).

Since the only indecomposable module which is not annihilated by the socle, is the algebra itself, the representation theory of (5') is identical to that of (5), and the representation theory of (6') is the same as that of (6).

(1.5) It follows from the preceding paragraphs that the only question which remains is to determine the representation type of (5) and (6). It is an interesting fact that these are "just" the group algebras of the generalised quaternion and the semi-dihedral groups. To be more precise: If k is an algebraically closed field of characteristic 2, and G is a generalised quaternion group, then the group algebra kG is of type (5'), and if G is semi-dihedral, then kG is of type (6').

It should be noted that for all other p-groups G, the representation type of kG is known: If char $k = p$ and G is a non-cyclic p-group, then kG is wild except in the case of a two-generator 2-group of maximal rank (Krugliak [9] and Brenner [1]), that is except in the case of dihedral, semi-dihedral, and generalised quaternion groups. Namely, in all the other cases, kG has a residue ring of type (a) or (c), and therefore is wild.

---

*) At the conference in Ottawa, theorem (1.3) was formulated by the author only with an additional hypothesis: that $kx_1+ky_1 = kx_2+ky_2$; the general case was conjectured. A complete proof will appear elsewhere.

2. The classification theorem

We want to prove theorem (1.4). Thus, we assume that k is algebraically closed. Let A be a complete local algebra, and let $J = \operatorname{rad} A$. We assume that A has no residue algebra of the form (a), (b), (b^o), (c) or (d). As a consequence, $\dim_k J/J^2 \leq 2$. If $\dim_k J/J^2 \leq 1$, then A is a homomorphic image of $\varprojlim k\langle X\rangle/(X^n)$, and this is a homomorphic image of the completion of the algebra (1). Thus, we may assume $\dim_k J/J^2 = 2$. Often we will denote by N a (suitable) k-subspace of A with $J = N \oplus J^2$.

(2.1) We may assume $\dim_k J^2/J^3 = 2$.

First, we show that for $\dim_k J^2/J^3 \geq 3$, there is a homomorphic image of one of the forms (a) – (d). This is obvious for dimension 4. We may assume $J^3 = 0$, and let $\dim_k J^2 = 3$. There is a non-trivial relation

$$\alpha x^2 + \beta xy + \gamma yx + \delta y^2 = 0,$$

where x, y is a basis of N. If $\alpha = \delta = 0$, then we use as additional relation $x^2 = 0$, and get as residue algebra an algebra of the form (b), (b^o) or (c). Thus, we may suppose $\alpha = 1$. Using $x' = x+\gamma y$ instead of x, we have a relation of the form

$$x'^2 + \beta' x'y + \delta' y = 0.$$

Adding the new relation $x'y = 0$, we get as residue algebra one of the form (b) or (d).

If $\dim_k J^2/J^3 = 1$, let A be the completion of some local algebra $k\langle X,Y\rangle/I$, where I is a twosided ideal. We want to construct an ideal $I' \subseteq I$ such that

$k\langle X,Y\rangle/I'$ again has no residue algebra of the form (a) – (d), but with $\dim_k J'^2/J'^3 = 2$, where $J' =$ rad $k\langle X,Y\rangle/I'$. It is fairly easy to see that $I+M_3$ contains elements x_2x_1 and y_2y_1 , where both x_1, y_1 as well as x_2, y_2 is a basis of a fixed N with $J = N \oplus J^2$. If x_2x_1+f and y_2y_1+g belong to I (with f, g in M_3), then let I' be generated by x_2x_1+f and y_2y_1+g .

(2.2) <u>There are elements</u> a, b <u>in</u> $J \setminus J^2$ <u>such that</u> ab <u>belongs to</u> J^3.

Again, we may assume $J^3 = 0$. Now A can be written in the form $k \oplus N \oplus N\otimes N/U$, where U is a subspace of $N\otimes N$ of dimension 2, and where the multiplication is given by the tensor product $\otimes$. Let x, y be a basis of N . We may assume that U intersects both $N\otimes x$ and $N\otimes y$ trivially, thus U is the graph of an isomorphism $N\otimes x \longrightarrow N\otimes y$, and therefore there is an automorphism $\varphi: N \longrightarrow N$ with $U = \{a\otimes x + \varphi(a)\otimes y \mid a \in N\}$. Let a be an eigenvector of φ with eigenvalue α . Then $0 \neq a\otimes(x+\alpha y)$ belongs to U.

(2.3) <u>There are elements</u> x_1, x_2, y_1, y_2 <u>with</u> $N = kx_1+ky_1 = kx_2+ky_2$ <u>and</u> x_2x_1, y_2y_1 in J.

Again, we may assume $J^3 = 0$. First, assume there is $x \in J \setminus J^2$ with $x^2 = 0$. Let x, y be a basis of N. There is another non-trivial relation

$$\alpha xy + \beta yx + \gamma y^2 = 0.$$

Now $\gamma \neq 0$, since otherwise we have one of the cases (b), (b^o) or (c). Thus, we may suppose $\gamma = 1$, and then

$$(y + \alpha x)(y + \beta x) = 0,$$

and we take $x_1 =, x_2 = x$ and $y_2 = y + \alpha x$, $y_1 = y + \beta x$.

Next, assume $x^2 \neq 0$ for all x in $J \setminus J^2$. By (2.2), there is now a basis x, y of N with $yx = 0$. As before, we consider another non-trivial relation, say

$$\alpha x^2 + \beta y^2 + \gamma xy = 0.$$

Again, $\gamma \neq 0$, since otherwise we are dealing with the algebra (d), thus assume $\gamma = 1$. Then

$$(x + \beta y)(\alpha x + y) = 0,$$

which shows that we may take $x_2 = y$, $x_1 = x$, $y_2 = x + \beta x$ and $y_1 = \alpha x + y$.

(2.4) A/J^3 <u>is residue algebra of one of the algebras (1) – (9)</u>.

Proof: Assume first, one of the elements x_2, y_2, say x_2, is linearly independent both from x_1 and from y_1. Using a suitable multiple of x_1 for x and of y_1 for y, we may assume $x_2 = y-x$. If y_2 is also linearly independent both from x and y, then a multiple of y_2 is of the form $x-\alpha y$, with $\alpha \neq 0,1$. Thus, A/J^3 is of the form (4). If y_2 is a multiple of x, then we have case (2) with $n=2$, if y_2 is a multiple of y, then we have case (8) with $n = 2$. In case both x_2 and y_2 are linearly dependent of x_1 or y_1, we get the cases (1) and (9).

(2.5) It remains to be shown: If, for $p \geq 3$, A/J^p is a residue ring of one of the algebras (1)–(9), then the same is true for A/J^{p+1}. Obviously, we may assume $J^{p+1} = 0$. As a by-product of our calculations, we also will determine a basis of the algebras (1)–(9).

Case (1). There are elements X, Y in rad A with YX and XY in $\mathrm{rad}^p A$. Now $\mathrm{rad}^p A$ is generated by X^p and Y^p, thus there are elements $\alpha, \beta, \gamma, \delta$ in k with

$$YX + \alpha X^p + \beta Y^p = 0 \quad \text{and} \quad XY + \gamma X^p + \delta Y^p = 0.$$

If we replace X by $X' = X + \beta Y^{p-1}$ and Y by $Y' = Y + \gamma X^{p-1}$, the new relations are

$$Y'X' + \alpha X'^p = 0 \quad \text{and} \quad X'Y' + \delta Y'^p = 0.$$

We show how to get rid of α and δ. If $\alpha = \delta = 0$, we are again in case (1). If $\alpha \neq 0$, and $\delta = 0$, we replace X' by $X'' = \sqrt[p-1]{-\alpha}\, X'$, and are in case (2). If $\alpha = 0$ and $\delta \neq 0$, then we interchange X' and Y', and are in the previous situation. Finally, assume $\alpha \neq 0 \neq \delta$. Consider $X' = \lambda X''$ and $Y' = \mu Y''$ where λ, μ are elements of k which we want to determine now, in order to have X'' and Y'' satisfying the relations (3). The old relations become

$$\mu\lambda Y''X'' + \alpha\lambda^p X''^p = 0 \quad \text{and} \quad \lambda\mu X''Y'' + \delta\mu^p Y''^p = 0.$$

This means that we have to find λ, μ such that

$$\alpha\lambda^{p-1}\mu^{-1} = -1 \quad \text{and} \quad \delta\mu^{p-1}\lambda^{-1} = -1,$$

in order to have

$$Y''X'' - X''^p = 0 \quad \text{and} \quad X''Y'' - Y''^p = 0.$$

Of course it is easy to write down λ and μ explicitly, and X'' and Y'' are generators of rad A, since X' and Y' had this property. Such a change of X' and Y' will be called a scalar transformation in the later part of the proof, and usually will be left to the reader.

Case (2). We can assume $n < p$. Now the elements XY and $YX-X^n$ both belong to J^p, therefore $X^{n+1} = XYX = 0$, and J^p is generated by the single element Y^p. Assume there is a relation

$$YX - X^n + \alpha Y^p = 0,$$

then we replace X by $X' = X + \beta Y^{p-1}$, and get that A is either residue ring of an algebra of type (2) or of one of type (3); in the latter case we use an obvious scalar transformation.

Case (3). We consider the case $n \leq m = p-1$, and we want to prove that $J^p = 0$. This then implies that the algebra of type (3) has dimension $n+m+1$. By assumption, there are elements X, Y in J with $YX - X^n$ and $XY - Y^m$ in J^p. As in case (2), J^p is generated by Y^p, but

$$Y^p = YXY = X^nY = X^{n-1}Y^m = X^{n-2}Y^{2m-1} = 0,$$

since $n+2m-3 \geq p+1$.

Case (4). We assume $J^4 = 0$ and show $J^3 = 0$. There are equalities

$$X^3 = XYX = \alpha Y^2X = \alpha YX^2 = \alpha X^3, \quad \text{and}$$

$$X^2Y = \alpha XY^2 = \alpha^2Y^3 = \alpha YXY = \alpha X^2Y.$$

Since $\alpha \neq 1$, the monomials X^3, XYX and X^2Y are zero. Since $\alpha \neq 0$, also all the other monomials vanish.

Case (9). Assume A/J^p is a rèsidue ring of the algebra of type (9). We distinguish two cases. First, let p be even, $p = 2q$. Then J^p is generated by the two elements $(YX)^q$ and $(XY)^q$, thus there are relations

$$X^2 + \alpha(YX)^q + \beta(XY)^q = 0, \quad Y^2 + \gamma(YX)^q + \delta(XY)^q.$$

If we replace X by $X' = X + \beta(YX)^{q-1}Y$ and $Y' = Y + \gamma(XY)^{q-1}X$, then the relations in X' and Y' (after some scalar transformation) have the form (7), (8) or (9). If p is odd, say $p = 2q+1$, then J^p is generated by the elements $(XY)^qX$ and $(YX)^qY$, and we have relations

$$X^2 + \alpha(XY)^qX + \beta(YX)^qY = 0, \quad Y^2 + \gamma(XY)^qX + \delta(YX)^qY = 0.$$

This time, we replace X by $X+\alpha(XY)^q$ and Y by $Y+\delta(YX)^q$, and, again after some scalar transformation, the newe relations are of the form (5), (6) or (9).

Case (8) Now, let $p = 2n+1$, and assume A/J^p is generated by two elements X and Y which satisfy the relations $X^2 - (YX)^n = 0$ and $Y^2 = 0$. Now J^p is generated by the elements $(XY)^nX$ and $(YX)^nY$, but

(+) $\quad (XY)^nX = X^3 = X(XY)^n = X^2Y(XY)^{n-1} = (XY)^nY(XY)^{n-1} = 0,$

therefore J^p is generated by the single element $(YX)^nY$. There are relations

$$X^2 - (YX)^n + \alpha(YX)^nY = 0, \quad Y^2 + \beta(YX)^nY = 0.$$

We replace X by $X' = X - \alpha YX + \alpha XY - \alpha^2 YXY$, and Y by $Y' = Y + \beta(YX)^n$. Then we get

$$X'^2 - (Y'X')^n = 0 \quad \text{and} \quad Y'^2 = 0.$$

To see the first, we note that

$$X'^2 = X^2 + \alpha X^2Y = X^2 + \alpha(YX)^nY,$$

where the first equality stems from the fact that all the other summands cancel each other, and the second follows from the fact that $X^2 - (YX)^n$ belongs to J^p. Thus, X' and Y' satisfy relations of the form (8).

Next, let $p = 2n+2$, and A/J^p be of type (8). Then, as we have seen above, $(XY)^nX$ belongs to J^p. But then $J^p = 0$, and therefore the algebra of type (8) has dimension $4n+2$.

Case (7). We assume $p = 2n+1$. We want to show that $J^p = 0$ in case A/J^p is residue algebra of the algebra (7). Using the calculation (+) of the previous case, we see that $(XY)^nX = 0$. Similarly, we have now also $(YX)^nY = 0$. This proves the assertion. As a consequence, we see that the algebra of type (7) has dimension $4n+1$.

Cases (5),(6). Finally, we have to consider the situation where A/J^p is residue algebra of an algebra of type (5) or (6). We first look at the case $p = 2n+2$. Since $X^2-(YX)^nY$ belongs to J^p, it follows that

$$(YX)^{n+1} = X^3 = (XY)^{n+1}.$$

Thus, if $J^p \neq 0$, then A is a Frobenius algebra, with socle generated by the element $(YX)^{n+1}$. This shows that A is of the form (5') of (6'). But if the characteristic of k is different of 2, then it is easy to bring (5') into the form (5),and (6') into the form (6).

If $p = 2m+3$, we know from the previous consideration that $(YX)^{n+1}-(XY)^{n+1}$ belongs to J^p, and therefore

$$(XY)^{n+1}X = (YX)^{n+1}X = (YX)^nYX^2 = (YX)^nY(YX)^nY = 0,$$

and then also $(YX)^{n+1}Y = 0$. As a consequence, the algebras of type (5), (5'), (6), (6') all are of dimension $4n+4$.

3. The wild algebras

In order to show that a given algebra A is wild, we will use the following procedure. We will start with a category $\underline{\underline{w}}$ which we know is wild, with a full subcategory $\underline{\underline{u}}$ of $_A\underline{\underline{m}}$, and with functors

$U: \underline{\underline{w}} \longrightarrow \underline{\underline{u}}$, and $P: \underline{\underline{u}} \longrightarrow \underline{\underline{w}}$,

such that the composition PU is the identity functor on $\underline{\underline{w}}$. Then, obviously, $\underline{\underline{w}}$ is representation equivalent to the full subcategory of $_A\underline{\underline{m}}$ of all modules which are images under U.

(3.1) <u>The algebra</u> $A = k\langle X,Y,Z\rangle/M_2$ <u>is wild</u>. Following Heller and Reiner [7], we embed the category $\underline{\underline{w}} = {}_{k\langle x,y\rangle}\underline{\underline{m}}$ into $_A\underline{\underline{m}}$. Let $\underline{\underline{u}}$ be the full subcategory of $_A\underline{\underline{m}}$ consisting of all $_AM$ with $Z^{-1}0 = ZM$ (that is, all A-modules which are free when considered as $K\langle Z\rangle/(Z^2)$-modules). The functor U associates with $_{k\langle x,y\rangle}V$ the module $_AM$ given by the diagram

$$V \;\underset{1}{\overset{x}{\underset{y}{\rightrightarrows}}}\; V ,$$

thus, as vectorspace, $_kM = V \oplus V$, and X operates on $V \oplus V$ by $\begin{bmatrix} 0 & x \\ 0 & 0 \end{bmatrix}$, and so on. Conversely, given $_AM$ in $\underline{\underline{u}}$, then $P(_AM)$ is the vector space ZM together with the two endomorphisms $x = XZ^{-1}$ and $y = YZ^{-1}$. Note that, for example, XZ^{-1} is well-defined, since $XZ^{-1}0 = XZM = 0$ according to the condition $Z^{-1}0 = ZM$, and that its image lies in ZM, using again the same condition.

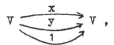

(3.2) <u>The algebra</u> $A = K\langle X,Y\rangle/(X^2,YX,XY^2,Y^3)$ <u>is</u>
<u>wild</u>. Again, we follow Heller–Reiner [7]. As $\underline{w}$, we
use the category

$$\bullet \hookrightarrow \bullet \circlearrowright$$

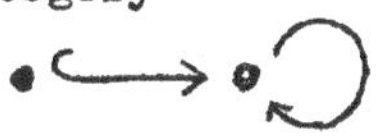

thus, an object of $\underline{w}$ is given by a tripel (W,V,φ)
with W a vector space, $V \subseteq W$ a subspace, and φ an
endomorphism of W . Let $\underline{u}$ be the full subcategory
of all $_AM$ in $_A\underline{m}$ with $XY^{-1}0 = 0$ and $Y^{-1}0 \subseteq YM$.
For (W,V,φ) in $\underline{w}$, define $_AM = U(W,V,\varphi)$ by the diagram

$$V \xrightarrow{\ Y\ } W \underset{Y=1}{\overset{X=\varphi}{\rightrightarrows}} W \ ,$$

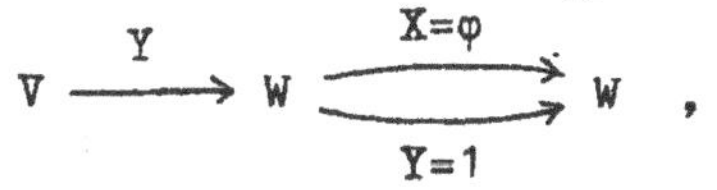

thus $M = V \oplus W \oplus W$, and X and Y operate on M as indicated.
Conversely, for $_AM$ in $\underline{u}$, let $P(_AM) = (Y^{-1}0, Y^2M, XY^{-1})$.
Obviously, Y^2M is a subspace of $Y^{-1}0$, and XY^{-1} is
well–defined, since we assume $XY^{-1}0 = 0$. Also, the image
of XY^{-1} lies in $Y^{-1}0$, since $YX = 0$.

(3.3) The algebra $A = k\langle X,Y\rangle/(X^2,XY,Y^2X,Y^3)$ is
wild, since it is just the opposite algebra to the
previously discussed one.

(3.4) <u>The algebra</u> $A = k\langle X,Y\rangle/(X^2,XY-\alpha YX,Y^2X,Y^3)$ <u>is</u>
<u>wild</u>. We may assume $\alpha \neq 0$, and give a construction due
to Drozd [4]. Again, $\underline{w}$ is the category $\bullet \hookrightarrow \bullet \circlearrowright$.
Let $\underline{u}$ be the full subcategory of all $_AM$ in $_A\underline{m}$ with
$YXY^{-2}XY^{-2}0 = 0$ and $YXM \subseteq Y^2M$, $XY^{-2}YXM \subseteq Y^2M$. For
(W,V,φ) in $\underline{w}$, define $_AM = U(W,V,\varphi)$ by the diagram

inclusion= X

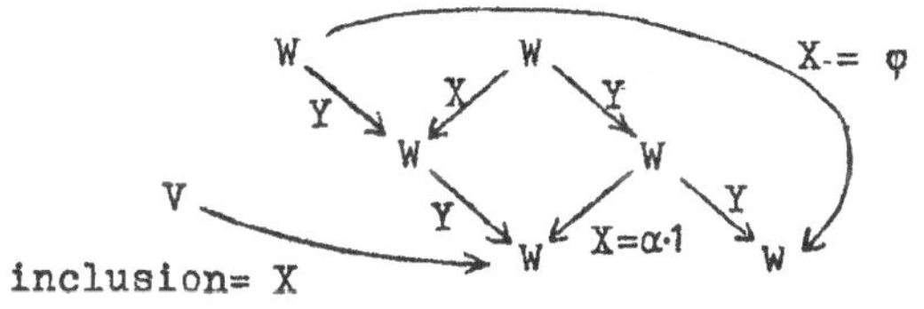

Thus, $_AM$ is the direct sum of six copies of W and one copy of V, and X and Y operate on it as indicated (where all but three maps are identity maps, one is given by φ, one is multiplication by α and one is the inclusion $V \subsetneq W$). It remains to define P. Given $_AM$ in $\underline{\underline{u}}$, let $P(_AM) = (YXM, XY^{-1}O\ YXM, YXY^{-2}XY^{-2})$. By the assumptions on $\underline{\underline{u}}$, $YXY^{-2}XY^{-2}$ is really an endomorphism of YXM, and it is easy to check that PU is the identity on $\underline{\underline{w}}$.

(3.5) <u>The algebra</u> $A=k\langle X,Y\rangle/(XY,X^2-Y^2)$ <u>is wild</u>. (Note that the ideal (XY,X^2-Y^2) contains M_3.)

We start with the category $\underline{\underline{w}}$ with objects V given as

$$\begin{array}{ccccccccccccccc} V_a & \twoheadleftarrow & V_b & \hookrightarrow & V_c & \twoheadleftarrow & V_d & \hookleftarrow & V_e & \twoheadrightarrow & V_f & \hookleftarrow & V_g & \twoheadrightarrow & V_h \\ & & & & & & \downarrow & & & & & & & & \\ & & & & & & V_i & & & & & & & & \end{array},$$

that is, we consider the category of representations of the corresponding quiver such that the maps are monomorphisms or epimorphisms as indicated. This is a well-known wild category. The functor $U\colon \underline{\underline{w}} \longrightarrow {}_A\underline{\underline{m}}$ maps the representation V onto the A-module $_AM$ given as

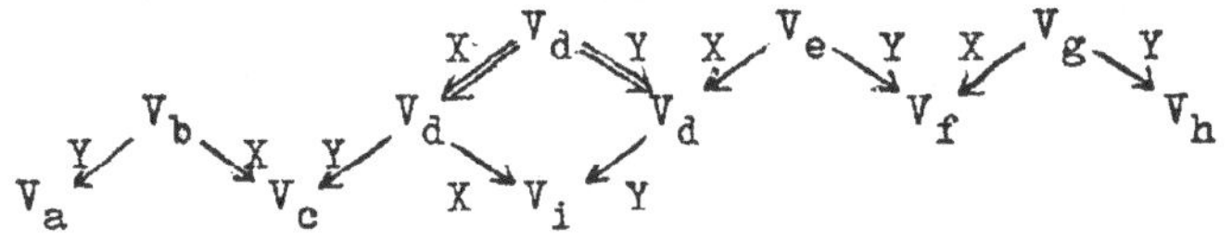

where (besides two identity maps) all maps are the ones given by V.

We define a functor $P\colon {}_A\underline{\underline{m}} \longrightarrow \underline{\underline{w}}'$, where $\underline{\underline{w}}'$ is the category of <u>all</u> representations of the quiver

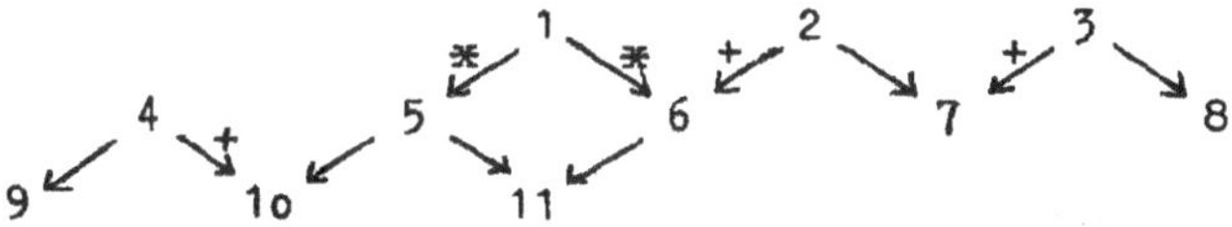

for which the square is commutative. The category $\underline{\underline{w}}$ is (equivalent to) the full subcategory of $\underline{\underline{w}}'$ of all representations for which the maps with $*$ are isomorphisms, those with $+$ are monomorphisms, and the remaining ones are epimorphisms. We will use as $\underline{\underline{u}}$ the full subcategory of all modules ${}_AM$ in ${}_A\underline{\underline{m}}$ with $P({}_AM)$ in $\underline{\underline{w}}$.

In order to define P , we note that there is a chain of subfunctors F_i $(o \leq i \leq 11)$ of the forget functor F_o from the category ${}_A\underline{\underline{m}}$ into the category of k-vector spaces, namely

$$F_o({}_AM) = M ,$$
$$F_1({}_AM) = X^{-1}YM ,$$
$$F_2({}_AM) = X^{-1}YX^{-1}YM ,$$
$$F_3({}_AM) = X^{-1}YXM + YM + XM ,$$
$$F_4({}_AM) = YM + XM ,$$
$$F_5({}_AM) = YM ,$$
$$F_6({}_AM) = YX^{-1}YM ,$$
$$F_7({}_AM) = YX^{-1}YX^{-1}YM ,$$
$$F_8({}_AM) = YX^{-1}YXM + YXM ,$$
$$F_9({}_AM) = YXM + X^2M ,$$
$$F_{1o}({}_AM) = X^2M ,$$
$$F_{11}({}_AM) = 0 .$$

Most of the inclusions $F_{i-1} \subseteq F_i$ are trivial, otherwise we use the relations $XM \subseteq X^{-1}YM$, $YM \subseteq X^{-1}0$ and $XM \subseteq X^{-1}YX^{-1}0$.

The functor $P: {}_A\underline{\underline{m}} \longrightarrow \underline{\underline{w}}'$ is now defined componentwise by $P_i = F_i/F_{i-1}$, and those natural transformations $P_i \longrightarrow P_j$ which we need, are the ones induced by multiplication by X or Y, respectively:

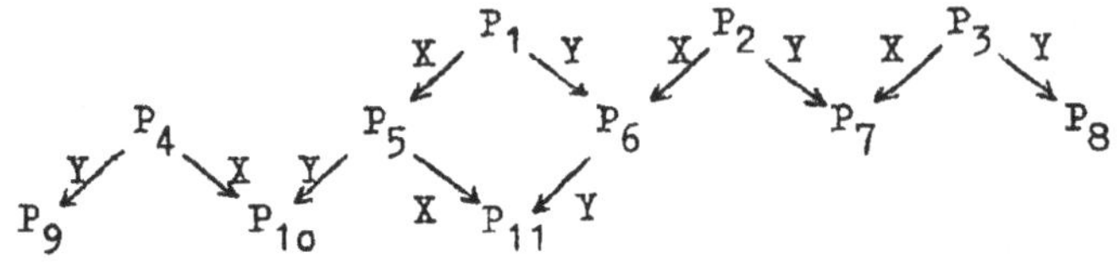

Again, in order to show that these maps are defined, we need only the relation $XY = 0$. Of course, the square is commutative, since we assume $X^2 = Y^2$.

It is easy to check that the composition PU is the identity functor on $\underline{\underline{w}}$.

4. Tame algebras

We want to give some indications about the proof of theorem (1.3). In order to show that a given algebra A is tame, it is reasonable to do two things: first to write down a list of certain indecomposable modules, and then to prove that every object of ${}_A\underline{\underline{m}}$ can be decomposed as a direct sum of copies of these modules. In our case, the decomposition will be achieved by using several functors and natural transformations.

We will start with an index set $\underline{W}$ on which a function $\underline{W} \longrightarrow \mathbb{N}$ is defined which associates to every D in $\underline{W}$ a natural number $|D| \geq 1$, the "length" of D. To every D in $\underline{W}$ we will define either one indecomposable module $M(D)$, or a whole set of indecomposable modules $M(D,\varphi)$ indexed by the set of (equivalence classes of) indecomposable automorphisms of k-vector spaces (thus, if k is algebraically closed, we may take as index set the set of Jordan matrices).

Then, we will consider the forget functor ${}_A\underline{\underline{m}} \longrightarrow {}_k\underline{\underline{m}}$, which associates to every A-module the underlying vector space. For every D in $\underline{W}$, we will construct $2\cdot|D|$ subfunctors of it, denoted by $F(D,i)^+$ and $F(D,i)^-$, where $1 \leq i \leq |D|$, such that $F(D,i)^- \leq F(D,i)^+$. We will denote by $F(D,i)$ the quotient functor $F(D,i)^+/F(D,i)^-$. Then, we will construct natural transformations

$$F(D,i) \longrightarrow F(D,i+1) \quad \text{or} \quad F(D,i) \longleftarrow F(D,i+1),$$

for $1 \leq i < |D|$, and for certain elements D in $\underline{W}$ also for $i = |D|$, calculating modulo $|D|$.

In this way, we will determine for every ${}_A M$ in ${}_A\underline{\underline{m}}$ a submodule of "type D" (that is, one which is a direct sum of copies either of M(D), or of some of the $M(D,\varphi)$.), such that ${}_A M$ is the direct sum of these submodules.

Obviously, the index set $\underline{W}$ will depend on the particular algebra A . The method will be easier to visualise, if we use a specific example. We have chosen the case of the algebra (4), that is $k\langle X,Y\rangle/(YX-X^2,XY-\alpha Y^2)$ with $\alpha \neq 0,1$, since, on the one hand, the algebra is rather small, and, on the other hand, the behaviour of the remaining algebras is somewhat intermediate between that of the algebra (4) and of the well-known cases (1) and (9).

Thus, let $A = k\langle X,Y\rangle/(YX-X^2,XY-\alpha Y^2)$, and $\alpha \neq 0,1$. We will denote the elements X, Y, $X-\alpha Y$ and $X-Y$ by a,b,c,d , in order to point out that these are just four elements of $kX+kY$ which are pairwise linearly independent. The set $\underline{W}$ will be the disjoint union of two subsets $\underline{W}_1$ and $\underline{W}_2$. Now, $\underline{W}_1$ is the set of all finite words in the letters a,b^{-1},c,d^{-1} (including the empty word 1), subject to the following rules: after c or b^{-1} follows either a or d^{-1}, after a follows only b^{-1}, and after d^{-1} follows only c . Thus, an example is the word

$$D = ab^{-1}d^{-1}cd^{-1},$$

and its length is defined to be 6 (= number of letters +1). If D and E are words, and DE is also a word, then we call DE the product of D and E . Of course, D^2

stands for DD, and so on. We call a word D non-periodic provided D^2 is also a word, and D is not of the form $D = E^n$ for some word E and $n>1$. Two non-periodic words D and E are called equivalent, provided one is a cyclic permutation of the other, and $\underline{W}_2$ will be the set of all equivalence classes of non-periodic words. Note that for elements of $\underline{W}_2$ the length is defined different: it is the precise number of letters of the corresponding word. (The word D above does not give rise to an element of $\underline{W}_2$, since D^2 is not an admissible word. An example of an element of $\underline{W}_2$ is the set of cyclic permutations of $ab^{-1}d^{-1}cd^{-1}c$.)

Next, we show how to define for D in $\underline{W}_1$ a module M(D). Namely, let M(D) be a $|D|$-dimensional vector space with base vectors $e_1, \dots, e_{|D|}$, such that X and Y operate on the base vectors according to the word D. Thus, for $D = ab^{-1}d^{-1}cd^{-1}$, we have the following schema

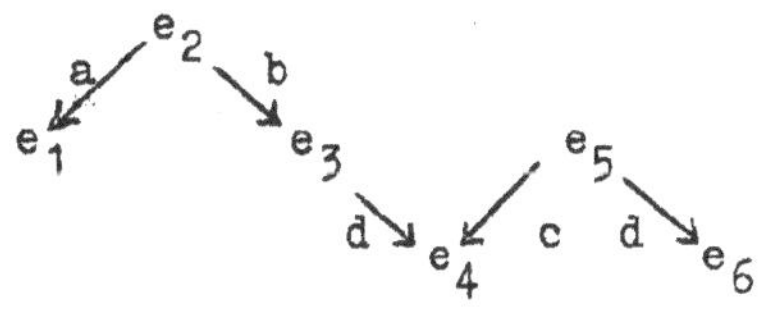

which means that $ae_2 = e_1$, $be_2 = e_3$, de_3 $(= (a-b)e_3) = e_4$, and so on. Note that in all but the terminal points e_1 and e_6 the action of a and b is uniquely defined. By definition this is true for e_2 . It is obvious for e_5, since the elements c and d are linearly independent. Since e_3 is image under b, we must have $ce_3 = 0$, thus also on e_3 the multiplication by two linearly independent elements (namely c and d) is given. Also,

e_4 is image both under a and b, thus we must have $ce_4 = de_4 = 0$. For the terminal points, we make the following convention. If, as in our case, also cD is a word, then we let $ce_1 = 0$, if aD is a word, we let $ae_1 = 0$, and if D starts with c, then we let $ae_1 = be_1 = 0$. Consequently, we define in our case also $ae_6 = be_6 = 0$.

In a similar way, we define for a word D in $\underline{W}_2$ and an automorphism φ of a vector space V, the module $M(D,\varphi)$. Namely, we take as underlying vector space the direct sum of $|D|$ copies of V, and define again the action of X and Y according to the word D, where all arrows but the last are taken as the identity map between the corresponding copies (as induced by the element of $kX+kY$ which correspond to the letter), and where the last letter gives just the map φ between the last and the first copy of V.

In order to define the subfunctors of the forget functor ${}_{A}\underline{\underline{m}} \longrightarrow {}_{k}\underline{\underline{m}}$ which are of interest to us, we note that the forget functor has two canonical filtrations, given by the equations $da = 0$ and $cb = 0$. Consider first the equation $da = 0$. We form finite and infinite words in the letters a and d^{-1}, and denote by $\underline{F}_a$ the set of all finite words together with those infinite words which are of the form $DE^{\infty} = DEEE\cdots$, where D and E are finite words. For every word D in $\underline{F}_a$, there are two obvious functors ${}_{A}\underline{\underline{m}} \longrightarrow {}_{k}\underline{\underline{m}}$, one defined by $M \longmapsto D(0_M)$, the other by

$M \longmapsto D(M)$. Here, we use the definition $E^{\infty}(0_M) = \bigcup E^n(0_M)$, and $E^{\infty}(M) = \bigcap E^n(M)$. It is easy to see that the set of all such functors is linearly ordered by inclusion, and we call this set the a-filtration. In a similar way, the equation $cb = 0$ gives rise to a set $\underline{F}_b$ of finite and infinite words in the letters b, c^{-1}, and then to the b-filtration.

If $F_2 \leq F_1$ are two subfunctors of the forget functor, we call $[{F_1 \atop F_2}]$ an intervall. The intersection of the two intervalls $[{F_1 \atop F_2}]$ and $[{G_1 \atop G_2}]$ is defined to be the intervall

$$[{F_1 \atop F_2}] \cap [{G_1 \atop G_2}] = [{F_1 \cap G_1 \atop (F_1 \cap G_2)+(F_2 \cap G_1)}]$$

For any word D in $\underline{W}$, the functors $F(D,i)^+$ and $F(D,i)^-$ are defined by intersecting suitable intervalls of the a-filtration with those of the b-filtration. We indicate the choice of the intervalls in the case of the word $D = ab^{-1}d^{-1}cd^{-1}$:

$$[{F(D,1)^+ \atop F(D,1)^-}] = [{ad^{-1}(d^{-1}a)\ M \atop ad^{-1}(d^{-1}a)\ 0}] \cap [{c^{-1}0 \atop bM}]$$

$$[{F(D,2)^+ \atop F(D,2)^-}] = [{d^{-1}(d^{-1}a)\ M \atop d^{-1}(d^{-1}a)\ 0}] \cap [{c^{-2}0 \atop c^{-1}bM}]$$

$$[{F(D,3)^+ \atop F(D,3)^-}] = [{(d^{-1}a)\ M \atop (d^{-1}a)\ 0}] \cap [{bc^{-2}0 \atop bc^{-1}bM}]$$

$$[{F(D,4)^+ \atop F(D,4)^-}] = [{(ad^{-1})\ M \atop (ad^{-1})\ 0}] \cap [{b^2c^{-2}0 \atop b^2c^{-1}bM}]$$

$$[{F(D,5)^+ \atop F(D,5)^-}] = [{(d^{-1}a)\ M \atop (d^{-1}a)\ 0}] \cap [{c^{-1}b^2c^{-2}0 \atop c^{-1}b^2c^{-1}bM}]$$

$$[{F(D,6)^+ \atop F(D,6)^-}] = [{(ad^{-1})\ M \atop (ad^{-1})\ 0}] \cap [{bc^{-1}b^2c^{-2}0 \atop bc^{-1}b^2c^{-1}bM}]$$

We now use the multiplication maps in order to define natural transformations between the quotient functors $F(D,i)$. Again, we use the word D as guide line. In our case, for example, we want to have the following transformations:

$$\begin{array}{ccccccccc} & & F(D,2) & & & & & & \\ & {\scriptstyle a}\swarrow & & \searrow{\scriptstyle b} & & & & & \\ F(D,1) & & & & F(D,3) & & F(D,5) & & \\ & & & & {\scriptstyle d}\searrow & \swarrow{\scriptstyle c} & & \searrow{\scriptstyle d} & \\ & & & & & F(D,4) & & & F(D,6), \end{array}$$

where the letter indicates the multiplying element. Of course, it has to be checked that the multiplication maps are well-defined and act as indicated, and that they induce even isomorphisms of the corresponding components.

It then only remains to be shown that the intervalls $[{F(D,i)^+ \atop F(D,i)^-}]$ cover the forget functor (that means, for every M and every $o \neq x \in M$, there is such an intervall with $x \in F(D,i)^+(M) \setminus F(D,i)^-(M)$.)

An outline of the background of the proof,may be found in Gabriel's paper [5] where he discusses the value of functor categories in order to determine all indecomposable objects of a given category.

Aknowledgement: The author is indepted to P.Gabriel for many fruitful discussions and helpful comments, and he would like to thank him.

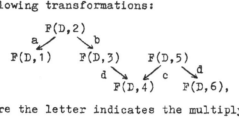

References

[1] S.Brenner. Modular representations of p-groups. J.Algebra 15, 89–1o2 (197o).

[2] S.Brenner. Decomposition properties of some small diagrams. Symposia Matematica (to appear).

[3] E.C.Dade. Une extension de la theorie de Hall et Higman. J.Algebra 2o, 57o–6o9 (1972).

[4] Yu.A.Drozd. Representations of commutative algebras. Funct.Analysis and its Appl. 6, (1972). Engl.transl. 286–288.

[5] P.Gabriel. Representations indecomposables. Seminaire Bourbaki 1973/74, Exp. 444.

[6] I.M.Gelfand, V.A.Ponomarev. Indecomposable representations of the Lorentz group. Usp. Mat.Nauk 23 3–6o (1968). Engl.Transl.: Russian Math.Surv.

[7] A.Heller, I.Reiner. Indecomposable representations. Ill.J.Math. 5, 314–323 (1961)

[8] G.J.Janusz. Faithful representations of p-groups at characteristic p. J.Algebra. I. 15, 335–351 (197o), II. 22, 137–16o (1972).

[9] S.A.Krugliak. Representations of the (p,p)-group over a field of characteristic p. Dokl.Acad. Nauk SSR 153, 1253–1256 (1963). Engl.Transl. Soviet Math Dokl. 4, 18o9–1813 (1964).

[1o] W.Müller. Gruppenalgebran über nicht-zyklischen p-Gruppen. J.Reine Ang.Math. I. 266, 1o–48, II. 267, 1–19 (1974).

[11] C.M.Ringel. The indecomposable representations of the dihedral 2-groups. (to appear).

[12] G.Szekeres. Determination of a certain family of finite metabelian groups. Trans.Amer.Math.Soc. 66, 1–43 (1949).

Mathematisches Institut
53 Bonn
Beringstr. 1
West-Germany

THE AUGMENTATION IDEAL OF A FINITE GROUP , AN INTERESTING MODULE

K. W. Roggenkamp

Representation theory of finite groups - both ordinary and modular - has become a useful tool in grouptheory. However, the connection between integral representation theory and group theory, apparently has not been exploited and applied to the extend it deserves. Usually one considers only the consequences,which special properties of the group have on its integral representation theory; one seldomly askes,which properties of the integral representation theory do reflect properties of the group. E.g. $\mathbb{Z}G$ reflects some more or less obvious properties of the finite group G:

(i) $\mathbb{Z}G$ is commutative iff G is abelian,

(ii) $\mathbb{Z}G$ has no twosided idempotent ideals iff G is solvable,

(iii) the augmentation ideal of $\mathbb{Z}G$ is idempotent iff G is perfect; similarly one can characterize nilpotent and simple groups by properties of its integral group ring.

A main reason that $\mathbb{Z}G$ reflects special properties of the finite group G is the existence of the canonical homomorphisms

$$\mathbb{Z}G \hookrightarrow \mathbb{Q}G \hookrightarrow \mathbb{R}G \hookrightarrow \mathbb{C}G$$

and

$$\mathbb{Z}G \twoheadrightarrow \mathbb{Z}/p^i\mathbb{Z}(G) \ , \quad p \text{ a prime.}$$

Consequently, $\mathbb{Z}G$ contains all informations one derives from ordinary as well as modular representation theory; but, what is more, it contains simultaneously informations on the modular representation theories for different primes linked together.

In the sequel I shall try to stress this point by explaining some of the results, which were obtained in joint work with K.W.Gruenberg [1,2] , and which hopefully will demonstrate the above philosophie.

§ 1 Groupextensions and augmentation ideals and some problems.

If S is a Dedekind domain, SG the groupring of the finite group G over S, then we have the augmentation sequence

(1) $$0 \to \mathfrak{g}_S \to SG \xrightarrow{\varepsilon} S \to 0 ,$$

which is an exact sequence of twosided SG-modules ($\varepsilon : g \mapsto 1, g \in G$), and $\mathfrak{g}_S$, which is the free S-module on $\{g-1 : g \in G\}$ is called the S-augmentation ideal of G.

There is a close and explicit connection between $\mathfrak{g}_S$ and G via extension categories:

By $\mathfrak{E}(G,S)$ we denote the category of S-module extensions of G:

The objects are short exact sequences of groups

$$\mathbb{E}_G : \qquad 1 \to K \xrightarrow{\varphi} E \xrightarrow{\psi} G \to 1 ,$$

where K is a finitely generated S-module, and E is a finitely generated group; morphisms are morphisms over G. Via conjugation K becomes a left SG-module.

$\mathbb{E}_G$ gives rise to the exact sequence of SE-modules

$0 \to SE \otimes_{SK} \mathfrak{k}_S \xrightarrow{\bar{\varphi}} SE \xrightarrow{\bar{\psi}} SG \to 0$, where $SE \otimes_{SK} \mathfrak{k}_S =: \mathfrak{k}_S^E$

is the augmentation ideal of K over S induced up to E. Passing to augmentation ideals we have the exact SE-sequence

$$0 \to \mathfrak{k}_S^E \xrightarrow{\tilde{\varphi}} \mathfrak{e}_S \xrightarrow{\tilde{\psi}} \mathfrak{g}_S \to 0 ;$$

which becomes an exact sequence of left SG-modules if one factors out $\mathfrak{e}_S \mathfrak{k}_S$:

$$0 \to \mathfrak{k}_S^E/\mathfrak{e}_S \mathfrak{k}_S \xrightarrow{\varphi''} \mathfrak{e}_S/\mathfrak{e}_S \mathfrak{k}_S \xrightarrow{\psi'} \mathfrak{g}_S \to 0 ;$$

however the map $K \to \mathfrak{k}_S^E/\mathfrak{e}_S \mathfrak{k}_S$, $k \mapsto (k-1) + \mathfrak{e}_S \mathfrak{k}_S$ is an SG-isomorphism, and so we finally obtain the exact sequence of left SG-modules

$$\mathbb{E}_{\mathfrak{g}} : \qquad 0 \to K \xrightarrow{\varphi'} \mathfrak{e}_S/\mathfrak{e}_S \mathfrak{k}_S \xrightarrow{\psi'} \mathfrak{g}_S \to 0 .$$

The correspondence $\mathbb{E}_G \longrightarrow \mathbb{E}_{\mathfrak{g}}$ induces a functor $\mathbb{F}$ from $\mathbb{G}(G,S)$ to $\mathbb{E}(\mathfrak{g}_S,S)$, the category of finitely generated left SG-module extensions, terminating in $\mathfrak{g}_S$. It turns out that $\mathbb{F}$ is an isomorphism of categories (It should be observed that the natural isomorphism $H^2(G,-) \cong \mathrm{Ext}^2_{\mathbb{Z}G}(\mathbb{Z},-) \cong \mathrm{Ext}^1_{\mathbb{Z}G}(\mathfrak{g}_{\mathbb{Z}},-)$ gives an isomorphism on the equivalence classes of extensions only.) The functor going back from $\mathbb{E}(\mathfrak{g}_S,S)$ to $\mathbb{G}(G,S)$ is given via a settheoretic pullback construction: Given an exact sequence of finitely generated left SG-modules

$$0 \longrightarrow K \longrightarrow M \xrightarrow{\psi'} \mathfrak{g}_S \longrightarrow 0 \quad ,$$

we form the settheoretic pullback of

$$\begin{array}{ccccc} M & \xrightarrow{\psi'} & \mathfrak{g}_S & & g-1 \\ \uparrow & & \uparrow \iota & & \uparrow \\ E & \xrightarrow{\psi} & G & & g \end{array} \quad ,$$

$$E = \{(m,g) \; : \; m\psi' = g-1 \; ; \; m \in M, \; g \in G\} \quad ;$$

then E becomes a group under $(m,g)(m',g') = (gm' + m, gg')$; moreover, ψ is a grouphomomorphism, and $\mathrm{Ker}\,\psi = K$, so we have constructed the desired group extension

$$1 \longrightarrow K \longrightarrow E \longrightarrow G \longrightarrow 1 \; .$$

By means of this close connection between G and $\mathfrak{g}_S$, module theoretic properties of $\mathfrak{g}_S$ should reflect group theoretic properties of G and conversely.

As an example we shall consider the Frattini extensions, which play an important rôle in group theory. Recall that an object in $\mathbb{G}(G,S)$,

$$1 \longrightarrow K \longrightarrow E \longrightarrow G \longrightarrow 1$$

is a Frattini extension, if, whenever $H \leq E$ with $E = KH$ and $H \cap K$ an SG-module, then $H = E$. Of particular interest are the maximal Frattini extensions and the question for uniqueness of these.

By means of the above correspondence one obtains immediately

Proposition 1: $\mathfrak{E}(G,S)$ has a unique maximal Frattini extension if and only if $\mathfrak{g}_S$ has a projective cover.

Hence the study of $\mathfrak{g}_S$ gives the answer of problems concerned with Frattini extensions.

For the study of $\mathfrak{g}_S$, some questions arise immediately :

I.) If G is generated by $\{g_i\}_{1 \leq i \leq n}$, then $\mathfrak{g}_{\mathbb{Z}}$ is generated as left $\mathbb{Z}G$-module by $\{g_i - 1\}_{1 \leq i \leq n}$. If $\{g_i\}_{1 \leq i \leq n}$ is a minimal set of generators for G, is then $\{g_i - 1\}_{1 \leq i \leq n}$ a minimal set of generators for $\mathfrak{g}_{\mathbb{Z}}$? Translated to extension theory we have the following problem: If

$$1 \longrightarrow R \longrightarrow F \xrightarrow{\psi} G \longrightarrow 1$$

is a minimal presentation of G by a free group F, then

$$1 \longrightarrow R/[R,R] \longrightarrow \overline{F} \longrightarrow G \longrightarrow 1$$

is an object in our category $\mathfrak{E}(G,\mathbb{Z})$. Can it happen that $R/[R,R]$ has a $\mathbb{Z}G$-projective direct summand. This in turn is equivalent to the existence of a epimorphism from F to $C \wr G$ - the wreath product of G with an infinite cyclic group C-induced by ψ .

II.) In connection with the Frattini extensions one may ask which semi-localized augmentation ideals can have projective covers. To be more precise: $\mathfrak{g} = \mathfrak{g}_{\mathbb{Z}}$ can not have a projective cover, since the Jacobsonradical of $\mathbb{Z}$ is zero. Put $\pi(G) = \{p,$ p a prime number, p divides $|G|\}$ and let $\pi \subset \pi(G)$; then

$$\mathbb{Z}_\pi = \{\tfrac{a}{b} \; ; \; a,b \in \mathbb{Z}, \text{ no prime of } \pi \text{ divides } b\}$$

is the semilocalization, and we write $\mathfrak{g}_\pi =: \mathfrak{g}_{Z_\pi}$. What does it mean for G, that $\mathfrak{g}_\pi$ has a projective cover ?

III.) In the integral augmentation sequence,

$$0 \longrightarrow \mathfrak{g} \longrightarrow \mathbb{Z}G \longrightarrow \mathbb{Z} \longrightarrow 0 ,$$

both $\mathbb{Z}G$ and $\mathbb{Z}$ are indecomposable as left $\mathbb{Z}G$-modules. Can it happen that $\mathfrak{g}$ decomposes as left $\mathbb{Z}G$-module ?

§ 2 Partial results to Problem III.

If $\mathfrak{g}$ decomposes, then there exists a finitely generated free left $\mathbb{Z}G$-module F and two non-projective left $\mathbb{Z}G$-modules A and B such that

$$\mathbb{Z} \oplus F = A \oplus B .$$

The following example was communicated to me by E.C.Dade. Let $G = \Sigma_3 = \{a,b: a^3 = 1, \; b^2 = 1, \; a^b = a^{-1}\}$ be the symmetric group on three letters. Then we have the group extension

$$1 \longrightarrow \langle a \rangle \longrightarrow G \longrightarrow \langle b \rangle \longrightarrow 1 ,$$

and we put $H = \langle a \rangle$, $C = \langle b \rangle \leq G$. We denote by $\mathbb{Z}[G/\!/C]$ the permutation representation on the cosets of G modulo C. We denote by $\mathfrak{f}$ the kernel of the canonical homomorphism $\mathbb{Z}[G/\!/C] \longrightarrow \mathbb{Z}$. As $\mathbb{Z}H$-module $\mathfrak{f}$ is the augmentation ideal of H, and C acts on it via conjugation. We then construct the exact sequence of left $\mathbb{Z}G$-modules

$$\mathbb{E} : \qquad 0 \longrightarrow \mathfrak{r}^G \longrightarrow \mathfrak{g} \longrightarrow \mathfrak{f} \longrightarrow 0 ,$$

and this sequence is split exact.

To see this,one observes that one only has to show the splitting locally At every prime different from 2 or 3 the sequence splits by a Maschke argument. At 3 $\mathbb{Z}_3 \otimes_{\mathbb{Z}} \mathfrak{r}$ is $\mathbb{Z}_3C$-projective, again by Maschke, and so $\mathbb{Z}_3 \otimes_{\mathbb{Z}} \mathfrak{r}^G$ is $\mathbb{Z}_3G$-projective. But then $\mathbb{Z}_3 \otimes_{\mathbb{Z}} \mathbb{E}$ is split exact, since in the category of integral representations, a module is projective iff it is injective. At 2 it suffices to show that $\mathbb{Z}_2 \otimes_{\mathbb{Z}} \mathfrak{f}$ viewed as $\mathbb{Z}_2C$-module is projective. But this is clear, since C acts fixpointfree on H. Hence $\mathfrak{g}$ decomposes.

Generalizing the above argument, one can show :

<u>Theorem 1</u>: Assume that G has a proper Hallsubgroup H with normalizer N such that for every $x \notin N$, $H \cap H^x = 1$. Suppose further that if $N > H$, then N is a Frobeniusgroup with kernel H. Then $\mathfrak{q}$ decomposes.

<u>Remarks</u>: Examples of such groups are

(i) the symmetric groups Σ_n for $n > 2$ and $n=p$ or $n=p+1$, p a prime number,

(ii) the alternating groups A_n for $n > 3$ and $n=p$ or $p+1$ or $p+2$, p a prime number,

(iii) all Zassenhaus groups,

(iv) the Frobenius groups,

(v) most of the known simple groups.

For solvable groups one can make a more precise statement:

<u>Theorem 2</u>: Let G be solvable. Then the following conditions are equivalent:

(i) $\mathfrak{q} = A \oplus B$, $0 \neq A, B$ left $\mathbb{Z}G$-modules,

(ii) $\mathbb{Z} \oplus \mathbb{Z}G = A_1 \oplus A_2$, A_i non-projective left $\mathbb{Z}G$-modules, $i = 1,2$,

(iii) $\pi(G) = \pi_1 \dot{\cup} \pi_2$, $\pi_i \neq \emptyset$, $i = 1,2$, and every element of G is either a π_1- or a π_2-element; i.e., if $g \in G$, then either $\pi(\langle g \rangle) \subset \pi_1$ or $\pi(\langle g \rangle) \subset \pi_2$,

(iv) G is either a solvable Frobenius group or a 2-Frobenius group; i.e.; $1 < H < T < G$ and T is a Frobenius group with kernel H and G/H is a Frobenius group with kernel T/H.

<u>Remarks</u>:

(i) If G is solvable, then $\mathfrak{q}$ decomposes into at most two summands, which are "essentially unique" and of the "same type" as in the above example,

(ii) for A_5 the augmentation ideal has a decomposition into three summands, all of which are of the "type" as in the example,

(iii) PSL(2,7) has a decomposition with $\pi_1 = \{2\}$, $\pi_2 = \{3,7\}$ but this decomposition does not arise "naturally";

(iv) if $\mathfrak{g}$ decomposes, then it never has a two-sided decomposition.

(v) If $\mathfrak{g} = A_1 \oplus A_2$, then for every prime p, $\mathfrak{g}/p\mathfrak{g} = A_1/pA_1 \oplus A_2/pA_2$ is a non-trivial decomposition, and since the Krull-Schmidt theorem holds for $(\mathbb{Z}/p\mathbb{Z})G$-representations, either A_1/pA_1 a A_2/pA_2 must be projective. This gives rise to the decomposition $\pi(G) = \pi_1 \cup \pi_2$ in (iii) of the theorem.

(vi) There are still many open questions in this connection:

a.) What structure do the possible summands of $\mathfrak{g}$ have (i.e. are they all connected with augmentation ideals of subgroups and permutation-representations.) ?

b.) Into how many direct summands can $\mathfrak{g}$ decompose, and is there an upper bound ?

c.) Are the conditions (i), (ii) and (iii) of Theorem 2 also equivalent for non-solvable groups ? (It seems to us that (iii) is equivalent to a non-projective decomposition of $\mathbb{Z} \oplus \mathbb{Z}G^{(n)}$.)

§ 3 Projective covers for augmentation ideals.

Let $\pi \subset \pi(G)$ and assume that $\mathfrak{g}_\pi$ has a projective coversequence

$$\mathbf{P}: \qquad 0 \to K_\pi \to P_\pi \to \mathfrak{g}_\pi \to 0 .$$

Then for every prime $p \in \pi$, the sequence

$$0 \to \mathbb{Z}_p \otimes_{\mathbb{Z}_\pi} K_\pi \to \mathbb{Z}_p \otimes_{\mathbb{Z}_\pi} P_\pi \to \mathfrak{g}_p \to 0$$

must be a projective cover , and for every $p \in \pi$ and every naturel number n the sequence

$$0 \to K_\pi/p^nK_\pi \to P_\pi/p^nP_\pi \to \mathfrak{g}/p^n\mathfrak{g} \to 0$$

must be a projective cover.

This means that the sequence $\mathbb{P}$ is in some sense universal, and so, if π is big enough, the existence of a projective cover should be a severe restriction.

Before formulating our results, we need some notation: Let L be an algebraic number field with ring of integers S and put

$$\Pi_S(G) = \{\mathfrak{p} \in \max(S) : \mathfrak{p} \supset |G|\, S\} .$$

For any subset Π of $\Pi_S(G)$ we put $\Pi\!\downarrow = \{p \in \max(\mathbb{Z}) : p\mathbb{Z} = \mathfrak{p} \cap \mathbb{Z}$ for some $\mathfrak{p} \in \Pi\}$. (Obeserve that Π need not contain all primes above $\Pi\!\downarrow$.)

Theorem 3: Let $\Pi \subset \Pi_S(G)$ and put $\pi = \Pi\!\downarrow$. Assume G is π-solvable and $|\pi| > 1$. Then $\mathfrak{g}_{S_\Pi}$ has a projective cover if and only if the following conditions are satisfied:

(i) $\bar{G} = G/O_{\pi'}$ is an extension

$$1 \to H \to \bar{G} \to C \to 1 ,$$

where H is a cyclic π-Hallsubgroup and C is cyclic of order n, ($O_{\pi'}$ is the largest normal subgroup, no prime divisor of the order of which lies in π),

(ii) L contains a primitive n-th root of unity,

(iii) there is an isomorphism of S_Π C-modules

$$A = : S_\Pi \otimes_{\mathbb{Z}} H \xrightarrow{\sim} S_\Pi / |H|\, S_\Pi : = B ,$$

where C acts on A via conjugation on H and on B via a faithful representation of C on S_Π.

Remarks:

(i) Unfortunately we do not have any conditions if $|\pi| = 1$.

(ii) The above conditions are of strongly arithmetical nature. For example, it may very well happen that for Π_1 and Π_2 with $\Pi_1\!\downarrow = \Pi_2\!\downarrow$, $\mathfrak{g}_{S_{\Pi_1}}$ has a projective cover, but $\mathfrak{g}_{S_{\Pi_2}}$ does not.

Corallary: Let $\pi \subset \pi(G)$, $|\pi| > 1$, and assume that G is π-solvable. Then $\mathfrak{g}_\pi$ has a projective cover if and only if the following conditions are satisfied:

(i) $\overline{G} = G/O_{\pi'}$ is an extension

$$1 \to H \to \overline{G} \to C \to 1 ,$$

with H a cyclic π-Hallsubgroup and $|C| \leq 2$,

(ii) if $|C| = 2$, then C operates as inversion on H.

§ 4 Generators for the augmentation ideal.

Let d(G) stand for the minimal number of generators of G and $d(\mathfrak{g})$ for the minimal number of generators of $\mathfrak{g}$ as left $\mathbb{Z}G$-modules.

Proposition 2: If either of the following conditions is satisfied

(i) G is solvable,

(ii) $d(G) \leq 2$,

(iii) $d(G) = d(G/[G,G])$,

then $d(G) = d(\mathfrak{g})$.

Remarks:

(i) If $G = A_5$ (19), then $d(G) < d(\mathfrak{g})$, and one can construct groups such that the difference $s = d(G) - d(\mathfrak{g})$ can be arbitrarily larg

(ii) $d(G) \leq \max_p d(G_p) + s + 1$,
where G_p denotes a p-Sylowsubgroup of G.

(iii) This number $s = d(G) - d(\mathfrak{g})$ is an invariant of G, and there should be a good grouptheoretic description of it.

References

[1] K.W.Gruenberg - K.W.Roggenkamp,
Decomposition of the augmentation ideal and of the relation modules of a finite group,
to appear in Proc.London Math.Soc. 1974

[2] ---
Projective covers for augmentation ideals of finite groups,
to appear in Journal for pure and applied algebra, 1974

[3] K.W.Roggenkamp,
Relation modules of finite groups and related topics,
Algebra i Logica 12 (1973), 351 - 359.

Anschrift:

Prof.Dr.K.W.Roggenkamp
Universität Stuttgart
Mathematisches Institut B
3.Lehrstuhl
7 S t u t t g a r t - 80
Pfaffenwaldring 57

REPRESENTATIONS OF DIFFERENTIAL GRADED CATEGORIES

A.V.Roiter, M.M.Kleiner

Introduction

Recently a series of papers has appeared considering the problems of linear algebra which can be interpreted as "matrix problems", i.e. the problems of equivalence of certain sets of matrices under a set of permissible transformations. In some cases these problems allow natural and beautiful formulation. The concept of a quiver representation introduced in [2] is an example. A quiver is a set of points connected together by some (directed) arrows. A quiver representation over a field K is given where a vector space is assigned to each point of the quiver and a linear mapping of the corresponding vector spaces is assigned to each arrow. The problem of describing the representations of some quiver (within the natural equivalence) can be interpreted as the problem of reducing a set of matrices $\{A_K\}$ by following transformations: $\bar{A}_K = C_i A_K C_j^{-1}$, where $\{C_i\}$ is a proper set of nonsingular matrices. Here the only permissible transformations are those of rays and columns within each matrix. But in other problems (cf. [4]) one may add rays and columns of some matrix to rays and columns of another. In this case the transformations themselves can be considered as certain linear operators of given vector spaces. So we can say that any matrix problem is based on a bigraph, i.e. on a graph consisting of arrows of two types. The transformated matrices are assigned to arrows of the first type and transformating ones to arrows of the second type.

But in general not every bigraph may be reasonably assigned to matrix problem and even if possible the way is not always unique. These considerations allow the authors to come to the idea of a possibility to interprete quite wide class of matrix problems as the problems of representations of differential graded categories. We mean here the category of all paths of bigraph, if "transformated" arrows have degree 0, "transformating" arrows have degree I, and the differential defines the way in which transformating matrices operate with transformated matrices.

I. Differential graded categories.

A category $\mathcal{U}$ is said to be a category over a commutative ring K if for any two objects A and B the set of morphisms $H(A,B)$ is a K -module, multiplication by a fixed morphism $a: A \to B$ being homomorphism of a K -module $H(X,A)$ into a K -module $H(X,B)$ and of a K -module $H(B,Y)$ into a K -module $H(A,Y)$ for any A, B, X, Y, a.

From the above definition we conclude that $\mathcal{U}$ is a preadditive but not necessarily additive category. Moreover the most interesting examples are those in which $\mathcal{U}$ contains a finite number of objects (an additive category contains, of course, an infinite number of objects). $\mathcal{U}$ is supposed to be a small category, i.e. the objects of $\mathcal{U}$ form a set. In this case the properties of $\mathcal{U}$ are similar to those of an algebra over the ring K. We shall construct the algebra $A(\mathcal{U})$ as a direct sum of all modules of morphisms of $\mathcal{U}$, the product in $A(\mathcal{U})$ assuming equal to 0 if it is not defined in $\mathcal{U}$.

A subset I of a set of all morphisms of $\mathcal{U}$ is called to be an ideal of the category $\mathcal{U}$ if:

a) for any two objects $A, B \in \mathcal{U}$ the set $I(A,B) = I \cap H(A,B)$ is a submodule of $H(A,B)$;

b) if $x \in I(A,B)$, $y \in H(B,C)$ $\big(z \in H(\mathcal{D},A)\big)$, then $xy \in I(A,C)$ $(zx \in I(\mathcal{D},B))$;

c) for any A $I(A,A) \neq H(A,A)$.

According to any ideal I of the category $\mathcal{U}$ we can simply construct a factorcategory $\mathcal{U}/_I$ with the same set of objects as that of $\mathcal{U}$.

A category $\mathcal{U}$ is called local if all the uninvertible morphisms form an ideal.

Let Γ be an oriented graph (a quiver by the terminology of [2]). The sequence of arrows $g_1,\dots,g_n$ is said to be a path of Γ if for each i the origin of the arrow g_i coincides with the end of the arrow g_{i+1}. Defining the multiplication of paths in a natural way and adding for every point A of Γ the "empty" path e_A we obtain the category in which objects are points of Γ and morphisms are paths. Introducing the linear combinations of paths over a commutative ring K, i.e. assuming for any $A,B \in \Gamma$ $H(A,B)$ equal to the free K-module with the paths from the point A to the point B as generators, we construct the category $\mathcal{U}(\Gamma, K)$ over a ring K and call it free over a ring K generated by a graph Γ.

It is clear that any category over a ring K can be obtained as a factorcategory of $\mathcal{U}(\Gamma, K)$ for an appropriate Γ.

Similar to a free product of algebras a free product of two categories can be introduced.

A category $\mathcal{U}$ is called a graded category over a ring K (abbreviated GC, cf. [I]) if any set of morphisms $H(A,B)$ is the union of a family of K-modules $H_i(A,B)$ where i runs all nonnegative integers, multiplication inducing a homomorphism

$$H(A,B) \otimes_K H(B,C) \longrightarrow H(A,C)$$

of degree 0 of graded K -modules. A number s is a degree of a morphism x $\deg x$ if $x \in H_s(A, B)$.

Strictly speaking, GC is not a "category over a ring K" and even not a preadditive category because $H(A,B)$ in GC is not a module but only a union of modules. We could introduce the interior graduation considering $\overline{H}(A,B) = \sum_{i=0}^{\infty} H_i(A,B)$ as a set of morphisms, but the exterior graduation is preferable because there are no nonhomogeneous elements.

The concepts of an ideal and a factorcategory are in an obvious way transfered onto GC. For each i the set I_i consisting of all morphisms of degree $> i$ is clearly an ideal in GC $\mathcal{U}$. Denote the factorcategory $\mathcal{U}/I_i$ by $\mathcal{U}_i$ which is obviously a graded category with the morphisms of degree $\leqslant i$. In particular, $\mathcal{U}_0$ is a (nongraded) category over a ring K. The factorcategory $\mathcal{U}_0$ is isomorphic to the subcategory of $\mathcal{U}$ consisting of all morphisms of degree 0 which we shall also denote by $\mathcal{U}_0$.

If $\mathcal{U}$ and $\mathcal{V}$ are two GC (over a ring K) then a functor φ of $\mathcal{U}$ into $\mathcal{V}$ is a functor which retains degrees of morphisms and defines for any $A, B \in \mathcal{U}$, $i \geqslant 0$ a homomorphism of K -module $H_i(A,B)$ into $H_i(\varphi(A), \varphi(B))$.

GC $\mathcal{U}$ is said to be semi-free over $\mathcal{U}_n$ if for any GC $\mathcal{V}$ each functor of the factorcategory $\mathcal{U}_n$ into $\mathcal{V}_n$ can be uniquely extended to a functor of $\mathcal{U}$ into $\mathcal{V}$. GC $\mathcal{U}$ can be semi-free over $\mathcal{U}_0$ only if $\mathcal{U} = \mathcal{U}_0$. GC which are semi-free over $\mathcal{U}_1$ are essential for us. Call them simply semi-free.

The category $\mathcal{U}(\Gamma, K)$ can be made graded if to each arrow of a graph Γ some nonnegative integer - a degree - is assigned (the "empty" arrows which are unit-morphisms must, of course, have degree 0). The GC obtained is semi-free over $\mathcal{U}_n$ if the degrees of arrows do not exceed n .

If $\mathcal{B}$ is a bigraph considered in Introduction we must assign the degree 0 to the "transformated" arrows and the degree I to the "transformating" arrows and form linear combinations of all the paths of a bigraph $\mathcal{B}$ with the coefficients from K . The GC obtained is denoted by $\mathcal{U}(\mathcal{B}, K)$ and called free GC. We shall depict a graph $\mathcal{B}$ with the arrows of degree 0 as the continuous arrows and the arrows of degree I as the dash arrows.

The degree (the sum of degrees of the arrows belonging to the path) is naturally assigned to each path on a bigraph $\mathcal{B}$. A path s is said to be a by-pass of an arrow g if the origin (end) of the path s coincides with the origin (end) of the arrow g and $\deg s = \deg g + 1$. Denote the set of all by-passes of an arrow g by $\mathcal{O}(g)$.

The free product of any (nongraded) category $\mathcal{S}$ over a ring K and a free category $\mathcal{U}(\Gamma, K)$ for some graph Γ is more general example for semi-free GC. We shall consider the morphisms of the first category to have degree 0 and the arrows of a graph to have degree 0. This GC is denoted by $\mathcal{U}(\mathcal{S}, \Gamma, K)$.

We say that a differential $\mathcal{D}$ of degree s is put in GC if for every morphism $a \in H_i(A, B)$ the morphism $\mathcal{D}(a) \in H_{i+s}(A, B)$ is assigned, and if A, B, i are fixed $\mathcal{D}$ is a homomorphism of a module $H_i(A, B)$ into a module $H_{i+s}(A, B)$ and Leibniz formula holds:

$$\mathcal{D}(ab) = \mathcal{D}(a)b + (-1)^{s \deg a} a\mathcal{D}(b).$$

Note that if $\mathcal{D}$ is a differential of an odd degree $\mathcal{D}^2$ is also a differential.

A differential $\mathcal{D}$ of degree s is called interior if for every object A there exists a morphism $h_A \in H_s(A, A)$ so that for every $x \in H(A,B)$ $\mathcal{D}(x) = h_A x - (-1)^{s \deg x} x h_B$.

A graded category $\mathcal{U}$ is called a differential graded category (DGC) if in $\mathcal{U}$ a differential $\mathcal{D}$ of degree s is defined so that $\mathcal{D}^2 = 0$.

A functor of DGC $\mathcal{U}$ into DGC $\mathcal{V}$ is called a DG-functor if it does not change a graduation and is commutable with elements of a ring K and with a differential.

An ideal I of DGC $\mathcal{U}$ is called differential if $\mathcal{D}(I) \subset I$. The corresponding factorcategory is obviously DGC.

It is trivial though quite convenient for us to use

Lemma I: If $\mathcal{D}$ is a differential and $\mathcal{D}(x) = 0$, $\mathcal{D}(y) = 0$, then $\mathcal{D}(xy) = 0$.

If we want to verify an equality $\mathcal{D}^2 = 0$ where $\mathcal{D}$ is a differential of degree I it is sufficient to verify this equality for the generators, in particular, in $\mathcal{U}(\mathcal{B}, K)$ for the arrows of a bigraph $\mathcal{B}$.

Consider an example.

Let $X = \{x_1, \dots, x_m\}$, $Y = \{y_1, \dots, y_n\}$ be two finite partially ordered sets. Construct the bigraph $\mathcal{B} = X \cup Y$ assuming that for any $x_i \in X$, $y_j \in Y$ there exists an arrow $\varphi(x_i, y_j)$ of degree 0 with the origin x_i and the end y_j and if $x_i < x_j$ $(y_k < y_\ell)$ there exists an arrow of degree I $\psi(x_j, x_i)(\psi(y_k, y_\ell))$ with the origin x_j (y_k) and the end x_i (y_ℓ).

Define the differential $\mathcal{D}$ for the arrows of the bigraph according to the formulas

$$\mathcal{D}(\varphi(x_i, y_k)) = -\sum_{y_s < y_k} \varphi(x_i, y_s)\psi(y_s, y_k) + \sum_{x_t < x_i} \psi(x_i, x_t)\varphi(x_t, y_k);$$

$$\mathcal{D}(\psi(x_i, x_t)) = \sum_{x_t < x_q < x_i} \psi(x_i, x_q)\psi(x_q, x_t);$$

$$\mathcal{D}(\psi(y_s, y_\kappa)) = \sum_{y_s < y_p < y_\kappa} \psi(y_s, y_p)\psi(y_p, y_\kappa).$$

We shall transfer the action of $\mathcal{D}$ to the rest elements of $\mathcal{U}(\mathcal{S}, K)$ using the Leibniz formula and the linearity. It is easy verified (using Lemma I) that $\mathcal{U}(\mathcal{S}, K)$ is DGC. As it will be seen later the category of the representations of DGC which we shall construct in the next section coincides in this case with that of a pair of partially ordered sets (cf. [IO]). In order to obtain the representations of one partially ordered set the other should be evidently considered to contain only one element.

In general if $\mathcal{U}(\mathcal{S}, K)$ is DGC then for any arrow g of degree 0 or I

$$\mathcal{D}(g) = \sum_{s \in \mathcal{O}(g)} k_{g,s}\, s \,, \quad k_{g,s} \in K.$$

Sometimes we use more generalization of DGC.

GC is said to be a quasidifferential graded category (QDGC) if a differential $\mathcal{D}$ of degree I is given such that $\mathcal{D}^2$ is an interior differential generated by a family of morphisms h_A where $\mathcal{D}(h_A) = 0$ for every A .

2. The category of the representations of QDGC.

The aim of this section is to show some construction which allows to construct a new category over a ring K according to any $\mathcal{U}$ semi-free QDGC (in particular DGC) and to any category L over a ring K . We shall call it the category of the representations of $\mathcal{U}$ into L and denote by $\mathcal{R}(\mathcal{U}, L)$.

Let us construct a graded category $\widehat{\mathcal{U}}$ (with the same objects) for any semi-free GC $\mathcal{U}$ introducing a new morphism d_A of degree I for every object A and generating $\widehat{\mathcal{U}}$ in a free (over K) way with the help of $\mathcal{U}$ and a family of d_A . If $\mathcal{U}$ is QDGC and $\mathcal{D}$ is a differential we define in $\widehat{\mathcal{U}}$

$$\widehat{\mathcal{D}}(d_A) = d_A^2 - h_A ;$$
$$\widehat{\mathcal{D}}(x) = \mathcal{D}(x) + d_A x - (-1)^{\deg x} x d_B ,$$

where $x \in H(A, B) \in \mathcal{U}$, $\{h_A\}$ is a family of morphisms mentioned in the definition of QDGC. Then we construct the operator $\widehat{\mathcal{D}}$ on $\widehat{\mathcal{U}}$ with the help of the Leibniz formula.

$\widehat{\mathcal{U}}$ is easily verified to be semi-free DGC if $\mathcal{U}$ is semi-free QDGC.

Construct now for any (nongraded) category L over a ring K the category of the triangular matrices $\widetilde{L}^{(2)}$ assuming the objects of $\widetilde{L}^{(2)}$ to be pairs of objects of L and the set of morphisms of $\widetilde{L}^{(2)}$ of an object $\bar{A} = (A_1, A_2)$ into an object $\bar{B} = (B_1, B_2)$ to be defined by formula

$$\widetilde{H}^{(2)}(\bar{A}, \bar{B}) = \left\{\begin{pmatrix} e_{11} & e_{12} \\ 0 & e_{22} \end{pmatrix}\right\}, \text{ where } A_i, B_j \in L,\ e_{ij} \in H(A_i, B_j) \in L,\ i,j = 1,2.$$

The morphisms in $\widetilde{L}^{(2)}$ are multiplied by the law of multiplication of matrices. Turn $\widetilde{L}^{(2)}$ into GC $L^{(2)}$ supposing

$$H_0^{(2)}(\bar{A}, \bar{B}) = \left\{\begin{pmatrix} e_{11} & 0 \\ 0 & e_{22} \end{pmatrix}\right\},\ H_1^{(2)}(\bar{A}, \bar{B}) = \left\{\begin{pmatrix} 0 & e_{12} \\ 0 & 0 \end{pmatrix}\right\},\ H_i^{(2)} = 0 \qquad \text{for } i > 1$$

(and eliminating from $H^{(2)}(\bar{A}, \bar{B})$ the morphisms not belonging to $H_0^{(2)}(\bar{A}, \bar{B}) \cup H_1^{(2)}(\bar{A}, \bar{B})$).

We shall consider $L^{(2)}$ to be DGC in which $\mathcal{D}(a) = 0$ for any a. Now define objects and morphisms of the category of the representations $\mathcal{R}(\mathcal{U}, L)$. The objects of $\mathcal{R}(\mathcal{U}, L)$ will be additive commutable with K functors of $\mathcal{U}_0$ into $L^{(2)}$ and the morphisms will be DG-functors of $\widehat{\mathcal{U}}$ into $L^{(2)}$. Let φ be a DG-functor from $\widehat{\mathcal{U}}$ into $L^{(2)}$. If a is a morphism of $\mathcal{U}_0$ then

$$\varphi(a) = \begin{pmatrix} \varphi_{11}(a) & 0 \\ 0 & \varphi_{22}(a) \end{pmatrix}, \text{ where } \varphi_{11}(a), \varphi_{22}(a) \in L.$$

It is possible to consider φ_{11} and φ_{22} as functors from $\mathcal{U}_o$ into L. The set of morphisms of $\mathcal{R}(\mathcal{U}, L)$ from a functor $\varphi_1 : \mathcal{U}_o \to L$ into a functor $\varphi_2 : \mathcal{U}_o \to L$ will be defined as the set of DG-functors $\overline{\varphi} : \widehat{\mathcal{U}} \to L^{(2)}$ for which $\varphi_{11} = \varphi_1$, $\varphi_{22} = \varphi_2$.

For the morphisms of the category $\mathcal{R}(\mathcal{U}, L)$ the addition and the multiplication by the elements of a ring K are introduced rather simply. The definition of the composition of the morphisms is not so trivial.

Let $X : \varphi_1 \to \varphi_2$ and $Y : \varphi_2 \to \varphi_3$ be two morphisms of the category. In order to define the morphism $XY : \varphi_1 \to \varphi_3$ it is sufficient to define the action of XY on the morphisms of degree I of DGC $\widehat{\mathcal{U}}$ (as it is clear that $(XY)_{11} = \varphi_1, (XY)_{22} = \varphi_3$). As $\widehat{\mathcal{U}}$ is semi-free we may claim that for any a of degree I from $\widehat{\mathcal{U}}$ $\mathcal{D}(a) = \sum_{i=1}^{m} a_i b_i$, where a_i and b_i are morphisms of degree I. Let $X(a_i) = \begin{pmatrix} 0 & X_{12}(a_i) \\ 0 & 0 \end{pmatrix}$, $Y(b_i) = \begin{pmatrix} 0 & Y_{12}(b_i) \\ 0 & 0 \end{pmatrix}$, $XY(a) = \begin{pmatrix} 0 & XY_{12}(a) \\ 0 & 0 \end{pmatrix}$. Suppose $XY_{12}(a) = \sum_{i=1}^{m} X_{12}(a_i) Y_{12}(b_i)$.

As $\mathcal{U}$ is semi-free it is not difficult to prove that the image $XY(a)$ does not depend on the choice of a_i, b_i in the formula $\mathcal{D}(a) = \sum_{i=1}^{m} a_i b_i$.

<u>Proposition I.</u> Under the introduced composition the functors of $\mathcal{U}_o$ into L and of $\widehat{\mathcal{U}}$ into $L^{(2)}$ form a category over the ring K. The unit-morphism corresponding to $\varphi : \mathcal{U}_o \to L$ is the functor $\overline{\varphi} : \widehat{\mathcal{U}} \to L^{(2)}$ such that for any object A of $\mathcal{U}$ and for morphism d_A introduced by constructing $\widehat{\mathcal{U}}$ from $\mathcal{U}$

$$\overline{\varphi}(d_A) = \begin{pmatrix} 0 & E_{\varphi(A)} \\ 0 & 0 \end{pmatrix}, \quad \overline{\varphi}(a) = \begin{pmatrix} \varphi(a) & 0 \\ 0 & \varphi(a) \end{pmatrix}, \text{ where } a \in \mathcal{U}_o,$$

and 0 is the image of morphisms of degree I of $\mathcal{U}$.

To prove it we must verify that:

a) XY is a functor;

b) XY is commutable with the differential;

c) for the introduced composition of morphisms the axioms of a category, in particular, the associative law hold.

Note that by proving XY as a DG-functor the equality $\hat{\mathcal{D}}^2(a)=0$ is used where a is a morphism of degree 0 while by proving the associative law the same equality is used but for morphisms of degree I from $\hat{\mathcal{U}}$.

It is obvious that if $\mathcal{U}$ is a free finitely generated trivially graded category and L is the category of all the finite dimensional vector spaces over a field K, then $\mathcal{R}(\mathcal{U},L)$ is the category of the representations of a quiver (generating $\mathcal{U}$) introduced in [2] (cf. also [3]). Ordinary representations of finite-dimensional algebras are also a special case of the representations of DGC (they contain one object and morphisms only of degree 0). Categories of the representations of partially ordered set and of pairs of partially ordered set have the form of $\mathcal{R}(\mathcal{U},L)$ where $\mathcal{U}$ is DGC constructed in I by the pair of partially ordered sets.

The scheme introduced covers all examples of matrix problems which are known to the authors (cf. [4], [5]). Here by studying the matrix problems over an algebraically closed field it is sufficient to take DGC as $\mathcal{U}$, but in the case of nonclosed field (in particular, for the problems considered in [6]) we have to use QDGC.

After constructing the category $\mathcal{R}(\mathcal{U},L)$ it is natural to introduce the terminology usual for the representation theory. We shall call objects of the category $\mathcal{R}(\mathcal{U},L)$ as representations of $\mathcal{U}$ (into L). A representation φ is faithful if $a=0$ follows from $\varphi(a) = 0$ (a is some morphism of degree 0 of DGC $\mathcal{U}$). The concepts of an indecomposable representation and of equivalent representations are introduced in an obvious way. The problems of finiteness or infinity of the number of indecomposable representations etc., that are usual in the representation theory also arise.

From now on we shall take a ring K (over which DGC $\mathcal{U}$ is defined) to be a field and L to be the category of all finite-dimensional vector spaces. Suppose also that $\mathcal{U}$ contains a finite number of objects $\{A_1, \dots, A_n\}$ and for any $A, B \in \mathcal{U}$ and non-negative integer i $H_i(A,B)$ is a finite - dimensional vector space. The dimension of a representation φ is the vector formed from the dimensions of vector spaces $\varphi(A_1), \dots, \varphi(A_n)$. Similar to the well known definitions for algebras by Brauer one can introduce the notions of DGC of finite type (if there exists a finite number of indecomposable representations), of bounded type (if the set of dimensions of indecomposable representations is bounded) and strongly unbounded type (if there exists an infinite number of dimensions each admitting infinitely many indecomposable representations).

One can apparently suppose statements similar to the Brauer-Thrall conjectures (cf. [II] , [5]) hold also for DGC or in any case for quite wide classes of DGC.

3. Operations over DGC.

Two DGC $\mathcal{U}$ and $\mathcal{U}'$ may be called similar if there exists a functor realizing an equivalence (cf. for instance [12]) of the categories $\mathcal{R}(\mathcal{U}, L)$ and $\mathcal{R}(\mathcal{U}', L)$. The method which is usual in matrix problem theory consists in reducing one or several matrices to some standard form. Then only those transformations are considered which do not change the form of matrices having been reduced and thus one gets a matrix problem being in a certain sense tantamount to the initial one. In our terminology this process is found to correspond to replacing DGC $\mathcal{U}$ by similar DGC $\mathcal{U}'$. Now it is naturally to construct for a given DGC the simplest category out of those similar to it. The following definitions are thus introduced. A category $\mathcal{V}$ over a field K is called simplest if $H(A,A)=K$, $H(A,B)=\{0\}$ where $A \neq B$ for any $A, B \in \mathcal{V}$. DGC $\mathcal{U}$ is called trivial if the subcategory $\mathcal{U}_o$ is simplest. For $\mathcal{U}= =\mathcal{U}(\mathcal{B},K)$ it means that a Digraph $\mathcal{B}$ contains no continuous arrows. It is evident that the structure of representations of trivial DGC is most simple and the number of the indecomposable objects of $\mathcal{R}(\mathcal{U}, L)$ is equal to the number of objects of $\mathcal{U}$.

While solving matrix problems one of matrices is often chosen the reducing of which is not affected by other matrices and is reduced by elementary transformations to the form $\begin{pmatrix} E & 0 \\ 0 & 0 \end{pmatrix}$. If an initial problem is of finite type then with the help of a finite number of such reducings the problem comes "to nothing" and in reducing process all its solutions are obtained.

In this section we describe a corresponding construction which for given DGC $\mathcal{U}$ and a morphism $a \in \mathcal{U}$ such that $\deg a = 0$ and $\mathcal{D}(a) = 0$ makes possible (under certain conditions) to construct DGC $\mathcal{U}_a$ similar to $\mathcal{U}$. If $\mathcal{U}$ is of finite type we construct trivial DGC which is similar to it with finite number of steps.

For simplicity we restrict ourselves to the case of free DGC $\mathcal{U} = \mathcal{U}(\mathcal{B}, K)$ although most results of this section are true under more general assumptions.

Introduce some more definitions.

Let I be an ideal of the category $\mathcal{U}_0$ generated by the continuous arrows of the bigraph $\mathcal{B}$ and $\bar{I}$ be an ideal (of the category $\mathcal{U}$) generated by the ideal I of the subcategory $\mathcal{U}_0$. DGC $\mathcal{U}$ is called regular if $\mathcal{D}(\mathcal{U}_0) \subset \bar{I}$.

Semi-free DGC $\mathcal{U}$ is called triangular if there exists a finite sequence of its subcategories $\mathcal{W}_i$: $\mathcal{U} = \mathcal{W}_0 \supset \mathcal{W}_1 \supset \ldots \supset \mathcal{W}_n$, where $\mathcal{W}_n$ is a simplest category, $\mathcal{D}(\mathcal{W}_{i-1}) \subset \mathcal{W}_i$ and $\mathcal{W}_i$ is generated by its intersections with $\mathcal{U}_0$ and $\mathcal{U}_1$, $i = 1, \ldots, n$. DGC corresponding to matrix problems which really arise, for instance, in representation theory of algebras, are triangular and regular (cf. [5]). As we see below nonregular DGC can appear in reducing process too.

Take an example of nontriangular but regular DGC. Let a bigraph $\mathcal{B}$ consist of one point and two arrows: continuous α and dash β and $\mathcal{D}(\alpha) = \alpha\beta\alpha$, $\mathcal{D}(\beta) = 0$. It is unknown to the authors whether these "pathological" DGC are of any interest. Apparently there is no simple method of reducing such DGC to triangular one. But we shall point out the way to reduce representations of nonregular DGC to those of regular DGC,.in any case, for triangular DGC of finite type.

<u>Proposition 2:</u> Let $\mathcal{U} = \mathcal{U}(\mathcal{B}, K)$ be a triangular DGC, $\mathcal{D}(f) =$

$= \alpha\psi + \sum_i c_i \psi_i h_i$ where f is a continuous arrow of $\mathcal{B}$; ψ, ψ_i are dash arrows of $\mathcal{B}$ and $\psi \neq \psi_i$; $c_i, h_i \in \mathcal{U}_o$, $\alpha \in K$, $\alpha \neq 0$. Then categories $\mathcal{U}$ and $\overline{\mathcal{U}}^1$ are similar where $\overline{\mathcal{U}}^1$ is the factorcategory of DGC $\mathcal{U}$ by the differential ideal generated by f and $\mathcal{D}(f)$.

Construct the actual form of $\overline{\mathcal{U}}^1 = \mathcal{U}(\overline{\mathcal{B}}^1, K)$. A bigraph $\overline{\mathcal{B}}^1$ is obtained from a bigraph $\mathcal{B}$ by eliminating the arrows f and ψ. In all expressions for differentials of arrows one must eliminate the members containing f and everywhere substitute $-\alpha^{-1}\sum_i c_i \psi_i h_i$ for ψ.

Since a graph $\overline{\mathcal{B}}^1$ contains less number of arrows than $\mathcal{B}$ and the categories $\mathcal{U}$ and $\overline{\mathcal{U}}^1$ are similar it is advisable, before examining the representations of some DGC, to use Proposition 2 as many times as possible. DGC obtained after that from $\mathcal{U}$ (for which Proposition 2 can no longer be applied) is denoted by $\overline{\mathcal{U}}$.

Remark that if $\mathcal{U}(\mathcal{B}, K)$ is nonregular DGC then there exists necessarily an arrow f of degree 0 such that $\mathcal{D}(f) = \alpha\psi + \sum_i c_i \psi_i h_i$ but in general the condition $\psi \neq \psi_i$ cannot be satisfied. This condition necessarily holds if there are no closed paths of degree 0 in a bigraph $\mathcal{B}$. Using this fact one can prove

<u>Proposition 3:</u> If a triangular DGC $\mathcal{U}$ is of finite type then $\overline{\mathcal{U}}$ is a regular DGC.

Before constructing $\mathcal{U}_a$ introduce two else operations with DGC: "bifurcation of an object" and "identification of a morphism".

Let $\mathcal{U} = \mathcal{U}(\mathcal{B}, K)$, A be a point of a bigraph $\mathcal{B}$. Construct a bigraph $T_A(\mathcal{B})$ from a bigraph $\mathcal{B}$ as follows:

a) substitute two points A_1 and A_2 for a point A;

b) substitute two arrows g_1, g_2 connecting respectively A_1 and A_2 with X (here the orientation and degree of arrows are preserved) for each arrow g connecting a point A with a point $X \neq A$;

c) substitute four arrows $p_{11}: A_1 \to A_1$, $p_{12}: A_1 \to A_2$, $p_{21}: A_2 \to A_1$, $p_{22}: A_2 \to A_2$ of the same degree for each arrow $p: A \to A$;

d) add one else arrow $\xi : A_2 \to A_1$ of degree I.

Construct GC $T_A(\mathcal{U})$ from a bigraph $T_A(\mathfrak{B})$. In order to turn $T_A(\mathcal{U})$ into DGC it is necessary (cf. I) to put coefficients $k_{g,x}$ for every $g \in T_A(\mathfrak{B})$ and $x \in \mathcal{O}(g)$. For each arrow $g \in T_A(\mathfrak{B})$ which is different from ξ we denote by $\varphi(g)$ a corresponding arrow of a bigraph $\mathfrak{B}$. If x is a path of a bigraph $T_A(\mathfrak{B})$ which contains no arrow ξ then we denote by $\varphi(x)$ a corresponding path of $\mathfrak{B}$. The set of paths of a bigraph $T_A(\mathfrak{B})$ which contain no arrow ξ is denoted by $\varphi^{-1}(\mathfrak{B})$.

Define:

a) $k_{g,x} = k_{\varphi(g),\varphi(x)}$ if $x, g \in \varphi^{-1}(\mathfrak{B})$;

b) $\mathcal{D}(\xi) = 0$.

If $g \neq \xi$ and $x \notin \varphi^{-1}(\mathfrak{B})$ then a coefficient $k_{g,x}$ differs from 0 if and only if the path x consists of two arrows ξ and g' such that $\varphi(g') = \varphi(g)$. In this case if $\deg g = 1$ then $k_{g,x} = 1$, and if $\deg g = 0$ then $k_{g,x}$ is equal to +I or -I depending on arrows g and g' having coinciding ends or coinciding origins.

It is not difficult to verify that the differential $\mathcal{D}$ defined with the help of coefficients $k_{g,x}$ satisfies the equation $\mathcal{D}^2 = 0$ and thus turns $T_A(\mathcal{U})$ into DGC. We shall call the operation of constructing $T_A(\mathcal{U})$ from $\mathcal{U}$ as "bifurcation of an object A".

Let again $\mathcal{U} = \mathcal{U}(\mathfrak{B}, K)$, a be an arrow of degree 0 of a bigraph $\mathfrak{B}$, $a \in H(B,C)$, $B \neq C$ and $\mathcal{D}(a) = 0$. Construct a bigraph $E_a(\mathfrak{B})$ uniting points B and C into one point X and eliminating the arrow a. In this case if the end or the origin of an arrow of $\mathfrak{B}$, different from a, coincides with one of the points B, C, then in $E_a(\mathfrak{B})$ the end or the origin of this arrow (respectively) coincides with the point X . Assign now a path $\psi(f) \in E_a(\mathcal{B})$ to any path $f \in \mathcal{B}, f \neq a$. If f contains no arrow a we do not change f . If f contains the arrow a then we eliminate a shortening respectively the path f . It is possible since the points B and C are united. Construct now DGC $E_a(\mathcal{U}) = \mathcal{U}(E_a(\mathcal{B}), K)$ defining $k_{\psi(g),\psi(f)} = k_{g,f}$ for every $g \in \mathcal{B}$, $f \in \mathcal{O}(g)$ and putting the rest coefficients $k_{x,y}$ equal to 0. We shall call an operation of constructing $E_a(\mathcal{U})$ from $\mathcal{U}$ as "identification of a morphism a ".

How does a category of representations behave when we perform the above operations with DGC? Under "bifurcation of an object" a category of representations becomes strongly complicated (under this operation DGC can turn its finite type into infinite) and under "identification of a morphism" a category of representations is considerably simplified (here DGC can turn its infinite type into finite).

We show that combining these two operations with factorisation by a differential ideal one can perform such transformation over DGC so DGC obtained becomes similar to the initial one.

Let again a be such a morphism of $\mathcal{U} = \mathcal{U}(\mathcal{B}, K)$ that $a \in H_0(B,C)$, $B \neq C$, $\mathcal{D}(a) = 0$. Use for $\mathcal{U}$ at first "bifurcation of B " and then "bifurcation of C ". Under the first bifurcation morphisms a_1, a_2 appear instead of a . Under the second bifurcation morphisms a_{11}, a_{12} and a_{21}, a_{22} appear instead of a_1 and a_2 respectively. The factorisation by the differential ideal generated by a_{11}, a_{12}, a_{22} is then applied to the results of the second bifurcation of DGC and finally, in DGC obtained the identification of the morphism a_{21} is performed. DGC resulting from all these operations is denoted by $\mathcal{U}'_a$.

<u>Proposition 4:</u> Categories $\mathcal{U}$ and $\mathcal{U}'_a$ are similar.

DGC $\mathcal{U}'_a$ is as a rule nonregular. Therefore it is advisable to use Proposition 2 for $\mathcal{U}'_a$ several times. Finally denote $\overline{\mathcal{U}}'_a$ by $\mathcal{U}_a$. We shall call the operation of constructing $\mathcal{U}_a$ from $\mathcal{U}$ as "reducing an arrow a ". Proposition 3 implies that if $\mathcal{U}$ is of finite type then $\mathcal{U}_a$ is a regular DGC.

<u>Proposition 5:</u> If $\mathcal{U}$ is a triangular DGC of finite type then $\mathcal{U}_a$ is also a triangular DGC of finite type.

Note that in constructing $\mathcal{U}_a$ from $\mathcal{U}$ with several steps some arrows at first appear and then are eliminated. Therefore it is advisable to show a direct algorithm for constructing a bigraph $\mathcal{B}_a$ corresponding to $\mathcal{U}_a$.

Thus again $a: A_1 \longrightarrow A_2$ is an arrow of degree 0 and $\mathcal{D}(a)=0$. A by-pass f of some continuous arrow g is called marked if f is equal to $a\psi$ or ψa (certainly, $deg\,\psi = 1$) and $k_{g,f} \neq 0$. A bigraph $\mathcal{B}$ is assumed to contain no closed paths of degree 0 (this assumption is always realized in DGC of finite type according to Lemma 2). To simplify the formulations we also assume that there exists not more than one marked by-pass for each arrow and every dash arrow is contained in not more than one marked by-pass (though "reducing an arrow" in the above way is also possible even if this condition is not realized). Continuous arrows which possess marked by-passes and dash arrows contained in such by-passes are called marked arrows.

For constructing a bigraph $\mathcal{B}_a$ we add to $\mathcal{B}$ a new point X , two new arrows of degree I $\xi_1: A_1 \to X$ and $\xi_2: X \to A_2$ and eliminate the arrow a . Besides, let h be a nonmarked arrow. Then if $h: A_i \to \mathcal{U}\ (h: \mathcal{V} \to A_j)\ (i,j = 1,2;\ \mathcal{U}, \mathcal{V} \neq A_i, A_j)$ we add a new arrow $\bar{h}: X \to \mathcal{U}\ (\bar{h}: \mathcal{V} \to X)$, and if $h: A_i \to A_j$ three new arrows $\bar{h}_{12} : A_i \to X$, $\bar{h}_{21} : X \to A_j$, $\bar{h}_{22} : X \to X$ are added. If g is a marked arrow of degree 0 or I then we add an arrow $\bar{g}$ if and only if $g : A_i \to A_j$. Here $\bar{g} : A_i \to X$ or $\bar{g} : X \to A_j$ is dependent on the fact whether an arrow a is included as the first or the second multiplier in marked by-pass corresponding to an arrow g . In all cases a generated arrow ($\bar{h}$, $\bar{h}_{ij}$ or $\bar{g}$) has the same degree as a generating arrow (h or g).

Thus a bigraph $\mathcal{B}_a$ has been constructed. To construct DGC we must also define the coefficients $k_{g,f}$.

Let $g_1, \ldots, g_s$ be a sequence of arrows of a bigraph $\mathcal{B}$ or $\mathcal{B}_a$ but not necessarily a path (i.e. the end of an arrow g_{i+1} may not coincide with the origin of an arrow g_i).

One of the following operation is called an elementary transformation of a sequence of arrows:

a) crossing out the arrow a from a sequence;

b) substitution of some arrow $\bar{g}$ or $\bar{g}_{ij}$ for a generating arrow g ;

c) if g is a marked dash arrow and a corresponding marked by-pass s is a summand of an expression $\mathcal{D}(h) = \alpha s + \sum_{f \in \mathcal{O}(h)} k_{h,f} f$ then one may substitute a sequence of arrows forming a path $f \in \mathcal{O}(h)$ for h if $k_{h,f} \neq 0$.

We shall say that the transformating coefficient is equal to 1 in cases a) and b) and is equal to $-\alpha^{-1} k_{h,f} \in K$ in c).

A path $\bar{f}$ of a bigraph $\mathcal{B}_a$ is said to be generated by a path f of a bigraph $\mathcal{B}$ if $\bar{f}$ can be obtained from f by superposition of several elementary transformations. A product of corresponding transformating coefficients is called a coefficient of generation $\pi(f, \bar{f})$.

Now let f be a path of a bigraph $\mathcal{B}_a$ and a by-pass of an arrow $w \in \mathcal{B}$, f not containing arrows ξ_1, ξ_2.

Put $k_{w,f} = \sum k_{w,h} \pi(h,f)$. If $\bar{w}$ is an arrow of $\mathcal{B}_a$, generated by an arrow w of a bigraph $\mathcal{B}$ then we put $k_{\bar{w},f} = \sum k_{w,h} \pi(h,f)$. In both cases the summation is performed over all paths h generating a path f.

Then put $\mathcal{D}(\xi_1)=0$, $\mathcal{D}(\xi_2)=0$. If however $g \neq \xi_i$ and a path f contains ξ_1 or ξ_2 then a coefficient $k_{g,f}$ differs from 0 if and only if f contains only one arrow g' besides one of ξ_i such that whether the arrows g and g' generate one another or arrows g and g' are generated by the same arrow of a bigraph $\mathcal{B}$. In this case $k_{g,f} = 1$ if $\deg g = 1$ but if $\deg g = 0$ then $k_{g,f}$ is equal to I or -I depending on arrows g and g' having common end and common origin.

Note that while reducing an arrow we can obtain arrows of degree I which are not included in the expression $\mathcal{D}(g)$ for any continuous arrow g. If we are interested only in objects of representation category we can of course eliminate such dash arrows from a bigraph $\mathcal{B}_a$. But if we are interested also in morphisms of representation category we must preserve such arrows.

Consider an example of "reducing" DGC correspondent to a problem of three subspaces. A reduced arrow is denoted by a double arrow.

Here we show only the form of bigraphs since the action of $\mathcal{D}$ (as in many similar cases) is naturally defined by a bigraph.

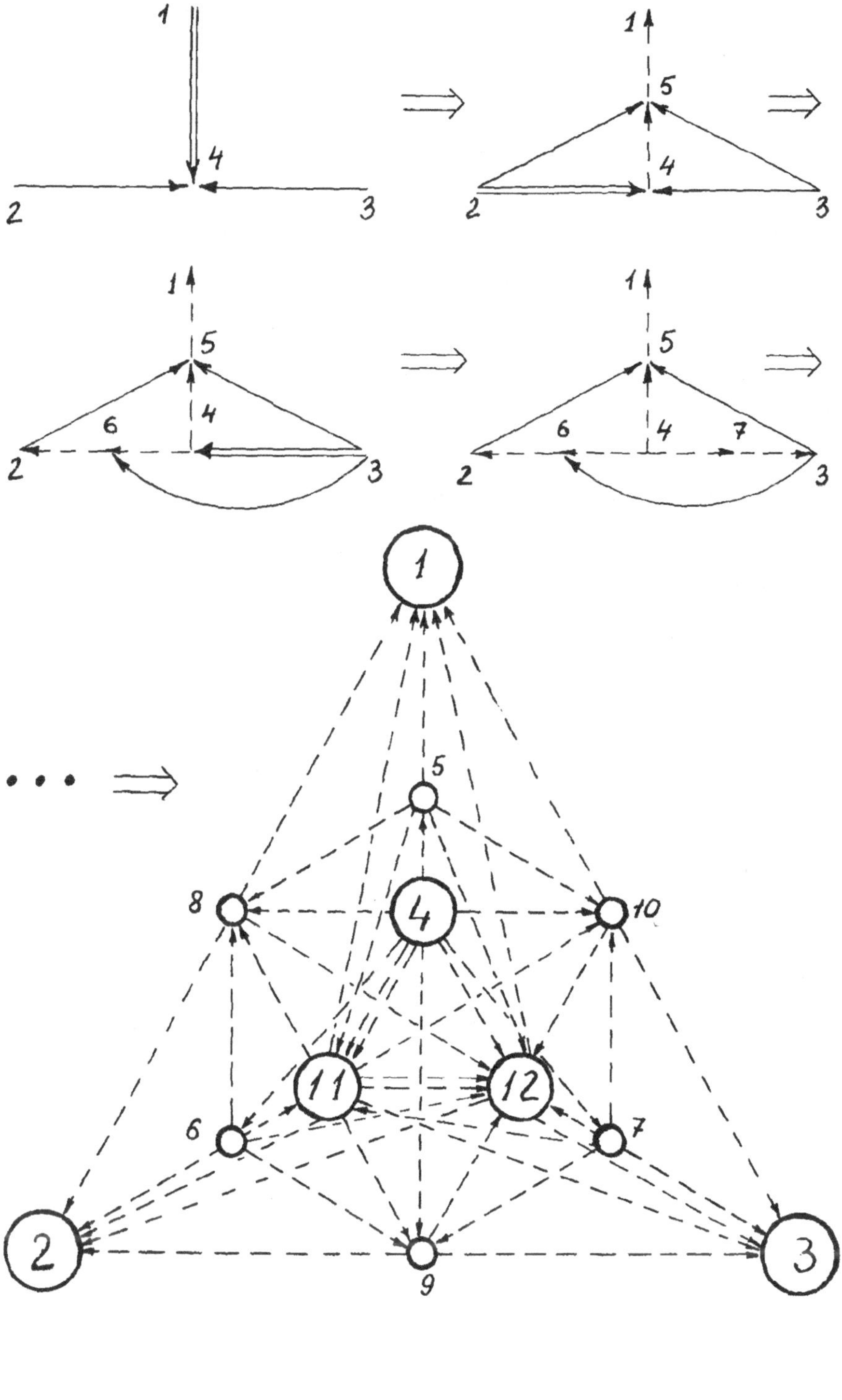
1
4
2
3
1
5
4
2
3
1
5
4
6
2
3
1
5
6
4
7
2
3
1
5
8
4
10
11
12
6
7
2
9
3

The above procedure may be obviously applied to any triangular free DGC of finite type. Namely there holds

Proposition 6: A free triangular DGC $\mathcal{U}$ is of finite type if and only if there exists a series of DGC $\mathcal{U}=\mathcal{U}^0, \mathcal{U}^1, \dots, \mathcal{U}^m$, where $\mathcal{U}^i$ is obtained from $\mathcal{U}^{i-1}$ by reducing some arrow, and DGC $\mathcal{U}^m$ is trivial. A number of the indecomposable representations of $\mathcal{U}$ is equal to a number of the objects of $\mathcal{U}^m$.

By reducing arrows it is not difficult to prove also

Proposition 7: For a free triangular DGC an analogue of the first Brauer-Thrall conjecture holds, i.e. a free triangular DGC is of finite type or of unbounded type.

Indeed, all the statements (from Lemma 2 to Proposition 6) are true if one substitutes "bounded type" for "finite type".

From Proposition 7 with the help of ordinary reducing the algebra representations to matrix problems (cf. for instance [5]) one can obtain a corresponding statement for algebras proved in [7]. But the authors don't know any statement in terms of DGC analogous to the second Brauer-Thrall conjecture which could imply the result of [5].

4. Schurian DGC and Tits form.

We show briefly a possibility to transfer considerations connected with the Tits form onto representations of DGC which were successfully used for representations of quivers and partially ordered sets (cf. [2], [3], [8], [9]).

The Tits form $f(x_1, \dots, x_n)$ of a bigraph $\mathcal{B}$ and DGC $\mathcal{U}(\mathcal{B}, K)$ is a quadratic form with a number of variables, equal to a number of points of a graph in which a coefficient of $x_i x_j$ $(i \neq j)$ is equal to $b_{ij} - a_{ij}$ and a coefficient of x_i^2 is equal to $1+b_{ii} - a_{ii}$, where b_{ij} (a_{ij}) is a number of the arrows of degree I (0) connecting corresponding points i and j of a bigraph $\mathcal{B}$.

Similar to [2], [8] one can prove

<u>Proposition 8:</u> If a free DGC is of finite type then the Tits form of a bigraph $\mathcal{B}$ is necessarily a positively definite form on the set of the vectors with nonnegative components.

DGC $\mathcal{U}$ is called Schurian if an algebra of the endomorphisms of every its indecomposable representation coincides with a field K. It is well known (cf. [4], [9]) that DGC corresponding to quivers and partially ordered sets of finite type are Schurian.

It is easy to see that if DGC $\mathcal{U}(\mathcal{B}, K)$ is Schurian then it is of finite type and therefore a corresponding Tits form is positively definite.

Following the idea stated in [9] one can prove a proposition which is well-known for quivers ([3], [9]) and partially ordered sets [8]. It is

<u>Proposition 9:</u> If a free DGC $\mathcal{U}$ is Schurian then there exists a natural bijection of the set of all its indecomposable representations and of the set of all integer nonnegative roots of equation $f(x_1, \dots, x_n) = 1$. Namely, let $X=(x_1,\dots,x_n)$, $x_i \geq 0$. If $f(X)=1$ then there exists precisely one indecomposable representation of $\mathcal{U}$ of dimension X. If however $f(X) \neq 1$ then there are no indecomposable representations of $\mathcal{U}$ of dimension X.

If a free DGC is Schurian then obviously there are no dash loops (i.e. the arrows of degree I with the origin and the end coinciding) in a bigraph. The authors have not succeeded in finding conditions under which DGC is Schurian. But the results of previous section imply obviously

Proposition IO: Let $\mathcal{U}$ be a triangular DGC of finite type and let $\mathcal{U}^0, \dots, \mathcal{U}^m$ be the series of DGC constructed in Proposition 6. Then $\mathcal{U}$ is Schurian if and only if the trivial DGC $\mathcal{U}^m$ contains no dash loops.

References.

I. S.MacLane, Homology, Berlin, I963.

2. P.Gabriel, Unzerlegbare Darstellungen. I, Manuscripta Math. 6(I972), 7I-I03.

3. I.N.Bernštein, I.M.Gel'fand and V.A.Ponomarev, Coxeter functors and Gabriel theorem, Uspehi Mat.Nauk 28 (I973), no.2 (I70), I9-33 (Russian).

4. L.A.Nazarova and A.V.Roiter, Representations of parially ordered sets, Zap.Naučn.Sem.Leningrad. Otdel.Mat.Inst.Steklov. (LOMI) 28 (I972), 5-3I (Russian).

5. L.A.Nazarova and A.V.Roiter, Category matrix problems and Brauer-Thrall problem, Preprint IM AN USSR, I973 (Russian).

6. V.Dlab, C.M.Ringel, Representations of graphs and algebras, Carleton Math.Lect.Notes, No.8, August I974.

7. A.V.Roiter, The unboundedness of the dimension of the indecomposable representations of algebras that have an infinite number of indecomposable representations. Izv.Akad.Nauk SSSR, 32 (I968), I275-I282 (Russian).

8. Yu.A.Drozd, Coxeter transformations and representations of partially ordered sets, Funct.Analyz, 8 (I974), no.3 (Russian).

9. P.Gabriel, Indecomposable representations - II. Istituto Nazionale di Alta Matematica, Symposia Mathematica, vol.XI (I973).

IO. M.M.Kleiner, Partially ordered sets of finite type, Zap.Naučn. Sem.Leningrad.Otdel.Mat.Inst.Steklov. (LOMI) 28 (I972),32-4I (Russian).

II. I.P.Jans, On indecomposable representations of algebras, Ann.of Math. (2) 66 (I957), 4I8-429.

I2. H.Bass, Algebraic K-theory, New York, I968.

AUTOMORPHISMS AND INVOLUTIONS OF INCIDENCE ALGEBRAS

Winfried Scharlau

In this note we study the automorphism group and involutions of the incidence algebra $A(I)$ of a finite partially ordered set I. Since a full matrix algebra $M(n,D)$ (D a skew-field) is a particular example of an incidence algebra our results will be generalizations of classical theorems like the Skolem-Noether theorem and results of Albert on the structure of involutions on simple algebras ([1], Chapter X). The automorphism group of an incidence algebra has been studied by R.P. Stanley [4] and our theorem 1.2 is a partial generalization of [4], theorem 2. Our proof, however, is different.

I am indebted to U. Hirzebruch and H. P. Petersson for useful discussions.

1. Incidence algebras and their automorphism groups

Let I be a finite partially ordered set, i.e. there is a relation $\prec$ on I satisfying

(01) $i \prec i$ for all $i \in I$

(02) $i \prec j$, $j \prec k$ implies $i \prec k$ for all $i,j,k \in I$.

(Warning: (03) $i \prec j, j \prec i$ implies $i=j$ is *not* required.)

Let K be a (commutative) field. The incidence algebra $A = A(I) = A_K(I)$ of I over K is defined as follows: For each pair $(i,j) \in I \times I$ such that $i \prec j$ there is a basis element e_{ij} of A and one has the multiplication table

$$e_{ij}\, e_{Kl} = \begin{cases} e_{il} & \text{if } j=k \\ 0 & \text{if } j \neq k . \end{cases}$$

It follows immedeately from this multiplication table that A can be identified with a subalgebra of the full matrix algebra $M(n,K)$ where $n = |I|$. For example, if $I = \{1,..,n\}$ with the usual ordering $\leqq$, then $A(I)$ is the algebra of triangular matrices. If $I = \{1,..,n\}$ with $i \prec j$ for all i,j, then $A(I)$ is the full matrix algebra $M(n,K)$. If D is a K-algebra define $A_D(I) := A_K(I) \otimes_K D$.

We introduce an equivalence relation on I as follows: $i \sim j$ if and only if $i \prec j$ and $j \prec i$. The set of equivalence classes will be denoted by S. This is an ordered set satisfying (O3).

1.1. Proposition. Let D be a skew-field with center K and let $A = A_D(I)$ be defined as above. Let

$$P := \sum_{i \sim j} D e_{ij}$$

$$R := \sum_{i \nsim j} D e_{ij} .$$

Then P *is a semisimple subalgebra of* A, *and* R *is the radical of* A. *Furthermore,* $A = P \oplus R$, *i.e.* P *is a Wedderburn complement of* R.

The proof is easy and will be left to the reader. In particular, one can show that $R^t=0$ where t is the maximal length of the chains in S, and that P is a product of s simple algebras with $s = |S|$.

If A is a partially ordered set, let $\mathrm{Aut}(I)$ be the group of order preserving permutations of I. If A is a K-algebra we mean by $\mathrm{Out}(A)$ the factor group $\mathrm{Aut}(A)/\mathrm{Inn}(A)$ of all K-automorphisms of A modulo the subgroup of inner automorphisms.

1.2. Theorem. Let I *be a finite partially ordered set which contains a minimal or a maximal element* i_o (*i.e.* $i_o < i$ *or* $i < i_o$ *for all* $i \in I$). *Let* D *be a finite-dimensional skew field with center* K *and* $A = A_D(I)$. *Then* $\mathrm{Out}(A)$ *is canonically isomorphic to the image of the canonical homomorphism* $\mathrm{Aut}(I) \to \mathrm{Aut}(S)$.

Proof: We consider the canonical homomorphism

$$\pi : \mathrm{Aut}(A) \longrightarrow \mathrm{Aut}(S)$$

defined as follows: S can be identified with the set of the indecomposable central idempotents of A/R

$$\bar{e}_t = e_t + R, \quad e_t = \sum_{i \in t} e_{ii} \quad , \quad t \in S \ .$$

Every automorphism α of A induces a permutation of the $\bar{e}_t$. This permutation is order preserving, because $r \prec t$ for $r,t \in S$ if and only if there is an $x \in A$ such that $e_r x e_t \neq 0$. (see [4], theorem 1). The inclusion Inn (A) $\subset$ Ker (π) is obvious.

Proof of Ker (π) $\subset$ Inn (A): Assume $\alpha \in$ Ker (π). We have $A = \alpha(P) \oplus \alpha(R) = \alpha(P) \oplus R$. By a theorem of Malcev [3] (see also [2]), there exists an <u>inner</u> automorphism β such that $\beta(\alpha(P)) = P$. Replacing α by $\beta\alpha$ we may assume that $\alpha(P) = P$. Let $P = P_1 x..x P_s$, $P_t = M(n_t, D)$ be the decomposition of P as a product of simple algebras. Since $\pi(\alpha) = \mathrm{id}$, $\alpha(P_i) = P_i$ and $\alpha|_P$ is an inner automorphism of P by the Skolem-Noether theorem. Therefore we may assume $\alpha|_P = \mathrm{id}$. Let $i \prec j$. Applying α to the equation

$$e_{ii}(de_{ij}) = de_{ij} \quad , d \in D$$

we get

$$e_{ii}\, \alpha(de_{ij}) = \alpha(de_{ij}) .$$

Similarly, we see

$$\alpha(de_{ij}) = \alpha(de_{ij})e_{jj} \quad ,$$

i.e.

$$\alpha(de_{ij}) = e_{ii}\alpha(de_{ij})e_{jj} \quad ,$$

hence

$$\alpha(De_{ij}) = De_{ij} \quad .$$

Hence $\alpha(de_{ij}) = \alpha_{ij}(d)e_{ij}$ for some map $\alpha_{ij} : D \to D$.

This map is right and left D-linear, since

$$(ae_{ii})(be_{ij}) = (ab)e_{ij}$$

implies

$$a\alpha_{ij}(b)e_{ij} = \alpha_{ij}(ab)e_{ij}$$

i.e. $\alpha_{ij}(ab) = a\alpha_{ij}(b)$, and similarly $\alpha_{ij}(ab) = \alpha_{ij}(a)b$. Hence, there exists a central element $0 \neq \lambda_{ij} \in K$ such that $\alpha_{ij}(d) = \lambda_{ij}d$, and therefore

$$\alpha(\sum d_{ij}e_{ij}) = \sum \lambda_{ij}d_{ij}e_{ij} .$$

If i_0 is e.g. a minimal element we compute

$$(\sum \lambda_{i_0 i}e_{ii})^{-1}(\sum d_{ij}e_{ij})\ (\sum \lambda_{i_0 j}e_{jj})$$

$$= \sum \lambda_{ij}d_{ij}e_{ij} ,$$

hence α is inner.

It is clear that $\pi(\mathrm{Aut}(A))$ is contained in the image of $\mathrm{Aut}(I) \to \mathrm{Aut}(S)$. Therefore it is sufficient to show that for every $\sigma \in \mathrm{Aut}(I)$ there exists an $\alpha \in \mathrm{Aut}(A)$ such that $\alpha(e_{ii}) = e_{\sigma i, \sigma i}$. However, α defined by $\alpha(e_{ij}) = \alpha_{\sigma i, \sigma j}$ is an automorphism. This concludes the proof of the theorem.

<u>1.3. Corollary</u>. <u>Let</u> A <u>be the algebra of matrices of the form</u>

$$\begin{pmatrix} * & & * \\ & \boxed{*} & \\ 0 & & \boxed{*} \end{pmatrix}$$

(s blocks of arbitrary size along the diagonal)

<u>where arbitrary elements of</u> D <u>are admitted in the</u>

diagonal blocks and above the diagonal and zeros are elsewhere. Then, every automorphism of D *is inner.*

Proof: This algebra is an incidence algebra with S totally ordered.

2. Involutions on incidence algebras

We use the same notations as in the last section, except that we write automorphisms and involutions exponentially. For convenience we assume char(K)$\neq$2. An involution on the partially ordered set I is a map $^*: I \to I$ such that $i \prec j$ implies $j^* \prec i^*$ and $i^{**} = i$ for all $i, j \in I$. An involution on the K-algebra A is a K-linear map $^\sigma: A \to A$ such that $(ab)^\sigma = b^\sigma a^\sigma$ for all $a, b \in A$ and $\sigma \circ \sigma = id$. We are interested in involutions on the incidence algebra $A = A_D(I)$.

2.1. Theorem. Let I satisfy the conditions of 1.2.

a) The incidence algebra $A = A_D(I)$ admits an involution if and only if I admits an involution and D admits an involution.

b) Two involutions σ, τ on A are conjugate by an automorphism of A if and only if the following conditions are satisfied:

(i) The induced involutions $\bar{\sigma}, \bar{\tau}$ on S are conjugate by an element of the image of $Aut(I) \to Aut(S)$.

(ii) The induced involutions on A/R are conjugate by an element of the image of $Aut(A) \to Aut(A/R)$.

The involutions σ, τ on A are conjugate by an inner automorphism of A if and only if the induced involutions on S are equal and (ii) is satisfied.

<u>Proof</u>: a) If $\sigma: D \to D$, $^*: I \to I$ are involutions, then

$$\sum d_{ij} e_{ij} \to \sum d_{ij}^{\sigma} e_{j^* i^*}$$

is an involution on A. If, on the other hand $\sigma: A \to A$ is an involution, σ induces a permutation $\bar{\sigma}$ of order 2 of S and it follows an the same way as in the proof of 1.2 that this permutation is order reversing. Furthermore, it is clear that σ can be lifted to an involution of I. To prove that D admits an involution, we note first that σ induces an involution on $A/R = \prod M(n_t, D)$. If one of the simple factors $M(n_t, D)$ is left invariant, then, by a theorem of Albert [1], D admits also an involution. If all simple factors of A/R are paired by the involution, then D is isomorphic to the opposite skew field D^o, and again by a theorem of Albert, this implies that D admits an involution.

b) If σ and τ are conjugate, then, obviously, (i) and (ii) are satisfied. To prove the other direction, we assume that σ and τ satisfy (i), (ii). By (i) and 1.2 there exists an $\alpha \in Aut(A)$ such that σ and $\alpha\tau\alpha^{-1}$ induce the same involutions on S. Therefore, we may assume from the beginning that σ and τ induce the same permutation of the simple factors of A/R. By a theorem of Taft [5], we can find involution invariant semisimple Wedderburn complements P_1, P_2 in A, i.e. $A = P_i \oplus R$, $P_1^{\sigma} = P_1$, $P_2^{\tau} = P_2$. Replacing σ and τ by conjugates obtained from the Malcev theorem we may assume $P^{\sigma} = P^{\tau} = P$. By (ii) we may assume $\sigma|_P = \tau|_P$. The rest of the proof is a rather typical (and well known) argument and will be isolated as a lemma.

<u>2.2. Lemma</u>. <u>If</u> σ, τ <u>are involutions on</u> A <u>such that</u> $\sigma|_P = \tau|_P = id_P$, <u>then</u> σ <u>and</u> τ <u>are conjugate.</u>

Proof: $\sigma\tau$ is an automorphism and by theorem 1.2 an inner automorphism. Hence, there exists a unit $a\in A$ such that

$$x^{\sigma\tau}=a^{\sigma}xa^{-\sigma} \qquad \text{or} \qquad x^{\tau}=a^{-1}x^{\sigma}a$$

for all $x\in A$. It follows from the equation

$$x^{\tau\tau}= x = a^{-1}a^{\sigma}xa^{-\sigma}a$$

that $a^{-1}a^{\sigma}$ lies in the center of A, i.e. $a^{\sigma}=\lambda a$ for some central unit λ. Since $\sigma|_P = \tau|_P$ we can write $a=1+a_1$ with $a_1\in R$. Since $\text{center}(A)\subset P$ it follows that $\lambda=1$. It is now sufficient to find a $b\in A$ such that $a=bb^{\sigma}$, because then $\tau=\gamma^{-1}\sigma\gamma$ where γ is the inner automorphism given by b. Since a_1 is symmetric and nilpotent, we obtain b from the Taylor series of $\sqrt{1+x}$

$$b=1 + \frac{1}{2}a_1 - \frac{1}{8}a_1^2 + \frac{1}{16}a_1^3 - \frac{5}{128}a_1^4 +-\ldots\ldots$$

(Here, we need $\text{char}(K)\neq 2$!). This concludes also the proof of 2.1.

We want to discuss condition b) (ii) and assume that we have a semisimple algebra $P=\prod P_i$, P_i simple, and two involutions σ,τ on P which induce the same permutation of the simple factors P_i. Then it is well known, that σ and τ are conjugate if their restrictions to those simple factors which are left invariant are conjugate. The simple factors which are paired by the involution can be disregarded. The conclusion of the theorem is therefore, that the classification problem for involutions on incidence algebras inducing a fixed involution of S can be reduced to the classification of involutions on the simple factors of A/R which correspond to the fixed elements of S. The classification of involutions on simple algebras is due to Albert [1].

2.3. Remark: In the proof of 2.1 we have already used that every involution of S obtained from an involution of A can be lifted to an involution of I. Similar as for automorphisms the converse is true also. To prove this it is sufficient to show that for every involution $*$ of I there exists an involution σ on A such that $e_{ii}^{\sigma}=e_{i^*i^*}$. We can define σ simply by

$$(\sum d_{ij}e_{ij})^{\sigma} = \sum d_{ij}^{\sigma}e_{j^*i^*} \text{ .}$$

2.4. Remark: There do not seem to be any examples of finite-dimensional algebras $A \cong A^{o}$ which do not admit an involution in the literature (probably because nobody ever looked into this problem). It is not difficult to construct incidence algebras satisfying these conditions. It is sufficient to construct a partially ordered set admitting an antiautomorphism but not admitting an involution. The simplest example I could find is the following one:

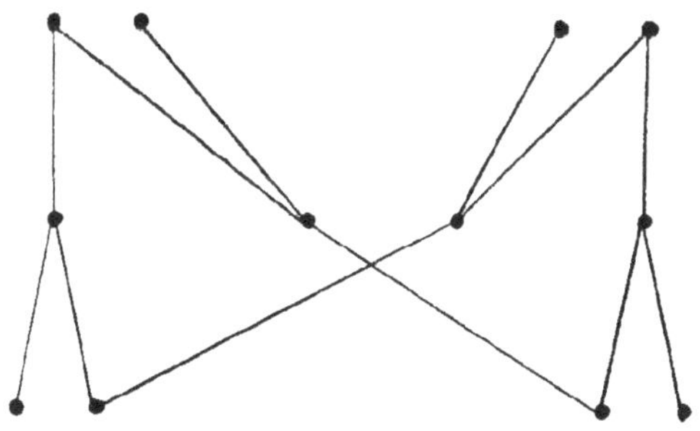

References

[1] A.A. Albert: Structure of algebras.
Am. Math. Soc. Coll. Publ. 24, Providence, R.I. (1961)

[2] C. Curtis, I. Reiner: Representation theory of finite groups and associative algebras
Interscience publishers, New York (1966)

[3] A. Malcev: On the representation of an algebra...,
Dokl. Akad. Nauk 36, 42-45 (1942)

[4] R.P. Stanley: Structure of incidence algebras and their automorphism groups.
Bull. Am. Math. Soc. (1970) 1236-1239

[5] E. Taft: Cleft algebras with operator groups.
Portugaliae math. 20, 195-198 (1961)

BALANCEDNESS AND LEFT SERIAL ALGEBRAS OF FINITE TYPE

Hiroyuki Tachikawa

Introduction. It is well known that serial rings (=generalized uniserial rings) are of finite (module) type. On the other hand, in [4] Janusz pointed out one-sided serial rings are not necessarily of finite type. Recently, C. M. Ringel [7] determined the structure of QF-1 algebras having square-zero radicals and he proved such algebras are of local-colocal type, so of finite type. The main purpose of this paper is to give an answer to a problem whether for one-sided serial algebras the balancedness on all faithful modules induces or not the finiteness of non-isomorphic indecomposable modules. The anwser is affirmative and we prove that left serial QF-1 algebras are of colocal type; i.e. every indecomposable left module has a unique minimal submodule [8].

§1. Notations and Preliminary.

Throughout this paper A is a finite dimensional associative algebra with unit 1 over a field P and by N the radical of A. For a left (right) A-module M call $T({}_RM) = M/NM$ $(T(M_R) = M/MN)$ the top of M and $S^i({}_RM) = \{m \in M \mid N^i m = 0\}$ $(S^i(M_R) = \{m \in M \mid mN^i = 0\})$ the i-th socle of M respectively. Especialy $S(M) = S^1(M)$ is the socle of M. If M is finitely generated, M has the composition series. By $|M|$ we denote the composition length of M. For another left (right) A-module M', put $T_r(M, M') = \{\Sigma \mathrm{Im}\,\sigma \mid \sigma \in \mathrm{Hom}_A(M, M')\}$ and we call it the trace submodule of M' from M. If L and L' are submodules of M and M' respectively such that there exists a left R-isomorphism θ of L onto L', then we can construct the factor module L of $M \oplus M'$ by the submodule consisting all elements of form $\{m, -\theta(m)\}$ for $m \in M$. Following [8] we call L the interlacing module of M and M'(recently it is called also the amalgamation) by using θ as the lacing isomorphism and denote by $\mathrm{Int}_\theta(M, M')$. In this paper homomorphisms will be always written on the side opposite the scalars. Let $B = \mathrm{End}({}_AM)$ and $C = \mathrm{End}(M_B)$. C is called the double centralizer of M and if $A \ni a \longmapsto (m \mapsto am) \in C$ is surjective, M is said to be balanced. Further, if all faithful modules are balanced, A is said to be QF-1. It was proved in [6] that the balancedness of modules is Morita invariant. So without loss of generality A may be assumed to be basic.

The next lemma, due to V. P. Camillo [1], is fundamental in the proof of our main theorem:

Lemma 1.1. If $M = \oplus_{i \in I} M_i$ is a direct sum of left A-modules, and $\{a_i \mid i \in I\} \subset A$, then the map given by $\Sigma m_i \longmapsto \Sigma a_i m_i \in \oplus_i M_i$ belongs to the double centralizer of M if and only if $(a_i - a_j) T_r(M_i, M_j) = 0$ for all i, $j \in I$.

In [7] C. M. Ringel proved that if e and f are primitive idempotents of QF-1 ring A with $f(S({}_A A) \cap S(A_A))e \neq$ $\neq 0$, then $|S(Ae)| \times |S(fA)| \leq 2$. Now, as $S(A_A)e \neq 0$ if and only if Ae appears as a direct summand of the minimum faithful projective left A-module (cf. [2]), we have

Lemma 1.2. Let A be a QF-1 algebra and e and f primitive idempotents of A. If Ae appears as a direct summand of the minimum faithful projective left A-module and Af/Nf appears as a submodule of Ae, then $|S(fA)| \leq 2$.

Here a faithful module M is said to be minimal if the deletion of any direct summand makes X unfaithful.

Lemma 1.3. Let $M = \oplus_{\lambda\in\Lambda} M_\lambda$ be a direct sum decomposition of a minimal faithful left A-module M into indecomposable direct summands. Then M_μ is not isomorphic to any subquotient of direct sum of copies of M_λ with $\lambda \neq \mu$.

Proof: Assume $M_\mu \cong L/L_o$, where $L_o \subset L \subset M_\lambda^I$ and $M_\lambda^I = \oplus_{i\in I} M_\lambda^{(i)}$ $(M_\lambda^{(i)} \cong M_\lambda)$ for some index set I and $\lambda(\neq\mu) \in \Lambda$. Then $\ell(M_\lambda) \subset \ell(M_\mu)$. Here $\ell(M_\lambda)$ and $\ell(M_\mu)$ are left annihilators of M_λ and M_μ in A respectively. Since M is faithful, we have

$$\begin{aligned}
0 &= \ell(\oplus_{\eta\in\Lambda} M_\eta) \\
&= \bigcap_{\substack{\eta\in\Lambda \\ \eta\neq\lambda \\ \eta\neq\mu}} \ell(M_\eta) \cap \ell(M_\lambda) \cap \ell(M_\mu) \\
&= \bigcap_{\substack{\eta\in\Lambda \\ \eta\neq\mu}} \ell(M_\eta) \\
&= \ell(\oplus_{\substack{\eta\in\Lambda \\ \eta\neq\mu}} M_\eta).
\end{aligned}$$

But this is a contradiction to minimal faithfulness of M.

Corollary 1.4. Let $\oplus_{\Lambda_o} Ae_\alpha$ be a direct sum decomposition of a minimal faithful projective module into primitive ideals Ae_α. Then Ae_α is not isomorphic to any subquotient of Ae_β, $\beta \neq \alpha$.

A left A-module M is said to be serial if the lattice of all left A-submodules of M, with respect to the inclusion, is linearly ordered.

In this paper an algebra A is said to be of left colocal type if the socle of each indecomposable left A-module is simple. By taking P-duals we know that A is of colocal type if and only if every indecomposable right A-module is isomorphic to a homomorphic image of a primitive ideal. So in [8] the author called such an algebras A strong right cyclic representation type (abbreviated by SRCRT) and established the following

Theorem 1.5. In order that an algebra A may be of left colocal type it is necessary and sufficient that the following conditions are satisfied:

I. A is a left serial algebra, i.e., all primitive left ideals are serial.

II. Let L_1 and L_2 be serial modules and M_1 and M_2 submodules of L_1 and L_2 respectively. If each of M_i, i=1,2, contains $S^2(L_i)$ and $\theta: M_1 \longrightarrow M_2$ is an isomorphism, then either θ or the inverse of θ is extendable to a monomorphism of L_1 into L_2 or of L_2 into L_1.

III. Assume that $Ne_\kappa/N^2e_\kappa \ncong Ne_\lambda/N^2e_\lambda \cong Ne_\mu/N^2e_\mu$ and

$Ae_\lambda \not\cong Ae_\mu$ for primitive idempotents e_κ, e_λ and e_μ, then Ae_κ is isomorphic to either Ae_λ or Ae_μ.

IV. Assume $Ne_\lambda/N^2e_\lambda \cong Ae_\kappa/Ne_\kappa$ for primitive idempotents e_λ and e_κ, then $e_\kappa Ne_\lambda/e_\kappa N^2e_\lambda$ is considered as a right vector space over a division ring $e_\lambda Ae_\lambda/e_\lambda Ne_\lambda$ and its dimension $(e_\kappa Ne_\lambda/e_\kappa N^2e_\lambda : e_\lambda Ae_\lambda/e_\lambda Ne_\lambda) \leq 2$. Particularly if the equality holds and $Ne_\lambda/N^2e_\lambda \cong Ne_\mu/N^2e_\mu$, then there results that $Ae_\lambda \cong Ae_\mu$.

§2. Maximal lacing length and quivers of serial submodules.

From now on we shall assume throughout that A is a left serial algebra. Further, as our investigation concerns exclusively with QF-1 algebras we restrict A to be basic and two sided indecomposable.

By Λ let us denote the totality of non-isomorphic primitive idempotents and let Λ_o be the subset of Λ consist of all primitive idempotents e_α such that $\oplus_{e_\alpha \in \Lambda_o} Ae_\alpha$ is a minimal faithful projective module. The meaning of Λ and Λ_o will be retained throughout.

For a pair of serial modules M_1 and M_2 with isomorphic socles L_1 and L_2 respectively, we consider the following commutative diagrams:

$$\begin{array}{ccc} M_1 & & M_2 \\ \uparrow & & \uparrow \\ N_1 & \longrightarrow & N_2 \\ \uparrow & & \uparrow \\ L_1 & \longrightarrow & L_2 \end{array},$$

where all vertical maps are inclusions and all row maps are isomorphisms.

If for given modules M_1 and M_2 N_1 is the largest submodule which appears in the above diagram, then we shall call $|N_1|$ the lacing length of M_1 and M_2. $\operatorname{Sup}|N_1|$ is called the maximal lacing length, where M_1 and M_2 go through all serial A-modules.

Lemma 2.1. Let ℓ be the maximal lacing length and let the lacing length of M_1 and M_2 equal to ℓ. Then at least one of the $\ell+1$-th socles of M_1 and M_2 is isomorphic to a primitive left ideal Af_1 and if $S^{\ell+1}(M_1) \cong Af_1$, there is a primitive ideal $Af^{(1)}$ such that $Af^{(1)} \supsetneq M_1$ and $f^{(1)} \in \Lambda_o$.

Proof: Since M_1 and M_2 are serial, there are primitive idempotents f_1 and f_2 such that $Af_1/K_1 \cong S^{\ell+1}(M_1)$ and $Af_2/K_2 \cong S^{\ell+1}(M_2)$. Suppose both K_1 and

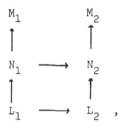

K_2 are nonzero. By the assumption we have $Nf_1/K_1 \cong S^{\ell}(M_1) \cong S^{\ell}(M_2) \cong Nf_2/K_2$, where N is the radical of A. Then, as A is left serial, $Nf_1/NK_1 \cong Nf_2/NK_2$ and $|Nf_1/NK_1| = \ell+1$. However, the isomorphism $Nf_1/NK_1 \longrightarrow Nf_2/NK_2$ is not extendable to a homomorphism: $Af_1/NK_1 \longrightarrow Af_2/NK_2$, for otherwise $Nf_1/K_1 \longrightarrow Nf_2/K_2$ is extendable. However this contradicts the maximality of ℓ. Thus, one of K_1 and K_2 must be zero. Assume $K_1 = 0$, $Af_1 \cong S^{\ell+1}(M_1)$, and since Af_1 is a left ideal, M_1 is also isomorphic to a left ideal. Take a left primitive ideal $Af^{(1)}$ which contains a copy of M_1 and have maximal length. As $Af^{(1)}$ must be embedded into a direct sum of copies of the minimal faithful projective module, we have $f^{(1)} \in \Lambda_0$.

Given a serial module M with the composition series $M = M_0 \supset M_1 \supset \cdots \supset M_n = (0)$. Then we shall define a quiver Q associated with M in the following way: As vertices of Q we take primitive idempotents e's which appear composition factors of M and edges are drawn between primitive idempotents e and f such that $\overline{Ae}$ and $\overline{Af}$ are adjacent composition factors of M, i.e., $\overline{Ae} \cong M_{i-1}/M_i$ and $\overline{Af} \cong M_i/M_{i+1}$ for some i $(1 \leq i \leq n-1)$.

Since A is a left serial algebra, we have the following three types of quivers:

(1)

(2)

(3)

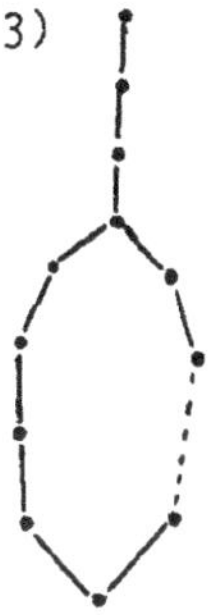

In the case (1) (resp. (2)) we say the quiver is a cycle (resp. straight line).

Lemma 2.2. Let Q be the quiver of a primitive ideal Ae. If Q is a cycle, then any automorphism of Ne can be extended to an automorphism of Ae.

If both Q and the quiver of Ne are not cycles, then in order that any automorphism of Ne may be extended to an automorphism of Ae it is necessary and sufficient that $\dim_{\bar{R}}(\bar{e}\bar{A}\bar{e}) = \dim_{\bar{R}}(\bar{e}_1\bar{A}\bar{e}_1)$ where $\bar{A}\bar{e}_1 \cong Ne/N^2e$.

Proof. By the assumption $\mathrm{End}_A(Ne/N^2e) \cong \bar{e}_1\bar{A}\bar{e}_1$ and an automorphism of Ae induces naturally an automorphism of Ne. It follows $\dim_{\bar{R}}(\bar{e}\bar{A}\bar{e}) \leq \dim_{\bar{R}}(\bar{e}_1\bar{A}\bar{e}_1)$. On the other hand, as Ae is serial, any automorphism of Ne/N^2e can be extended to an automorphism of Ne. Hence, in order that any automorphism of Ne may be extended to one of

Ae it is necessary $\dim_F(\bar{e}\bar{A}\bar{e}) = \dim_F(\bar{e}_1\bar{A}\bar{e}_1)$.

Let $N^j e/N^{j+1}e \cong \bar{A}\bar{e}_j$ $j=1,2,\cdots$.

In case where Q and the quiver of Ne are not cycle, it is clear that eNe and $e_1Ne_1 = 0$. Hence $\dim_F(eAe) = \dim_F(\bar{e}\bar{A}\bar{e}) = \dim_F(\bar{e}_1\bar{A}\bar{e}_1) = \dim_F(e_1Ae_1)$. It follows that any automorphism of Ne can be extended to an automorphism of Ae.

In case of Q being a cycle, let e_r be the first primitive idempotent isomorphic to e.

Then, the condition that $\dim_F \bar{e}\bar{A}\bar{e} = \dim_F \bar{e}_1\bar{A}\bar{e}_1$ is automatically satisfied because

$$\dim_F \bar{e}\bar{A}\bar{e} \leq \dim_F \bar{e}_1\bar{A}\bar{e}_1 \leq \dim_F \bar{e}_2\bar{A}\bar{e}_2$$

$$\cdots \qquad \leq \dim_F \bar{e}_r\bar{A}\bar{e}_r = \dim_F\bar{e}\bar{A}\bar{e},$$

and $e_1 = e_{r+1}$, $eNe = eN^{ir}e$, $e_1Ne_1 = e_1N^{ir}e_1$, $i=1.2.\cdots$. Take $x_i \in eN^{ir}e \setminus eN^{ir+1}e$, $y_i \in e_1N^{ir}e_1 \setminus e_1N^{ir+1}e_1$ and $e_1ue \in Ne \setminus N^2e$. It holds $\dim_F N^{ir+1}e/N^{ir+2}e = \dim_F \bar{e}_1\bar{A}\bar{e}_1 = \dim_F \bar{e}\bar{A}\bar{e}$. Hence for any $a \in e_1Ae_1 \setminus e_1Ne_1$ there exists an element $b \in eAe \setminus eNe$ such that $e_1ae_1y_ie_1ue \equiv e_1uex_iebe$ mod. N^{ir+2}. More simply for any $\sigma \in e_1N^{ir}e_1$ there exists an element $\tau \in eN^{ir}e$ such that $\sigma e_1ue_1 \equiv e_1ue\tau$ mod. N^{ir+2}.

Now, let θ be an automorphism of Ne, and $\theta(e_1ue) = \alpha e_1ue$, $\alpha \in e_1Ae_1$. Then there exists $\beta \in eAe$ such that

$$\alpha e_1 ue \equiv eue\beta \quad \mod N^2e.$$

We can put $e\alpha e_1ue - e_1ue\beta = \alpha_1 e_1 ue$ for $\alpha_1 \in e_1N^r e_1$ and find an element $\beta_1 \in eN^r e$ such that $\alpha_1 e_1 ue \equiv e_1 ue\beta_1$ mod $N^{r+2}e$. And successively we obtains $\alpha_i \in e_1 N^{ir} e_1$ and $\beta_i \in eN^{ir}e$ as follows: $\alpha_{i-1}e_1ue_1 - e_1ue\beta_{i-1} = \alpha_i e_1 ue$ and

$$\alpha_i e_1 ue \equiv e_1 ue\beta_i \quad \mod N^{ir+2}e.$$

Since the radical of A is nilpotent, we obtain finally $\gamma = \beta+\beta_1+\cdots$, such that $\alpha e_1 ue = e_1 ue\gamma$, where $\beta \in eAe \setminus eNe$ and $\beta_i \in eN^{ir}e \setminus eN^{ir+1}e$. Thus a map $Ae \ni ae \longmapsto ae\gamma \in Ae$ defines an automorphism of Ae which is an extension of θ. Since a serial module may be considered as a primitive ideal of a factor ring of A by a power of the radical,we have

Corollary 2.3. Let Q be the quiver of a serial module L with $NL \neq 0$. If Q is a cycle, then any automorphism of NL is extendable to an automorphism of L.

Lemma 2.4. Let A be an algebra (not necessarily left serial). Let J be a homomorphic image of a primitive left ideal and M a left A-module. If a monomorphism ϕ of a proper submodule L of J into M has no extension which is a monomorphism, then

$$|T(\mathrm{Int}_\phi(J, M))| \geq 1+|T(M)|-|\mathrm{Im}\phi+NM/NM|.$$

Especially, if J is serial,

$$|T(\mathrm{Int}_\phi(J, M))| \geq |T(M)|.$$

Proof. For a suitable primitive idempotent e and a left ideal $I \subseteq Ne$ we can put $J = Aeu$ and $Iu = L$ for $u \in J$. Denote by α and β the canonical injections of J and M into $\mathrm{Int}_\phi(J, M)$. $Neu\alpha+NM\beta$ is the radical of $\mathrm{Int}_\phi(J, M)$. If $aeu\alpha \in Neu\alpha+NM_\beta$, then $aeu\alpha = neu\alpha +$ $+\Sigma n_i'm_i\beta$ for $n, n_i' \in N$ and $m_i \in M$. This implies $(ae-ne)u \in Iu$. Hence if $be \notin Ne$, then $beu\alpha \notin Neu\alpha+NM_\beta$. Similarly if $m\beta \in Neu\alpha+NM\beta$ and $m\beta = neu\alpha+\Sigma n_j'm_j'\beta$ for $n, n' \in N$ and $m_j' \in M$, then $neu \in Iu$ and $\phi(neu) =$ $= \Sigma n_j'm_j'-m$. Hence $m \in \mathrm{Im}\,\phi+NM$. Since M/NM is completely reducible, we can choose $m_i \in M$, $i=1,2,\cdots,n$ such that $\mathrm{Im}\,\phi+NM/NM \cong \oplus_{i=k+1}^{n} A(m_i+NM)$, $k+1 \leq n$, and $\oplus_{i=1}^{n} A(m_i + NM)$ $\cong M/NM$. Let $ybeu\alpha+ \Sigma_{i=1}^{k} x_im_i\beta \equiv 0 \mod Neu\alpha+NM\beta$ and $ybeu\alpha+\Sigma_{i=1}^{k} x_im_i\beta = neu\alpha+\Sigma_{i=1}^{n} n_im_i$. Then $(ybe-ne)u\alpha =$ $= \Sigma_{i=1}^{k} -(x_i-n_i)m_i\beta+\Sigma_{i=k+1}^{n} n_im_i\beta$, and hence $(ybe-ne)u \in Iu$ $\subset Nu$. So it holds

$$ybeu\alpha \equiv 0 \qquad \mathrm{mod.}\ Neu\alpha + NM\beta$$

$$\text{and}\quad \Sigma_{i=1}^{k} x_im_i\beta \equiv 0 \quad \mathrm{mod.}\ Neu\alpha + NM\beta.$$

Hence $\Sigma_{i=1}^{k} x_i m_i \equiv 0$, mod. Im ϕ+NMβ, and It follows $x_i m_i \equiv 0$ mod. Im ϕ+NM for $i=1,2,\cdots,k$.

This implies $x_i m_i \beta \equiv 0$ mod. Neuα+NMβ for $i=1,2,\cdots k$. Thus, beuα and $m_i\beta$, $i=1,2,\cdots,k$, constitute a part of minimal generators of $\mathrm{Int}_\phi(J, M)$. Hence, we have

$$|T(\mathrm{Int}_\phi(J, M))| \geq 1+|T(M)|-|\mathrm{Im}\ \phi+NM/NM|.$$

The last statement now is obvious.

Proposition 2.5. Let A be a left serial QF-1 algebra and Agu homomorphic image of primitive ideal Ag. Let $S^n(Af)$ be the n-th socle of Af with $f \in \Lambda_o$. If a monomorphism ϕ: $S^n(Af) \longrightarrow$ Agu has no extension, then the quiver of $S^{n+1}(Af)/S^1(Af)$ is not a cycle.

Proof: Let $S^{i+1}(Af)/S^i(Af) \cong \overline{A}\overline{e}_{i+1}$, $0 \leq i \leq n$. Suppose the quiver of $S^{n+1}(Af)/S^1(Af)$ is a cycle and i is the first integer such that $\overline{A}\overline{e}_{i+1} \cong \overline{A}\overline{e}_{n+1}$, $i \geq 1$. By Corollary 2.3 and the maximality of ϕ, $S^{n+1}(Agu)/S^n(Agu) \not\cong S^{n+1}(Af)/S^n(Af)$. According to the maximality of ϕ we can conclude

$$S^n(\mathrm{Int}_\phi(Af, Agu)) = S^n(Af\alpha),$$

where α is the canonical injection of Af into Int_ϕ(Af, Agu). On the other hand,

$$S^{i+1}(Agu/S^{n-i}(Agu)) \cong S^{n+1}(Agu)/S^{n-i}(Agu)$$

$$\begin{aligned} S^{i}(Agu/S^{n-i}(Agu)) &\cong S^{i}(Af/S^{n-i}(Af)) \\ &\cong S^{n}(Af)/S^{n-i}(Af) \\ &\cong S^{i}(Af). \end{aligned}$$

Since $S^{i+1}(Agu/S^{n-i}(Agu))/S^{i}(Agu/S^{n-i}(Agu)) \not\cong S^{i+1}(Af)/S^{i}(Af)$, we have a monomorphism ψ of $S^{i}(Agu/S^{n-i}(Agu))$ into $Af\alpha(\subset \mathrm{Int}_{\phi}(Af, Agu))$.

Now, by Lemma 2.4 we have

$$|T(\mathrm{Int}_{\psi}(Agu/S^{n-i}(Agu), \mathrm{Int}_{\phi}(Af,Agu)))| = 3$$

and $$S^{1}(\mathrm{Int}_{\psi}(Agu/S^{n-i}(Agu), \mathrm{Int}_{\phi}(Af, Agu))) \cong \bar{A}\bar{e}_1,$$

because $$|T(\mathrm{Int}_{\phi}(Af, Agu))| = 2.$$

Since A is left serial, the injective hull $E(\bar{A}\bar{e}_1)$ is obtained by interlacings of serial modules with socles isomorphic to $\bar{A}\bar{e}_1$ and it follows from Lemma 2.4, $|T(E(\bar{A}\bar{e}_1))| \geq 3$. By taking K-dual we have $|S(e_1A)| \geq 3$, but this contradicts Lemma 1.2.

Corollary 2.6. Under the same assumption of Proposition 2.5, let $S^{n}(Af)/S^{n-1}(Af) \cong \bar{A}\bar{e}_n$ and $S^{n+1}(Af)/S^{n}(Af) \cong \bar{A}\bar{e}_{n+1}$. Then $e_n \neq e_{n+1}$.

Lemma 2.7. Let A be a basic, left serial algebra.

Let θ be a monomorphism of $S^{\ell_1}(Ae_1)$ into $S^{\ell_1}(Ae'/J)$ such that θ has no extension, where $1 \leq \ell_1$, $e_1 \in \Lambda_0$, $e' \in \Lambda$, and J is a left ideal contained in Ae'.
According as

1) $J \neq 0$ or $J = 0$, $e' \notin \Lambda_0$

or 2) $J = 0$, $e' = e_j \in \Lambda_0$, we put

$$L = M \oplus \oplus_{e_i \in \Lambda_0 - \{e_1\}} Ae_i$$

or

$$L = M \oplus \oplus_{e_i \in \Lambda_0 - \{e_1, e_j\}} Ae_i \text{ , where}$$

$$M = \mathrm{Int}_\theta(Ae_1, Ae'/J).$$

Then L is a minimal faithful module.

Proof: It is obvious that L is faithful. In the case 1) suppose

$$L' = M \oplus \oplus_{e_i \in \Lambda_0 - \{e_1, e_k\}} Ae_i \text{ ,} \quad k \neq 1,$$

is faithful. Then $Ae_k \subset L'^{(n)}$ (= a direct sum of n-copies of L'). Since Ae_k is serial and $e_i \in \Lambda_0 - \{e_1\} - \{e_k\}$, Ae_k must be embedded into a indecomposable direct summand of L'. In fact, by Corollary 1.4 we would have $Ae_k \subset M$. Put $e_k = e_k x_1 e_1 \alpha + (e_k x_2 e' + J)\beta$. However, as e_1, $e_k \in \Lambda_0$, $S(Ae_k) e_k x e_1 = 0$, and consequently $(e_k x_2 e' + J)\beta \neq 0$. This implies that $Ae_k \subset Ae'/J$. It follows $J = 0$ and $e_k = e'$, but this is not the case 1).

In case 2) suppose also

$$L' = M \oplus \oplus_{e_i \in \Lambda_0 - \{e_1, e_j, e_k\}} Ae_i, \; k \neq 1 \text{ and } j,$$

is faithful. Then, similarly as above, $Ae_k \subset M$ and $k = 1$ or $k = j$, but this is a contradiction.

Lemma 2.8. Let M_1, M_2, $f^{(1)}$ and f_1 be left A-modules and primitive idempotents as in Lemma 2.1. If $S^{\ell+1}(M_2) \cong Ah/N^{\ell+1}h$, $h \in \Lambda$, then canonical isomorphisms

$$\theta: Nh/N^{\ell-j+1}h \longrightarrow S^{\ell}(Af^{(1)})/S^{j}(Af^{(1)}) \quad 0 \leq j \leq \ell-1,$$

have no extensions: $Ah/N^{\ell-j+1}h \longrightarrow S^{\ell+1}(Af^{(1)})/S^{j}(Af^{(1)})$ respectively.

Proof: It is clear this lemma is true if $f_1 \neq h$. Assume $f_1 = h$. Then $Ah/N^{\ell+1}h \cong Af_1$. According to the selection of $Af^{(1)}$ we know immediately our conclusion is true if $j = 0$. Then, by Corollary 2.3 the quiver of Af_1 is not a cycle.

So, at first we shall consider the case where the quiver of Af_1 is not a cycle, but the quiver of Nf_1 is a cycle. Let $S^{j}(Af^{(1)})/S^{j-1}(Af^{(1)}) \cong \overline{A}\overline{e}_j$, $1 \leq j \leq \ell$. If ℓ' is the least positive integer such that $e_{\ell'} = e_{\ell}$, then $e_{\ell'+1} \neq f_1$, because the quiver of Af_1 is not a cycle. Hence we have an isomorphism $\theta'\colon Nf_1/N^{\ell'+1}f_1 \longrightarrow S^{\ell'}(Af^{(1)})$ which has no extension: $Af_1/N^{\ell'+1}f_1 \longrightarrow S^{\ell'+1}(Af^{(1)})$. Denote by L $\mathrm{Int}_{\theta'}(Af_1/N^{\ell'+1}f_1, Af^{(1)})$ and by β the canonical injection of $Af^{(1)}$ into L. Then, similarly as in the proof of Proposition 2.5 $S^1(\mathrm{Int}_{\beta\theta}(Af_1, L)) \cong \bar{A}\bar{e}_1$ and $|T(\mathrm{Int}_{\beta\theta}(Af_1, L))| = 3$ and by Lemma 2.4 it follows $|S(e_1A)| \geq 3$. But, since $f^{(1)} \in \Lambda_0$, this contradicts Lemma 1.2.

Now it remains to prove this lemma for the case where both the quivers of Af_1 anf Nf_1 are not cycles. At beginning we remark, if $j = 0$ then our conclusion is true and in this case it follows by Lemma 2.2 that $\dim_{\bar{F}}(\bar{e}_\ell\bar{A}\bar{e}_\ell) > \dim_{\bar{F}}(\bar{f}_1\bar{A}\bar{f}_1)$. But, since both the quivers of $Af_1/N^{\ell-j+1}f_1$ (= $Ah/N^{\ell-j+1}h$) and $Nf_1/N^{\ell-j+1}f_1$ are not cycles, θ has no extension.

Lemma 2.9. Let $f^{(1)}$ be the primitive idempotent selected in Lemma 2.1. If $Af^{(1)}/S^k(Af^{(1)})$ is embedded into $Af^{(j)}$, $f^{(j)} \in \Lambda_0$ and $\ell-k \geq 2$, then there exists a primitive idempotent $f^{(i)} \in \Lambda_0$ such that $Af^{(j)}/S(Af^{(j)})$ is embedded into $Af^{(i)}$.

Proof: In Lemma 2.1, let $S^{\ell+1}(M_2)$ be a homomorphic image Ah. Since by Lemma 2.8 $S^{\ell-k}(Af^{(j)})(\cong S^\ell(Af^{(1)})/S^k(Af^{(1)})) \xrightarrow{\theta} Nh/N^{\ell-k+1}h$ has no extension, we can use

$f^{(j)}$ and h as e_1 and e' in Lemma 2.7 respectively. Putting $M = \mathrm{Int}_\theta(Af^{(j)}, Ah/N^{\ell-k+1}h)$ we have a minimum faithful module

$$L = M \oplus \bigoplus_{f^{(k)} \in \Lambda_0'} Af^{(k)}, \quad f^{(j)} \in \Lambda_8',$$

where Λ_0' is a suitable subset of Λ_0.

Suppose

$$L^{(n)} \longrightarrow Af^{(j)}/S(Af^{(j)}) \longrightarrow 0$$

is exact. Then since $Af^{(j)}$ is serial we have

$$M \xrightarrow{\psi} Af^{(j)}/S(Af^{(j)}) \longrightarrow 0$$

is exact. Since $|M| = |Af^{(j)}|+1$, $|\mathrm{Ker}\ \psi| = 2$ and since $S(M)$ is simple, $\mathrm{Ker}\ \psi = S^2(\mathrm{Ker}\ \psi) \subset S^2(M) \subset Nh/N^{\ell-k+1}h\beta$ where β is the canonical injection of $Ah/N^{\ell+k+1}h$ into M. It follows

$$Af^{(j)}/S(Af^{(j)})(\cong M/\mathrm{Ker}\ \psi) \longrightarrow M/(Nh/N^{\ell-k+1}h)\beta \longrightarrow 0$$

is exact. But it is clear $M/(Nh/N^{\ell+k+1}h)\beta$ is not serial and this is a contradiction.

Hence, by Morita's Criterion [5] we may assume $Af^{(j)}/S(Af^{(j)})$ is embedded into $L^{(m)}$ for some positive integer m. Suppose $Af^{(j)}/S(Af^{(j)})$ is embedded into M.

Then, putting $\ell_1 = \ell-k$ we have $Af_1 \cong S^{\ell+1}(Af^{(1)})/S^{\ell}(Af^{(1)})$ $\cong S^{\ell_1+1}(Af^{(j)})/S^{\ell_1}(Af^{(j)}) \cong S^{\ell_1}(Af^{(j)})/S^{\ell_1-1}(Af^{(j)})$ $\cong S^{\ell}(Af^{(1)})/S^{\ell-1}(Af^{(1)})$. However this contradicts to Corollary 2.6. Hence there must exist a primitive idempotent $f^{(i)} \in \Lambda_o$ such that $Af^{(j)}/S(Af^{(j)})$ is embedded into $Af^{(i)}$.

Corollary 2.10. Let $f^{(1)}$ be the primitive idempotent selected in Lemma 2.1. Then there are primitive idempotents $f^{(j)} \in \Lambda_o$, $j=2,\cdots,\ell$, such that $Af^{(j-1)}/S^1(Af^{(j-1)})$ is embedded into $Af^{(j)}$.

Proposition 2.11. In Corollary 2.10 each $f^{(i)}$ $(i=1,2,\cdots,\ell)$ is not isomorphic to the others.

Proof: We shall number the composition factors of $Af^{(i)}$'s as follows:

$$S^1(Af^{(1)}) \cong \bar{A}\bar{e}_1 ,$$

$$S^2(Af^{(1)})/S^1(Af^{(1)}) \cong \bar{A}\bar{e}_2 \cong S(Af^{(2)}),$$

$$\vdots$$

$$S^i(Af^{(1)})/S^{i-1}(Af^{(1)}) \cong \bar{A}\bar{e}_i \cong S^{i-1}(Af^{(2)})/S^{i-2}(Af^{(2)}) \cong \cdots \cong S(Af^{(i)}),$$

$$S^{\ell}(Af^{(1)})/S^{\ell-1}(Af^{(1)}) \cong \bar{A}\bar{e}_{\ell} \cong S^{\ell-1}(Af^{(2)})/S^{\ell-2}(Af^{(2)}) \cong \cdots \cong S(Af^{(\ell)}) ,$$

$$S^{\ell+1}(Af^{(1)})/S^{\ell}(Af^{(1)}) \cong \overline{\overline{Af}}_1 \cong S^{\ell+1}(Af^{(2)})/S^{\ell}(Af^{(2)}) \cong$$
$$\cdots \cong S^2(Af^{(\ell)})/S(Af^{(\ell)}),$$
$$S^{\ell+2}(Af^{(1)})/S^{\ell+1}(Af^{(1)}) \cong \overline{\overline{Af}}_2 \cong \cdots$$
$$\cong S^3(Af^{(\ell)})/S^2(Af^{(\ell)}),$$
$$\vdots$$
$$Af^{(1)}/Nf^{(1)} \cong \overline{\overline{Af}}_{\kappa_1},$$
$$S^{\kappa_1+1}(Af^{(2)})/S^{\kappa_1}(Af^{(2)}) \cong \overline{\overline{Af}}_{\kappa_1+1},$$
$$S^{\kappa_1+2}(Af^{(2)})/S^{\kappa_1+1}(Af^{(2)}) \cong \overline{\overline{Af}}_{\kappa_1+2},$$
$$\vdots$$
$$Af^{(2)}/Nf^{(2)} \cong \overline{\overline{Af}}_{\kappa_2},$$
$$S^{\kappa_2+1}(Af^{(3)})/S^{\kappa_2}(Af^{(3)}) \cong \overline{\overline{Af}}_{\kappa_2+1},$$
$$\vdots$$
$$Af^{(j)}/Nf^{(j)} \cong \overline{\overline{Af}}_{\kappa_j},$$
$$S^{\kappa_j+1}(Af^{(j+1)})/S^{\kappa_j}(Af^{(j+1)}) \cong \overline{\overline{Af}}_{\kappa_j+1},$$
$$\vdots$$
$$Af^{(\ell)}/Nf^{(\ell)} \cong \overline{\overline{Af}}_{\kappa_\ell}.$$

Now, suppose $f^{(i)} = f^{(j)}$, $j > i$. Then there is a positive integer r such that $\kappa_i + r = \kappa_j$.

Since $|Af^{(k)}| \leq |Af^{(k+1)}|$, $k=1,2,\cdots,\ell$, from our assumption it follows that $|Af^{(i)}| = |Af^{(i+1)}| = \cdots = |Af^{(j)}|$

$$f^{(i+t)} = f_{\kappa_i+t} \quad , \quad 0 \leq t \leq \ell.$$

This implies the quiver of $Af^{(j)}$ $(= Af^{(i)})$ is a cycle with period at most r. However from $S(Af^{(j)}) \cong \overline{\overline{Ae}}_{i+r}$, we know $i+r \leq \ell$ and hence $r < \ell-i \leq \ell-1$. As $\overline{\overline{Af}}_1$ appears as a composition factor of $Af^{(j)}$, the quiver of Af_1 is a cycle with period at most $\ell-1$. This contradicts Proposition 2.5.

Proposition 2.12. f_{κ_1} $(= f^{(1)})$ is different from each of $f_{\kappa_1+1}, f_{\kappa_1+2},\ldots,f_{\kappa_2}, f_{\kappa_2+1},\ldots, f_{\kappa_\ell}$ $(= f^{(\ell)})$.

Proof: Suppose $f_{\kappa_1} = f_{\kappa_i+j}$, $1 < i \leq \ell-1$, $\kappa_i \leq \kappa_i+j \leq \kappa_{(i+1)}$. Put $k = (\kappa_i-\kappa_1)+j+1$. By assumption the quiver of Af_{κ_1} $(= Af_{\kappa_i+j})$ is a cycle with the period at most $k-1$. So $S^{k+\ell-i}(Af_{\kappa_{(i+1)}})/S^{k-1+(\ell-i)}(Af_{\kappa_{(i+1)}})$ is isomorphic to $\overline{\overline{Af}}_1$, because $S^{1+\ell-i}(Af_{\kappa_{(i+1)}})/S^{\ell-i}(Af_{\kappa_{(i+1)}}) \cong \overline{\overline{Af}}_1$. If $k+\ell-i > \ell+1$, then $|Af_1| > \ell+1$ and this contradicts the selection of f_1. Hence we may assume $\ell-i+k \leq \ell+1$. On the other hand, $k = (\kappa_i-\kappa_1)+j+1 \geq i-1+j+1 = i+j$ and hence $\ell-i+k \geq \ell+j$. So j must be 0 or 1. If $j = 0$, then $f_{\kappa_i} = f_{\kappa_1}$ but by Proposition 2.11 this is impossible. If $j = 1$, then from the assumption $\ell-i+k \leq \ell+1$ we have $\kappa_i-\kappa_1 = i-1$. If follows that $f_{\kappa_2} = f_{\kappa_1+1}, f_{\kappa_3} = f_{\kappa_1+2},\ldots, f_{\kappa_i} = f_{\kappa_1+i-1}, f_{\kappa_i+1} = f_{\kappa_i+j} = f_{\kappa_1}$.

Hence the quiver of Af_{κ_1} $(= Af^{(1)})$ is a cycle of period at most i. This implies the quiver of Af_1 is a cycle of period at most $\ell-1$, but this is again a contradiction to Proposition 2.5.

§3. Main theorems

Theorem 3.1. Let A be a left serial QF-1 algebra. Then the largest lacing length ℓ of serial modules is at most one.

Proof: Assume $\ell \geq 2$. By Propositions 2.10 and 2.11 there are primitive idempotents $f_1, f^{(1)}, f^{(2)}, \cdots, f^{(\ell)}$ and h such that $Af_1 \subsetneq Af^{(1)}$, $Af^{(i)}/S(Af^{(i)}) \subsetneq Af^{(i+1)}$, $1 \leq i \leq \ell-1$ and $Nf_1 \cong Nh/N^{\ell+1}h$. Without loss of generality we can number the factors of composition series of $Af^{(\ell)}$ and Af_1 as in Proposition 2.11.

Denote by ρ_i and σ_i canonical homomorphisms $Af_{\kappa_i} \longrightarrow Af_{\kappa_i+1}$ and $Ah/N^{\ell-i+2}h \longrightarrow Ah/N^{\ell-i+1}h$ $(1 \leq i \leq \ell)$ respectively. Then, by Lemma 2.8 in the following commutative diagram θ_i are not extendable to monomorphisms $Ah/N^{\ell-i+2}h$ into Af_{κ_i} :

$$
\begin{array}{ccccccccccc}
Af_{\kappa_1} & \xrightarrow{\rho_1} & Af_{\kappa_2} & \xrightarrow{\rho_2} \cdots \rightarrow & Af_{\kappa_i} & \xrightarrow{\rho_i} & Af_{\kappa_{i+1}} & \xrightarrow{\rho_{i+1}} \cdots & \xrightarrow{\rho_{\ell-1}} & Af_{\kappa_\ell} \\
\uparrow & & \uparrow & & \uparrow & & \uparrow & & & \uparrow \\
Nf_1 & \longrightarrow & (Nf_1)\rho_1 & \rightarrow \cdots \rightarrow & (Nf_1)\rho_1 \cdots \rho_{i-1} & \rightarrow & (Nf_1)\rho_1 \cdots \rho_i & \rightarrow \cdots & \rightarrow & (Nf_1)\rho_1 \cdots \rho_{\ell-1} \\
\uparrow \theta_1 & & \uparrow \theta_2 & & \uparrow \theta_i & & \uparrow \theta_{i+1} & & & \uparrow \theta_\ell \\
Nh/N^{\ell+1}h & \longrightarrow & Nh/N^{\ell}h & \rightarrow \cdots \rightarrow & Nh/N^{\ell-i+2}h & \longrightarrow & Nh/N^{\ell-i+1}h & \rightarrow \cdots & \longrightarrow & Nh/N^2h \\
\downarrow & & \downarrow & & \downarrow & & \downarrow & & & \downarrow \\
Ah/N^{\ell+1}h & \xrightarrow{\sigma_1} & Ah/N\,h & \xrightarrow{\sigma_2} \cdots \rightarrow & Ah/N^{\ell-i+2}h & \xrightarrow{\sigma_i} & Ah/N^{\ell-i+1}h & \xrightarrow{\sigma_{i+1}} \cdots & \xrightarrow{\sigma_{\ell+1}} & Ah/N^2h ,
\end{array}
$$

where vertical maps except θ_i are all inclusions. Put $X_i = \mathrm{Int}_{\theta_i}(Ah/N^{\ell-i+2}h,\ Af_{\kappa_i})$ for $i=1,2,\cdots,\ell-1$. Then X_i have simple socles and are indecomposable. Complementing primitive idempotents g_j, $1 \leq j \leq t$, of Λ_o to $\{f_{\kappa_i}\}$ we obtain a faithful left A-module

$$X = X_1 \oplus X_2 \oplus \cdots \oplus X_{\ell-1} \oplus Af_{\kappa_\ell} \oplus Ag_1 \oplus \cdots \oplus Ag_t .$$

Here $Af_{\kappa_1} \oplus \cdots \oplus Af_{\kappa_{\ell-1}} \oplus Af_{\kappa_\ell} \oplus Ag_1 \oplus \cdots \oplus Ag_t$ is the minimal faithful projective module and $f_{\kappa_i} = f^{(i)}$ $i=1,2,\cdots,\ell$. Now we shall prove that X is not balanced. Since Af_{κ_1} is serial, there is an element $e_\ell a f_{\kappa_1} \in A$ such that $Ae_\ell a f_{\kappa_1} = S^{\ell}(Af_{\kappa_1})$. At first we prove

$$(1)\quad \begin{aligned} &e_\ell a f_{\kappa_1} T_r(Af_{\kappa_\ell}, X_i) = 0, && 1 \le i \le \ell-1,\\ &e_\ell a f_{\kappa_1} T_r(Af_{\kappa_\ell}, Ag_j) = 0, && 1 \le j \le t. \end{aligned}$$

If there exists a monomorphism of Af_{κ_ℓ} into X_i, then $\overline{\overline{Af}}_1 \cong \overline{S^2(Af_{\kappa_\ell})} \cong \overline{S^2(X_i)} \cong \overline{\overline{Ae}}_{i+1}$ and it follows that the quiver of Af_1 is a cycle. This contradicts Proposition 2.5. So, we may assume $\operatorname{Ker}\phi \ne 0$ for all $\phi \in \operatorname{Hom}_A(Af_{\kappa_\ell}, X_i)$. On the other hand, as is proved in Proposition 2.12, idempotents $f_{\kappa_1+1}, f_{\kappa_1+2}, \dots, f_{\kappa_2}, \dots, f_{\kappa_\ell}$ are different from f_{κ_1} and hence $Ae_\ell a f_{\kappa_1} Af_{\kappa_\ell} = S^1(Af_{\kappa_\ell}) \subset \operatorname{Ker}\phi$. Since $T_r(Af_{\kappa_\ell}, X_i) = \Sigma_{\phi \in \operatorname{Hom}_A(Af_{\kappa}, X_i)} \operatorname{Im}\phi$ and $\operatorname{Im}\phi \simeq Af_{\kappa_\ell}/\operatorname{Ker}\phi$, we have

$$e_\ell a f_{\kappa_1} \cdot T_r(Af_{\kappa_\ell}, X_i) = 0, \qquad 1 \le i \le \ell-1.$$

For all $\psi \in \operatorname{Hom}_A(Af_{\kappa_\ell}, Ag_j)$ we may assume also that $\operatorname{Ker}\psi \ne 0$, because both f_{κ_ℓ} and g_j belong to Λ_o. Then, similarly as above we have

$$e_\ell a f_{\kappa_1} \cdot T_r(Af_{\kappa_\ell}, Ag_j) = 0, \qquad 1 \le j \le t.$$

Next we prove

(2)
$$e_\ell af_{\kappa_1}\, T_r(X_i,\ Af_{\kappa_\ell}) = 0, \quad 1 \le i \le \ell-1,$$
$$e_\ell af_{\kappa_1}\, T_r(Ag_j,\ Af_{\kappa_\ell}) = 0 \quad 1 \le j \le t.$$

Since Af_{κ_ℓ} is serial, $T_r(X_i, Af_{\kappa_\ell})$ is isomorphic to either $Af_{\kappa_i}/S^{\ell-i+1}(Af_{\kappa_i})$ or Ah/Nh. However the factors of composition series of the former is Af_{κ_i}, $Af_{\kappa_i-1}, \cdots,$ Af_1 and Ah/Nh is simple. Hence $e_\ell af_{\kappa_1}(Af_{\kappa_i}/S^{\ell-i+1}(Af_{\kappa_i})) =$ $= 0$ and

$$e_\ell af_{\kappa_1}(Ah/Nh) = 0.$$

It follows that $e_\ell af_{\kappa_1} T_r(X_i, Af_{\kappa_\ell}) = 0,\ 1 \le i \le \ell-1.$

Now, suppose $T_r(Ag_j, Af_{\kappa_\ell}) \neq 0$. Since both g_j and f_{κ_ℓ} belong to Λ_o, any homomorphism of Ag_j into Af_{κ_ℓ} is not a monomorphism. Assume there exist a homomorphism ϕ of Ag_j into Af_{κ_ℓ} such that $|\mathrm{Ker}\ \phi| = 1$. Then $\mathrm{Ker}\ \phi \cong \overline{\overline{Ae}}_{\ell-1}$ and $g_j \in \{f_1, f_2, \cdots, f_{\kappa_\ell}\}$. And g_j must be $f_{\kappa_{\ell-1}}$, for otherwise it happens either $g_j \notin \Lambda_o$ or $f_{\kappa_{\ell-1}} \notin \Lambda_o$. However it does not happen because g_j was a complemented primitive idempotent of Λ_o to $\{f_{\kappa_1}, \cdots, f_{\kappa_{\ell-1}}, f_{\kappa_\ell}\}$. Hence $|\mathrm{Ker}\ \phi| > 1$. However from the assumption that $|\mathrm{Ker}\ \phi| = 2$ we arrive at a similar contradiction. In this case g_j must be $f_{\kappa_{\ell-2}}$ and this is again impossible. It will be easily seen that we can repeat similar agruments for the cases $|\mathrm{Ker}\ \phi| = 3, 4, \cdots, \ell-1$.

Finally, assume $|\mathrm{Ker}\ \phi| \geq \ell$. Then $|Af_1| \geq |\mathrm{Ker}\ \phi|+2 = \ell+2$, but this contradicts the fact that Af_1 is a left ideal with composition length $\ell+1$. Consequently $T_r(Ag_j, Af_{\kappa_\ell}) = 0$, and $e_\ell af_{\kappa_1} \cdot T_r(Ag_j, Af_{\kappa_\ell}) = 0$ for $1 \leq j \leq t$.

Now, consider the map δ given by $X \ni (x_1, \cdots, x_{\ell-1}, z_\ell, y_1, \cdots, y_t) \longmapsto (0, \cdots, 0, e_\ell af_{\kappa_1} z_\ell, 0, \cdots, 0) \in X$, where $x_i \in X_i$, $y_i \in Ag_j$ and $z_\ell \in Af_{\kappa_\ell}$. Then by Lemma 1.1, (1) and (2) assure us δ belongs to the double centralizer of X.

Suppose δ is obtained by the left multiplication with an element b of A. Then, as there exists an element z_ℓ of Af_{κ_ℓ} such that $e_\ell af_{\kappa_1} \cdot f_{\kappa_1} z_\ell f_{\kappa_\ell} \neq 0$, $bf_{\kappa_1} z_\ell f_{\kappa_\ell} \neq 0$ but $b \cdot f_{\kappa_1} = 0$ because $b \cdot X_1 = 0$. This completes the proof of unbalanceness of X. This contradiction follows from $\ell \geq 2$. Thus we can conclude $\ell \leq 1$.

Theorem 3.2. Let A be a left serial algebra. If A is QF-1, then A is of left colocal type.

Proof: The condition I of Theorem 1.5 is satisfied

by the assumption. By Theorem 3.1 we know immediately the condition II of Theorem 1.3 is also satisfied.

For different primitive idempotents e_κ, e_λ and e_μ assume $Ne_\kappa/N^2e_\kappa \cong Ne_\lambda/N^2e_\lambda \cong Ne_\mu/N^2e_\mu \cong Af/Nf$. Then, either N^2e_κ, N^2e_λ or N^2e_μ is zero, for otherwise the maximal lacing length $\ell \geq 2$, contradicting Theorem 3.1 again. So, let $N^2e_\kappa = 0$. Then there exists a primitive idempotent $e_\eta \in \Lambda_0$ such that $Ae_\eta \supsetneq Ae_\kappa$. Hence by Lemma 1.2 we have $|S(fA)| \leq 2$. On the other hand, by the assumption for e_κ, e_λ and e_μ it follows $|S(fA/fN^2)| = |fN/fN^2| \geq 3$. Denote by $(fA/fN^2)^*$ the P-dual of fA/fN^2. Then we have $|T(fA/fN^2)^*| = |S(fA/fN^2)| \geq 3$. However $S((fA/fN^2)^*) \cong Af/Nf = \bar{\bar{A}}\bar{\bar{f}}$ and

$$(fA/fN^2)^* \subsetneq E(\bar{\bar{A}}\bar{\bar{f}}).$$

Hence, $|T(E(\bar{\bar{A}}\bar{\bar{f}}))| \geq 3$ by Lemma 2.4. Take P-dual of $E(\bar{\bar{A}}\bar{\bar{f}})$. Then we have $|S(fA)| \geq 3$ and this is a contradiction. Consequently, the condition III of Theorem 1.3 is satisfied.

Next, for a primitive idempotent e_κ assume $Ne_\mu/N^2e_\mu \cong Af/Nf$ and the dimension $(fNe_\kappa/fN^2e_\kappa : e_\kappa Ae_\kappa/e_\kappa Ne_\kappa) \geq 3$. Then similarly as above by Theorem 3.1, it follows $N^2e_\mu = 0$ and there exists a primitive idempotent e_η such that $Ae_\eta \supsetneq Ae_\kappa$. And by Lemma 1.2 $S(fA) \leq 2$. However, by our assumption $|S(fN/fN^2)| \geq 3$ and quite similarly, we have

$|S(fA)| \geq 3$. This contradiction implies that the condition IV of Theorem 1.3 are satisfied.

References

[1] V. P. Camillo, Balanced rings and a problem of Thrall, Trans. Amer. Math. Soc. 149 (1970), 143-153.

[2] K. R. Fuller, Double centralizers of injectives and projectives over artinian rings, Illinois J. Math. 14 (1970), 658-664.

[3] ___________, Generalized uniserial rings and their Kupish series, Math. Z. 106 (1968), 248-260.

[4] G. J. Janusz, Some left serial algebras of finite type, J. of Algebra 23 (1972), 404-411.

[5] K. Morita, On algebras for which every faithful representation is its own second commutator, Math. Z. 69 (1958), 429-434.

[6] K. Morita and H. Tachikawa, QF-3 rings (unpublished).

[7] C. M. Ringel, Socle conditions for QF-1 rings, Pacific J. Math. 41 (1973), 309-336.

[8] H. Tachikawa, On rings for which every indecomposable right module has a unique maximal submodule, Math. Z. 71 (1959), 200-222.

[9] R. M. Thrall, Some generalizations of quasi-Frobenius algebras, Trans. Amer. Math. Soc. 64 (1948), 173-183.

Tokyo Kyoiku Daigaku,
Tokyo, Japan

Vol. 342: Algebraic K-Theory II, "Classical" Algebraic K-Theory, and Connections with Arithmetic. Edited by H. Bass. XV, 527 pages. 1973. DM 40,-

Vol. 343: Algebraic K-Theory III, Hermitian K-Theory and Geometric Applications. Edited by H. Bass. XV, 572 pages. 1973. DM 40,-

Vol. 344: A. S. Troelstra (Editor), Metamathematical Investigation of Intuitionistic Arithmetic and Analysis. XVII, 485 pages. 1973. DM 38,-

Vol. 345: Proceedings of a Conference on Operator Theory. Edited by P. A. Fillmore. VI, 228 pages. 1973. DM 22,-

Vol. 346: Fučík et al., Spectral Analysis of Nonlinear Operators. I, 287 pages. 1973. DM 26,-

Vol. 347: J. M. Boardman and R. M. Vogt, Homotopy Invariant Algebraic Structures on Topological Spaces. X, 257 pages. 1973. DM 24,-

Vol. 348: A. M. Mathai and R. K. Saxena, Generalized Hypergeometric Functions with Applications in Statistics and Physical Sciences. VII, 314 pages. 1973. DM 26,-

Vol. 349: Modular Functions of One Variable II. Edited by W. Kuyk and P. Deligne. V, 598 pages. 1973. DM 38,-

Vol. 350: Modular Functions of One Variable III. Edited by W. Kuyk and J.-P. Serre. V, 350 pages. 1973. DM 26,-

Vol. 351: H. Tachikawa, Quasi-Frobenius Rings and Generalizations. XI, 172 pages. 1973. DM 20,-

Vol. 352: J. D. Fay, Theta Functions on Riemann Surfaces. V, 137 pages. 1973. DM 18,-

Vol. 353: Proceedings of the Conference on Orders, Group Rings and Related Topics. Organized by J. S. Hsia, M. L. Madan and T. G. Ralley. X, 224 pages. 1973. DM 22,-

Vol. 354: K. J. Devlin, Aspects of Constructibility. XII, 240 pages. 1973. DM 24,-

Vol. 355: M. Sion, A Theory of Semigroup Valued Measures. V, 140 pages. 1973. DM 18,-

Vol. 356: W. L. J. van der Kallen, Infinitesimally Central-Extensions of Chevalley Groups. VII, 147 pages. 1973. DM 18,-

Vol. 357: W. Borho, P. Gabriel und R. Rentschler, Primideale in Einhüllenden auflösbarer Lie-Algebren. V, 182 Seiten. 1973. DM 20,-

Vol. 358: F. L. Williams, Tensor Products of Principal Series Representations. VI, 132 pages. 1973. DM 18,-

Vol. 359: U. Stammbach, Homology in Group Theory. VIII, 183 pages. 1973. DM 20,-

Vol. 360: W. J. Padgett and R. L. Taylor, Laws of Large Numbers for Normed Linear Spaces and Certain Fréchet Spaces. VI, 111 pages. 1973. DM 18,-

Vol. 361: J. W. Schutz, Foundations of Special Relativity: Kinematic Axioms for Minkowski Space Time. XX, 314 pages. 1973. DM 26,-

Vol. 362: Proceedings of the Conference on Numerical Solution of Ordinary Differential Equations. Edited by D. Bettis. VIII, 490 pages. 1974. DM 34,-

Vol. 363: Conference on the Numerical Solution of Differential Equations. Edited by G. A. Watson. IX, 221 pages. 1974. DM 20,-

Vol. 364: Proceedings on Infinite Dimensional Holomorphy. Edited by T. L. Hayden and T. J. Suffridge. VII, 212 pages. 1974. DM 20,-

Vol. 365: R. P. Gilbert, Constructive Methods for Elliptic Equations. VII, 397 pages. 1974. DM 26,-

Vol. 366: R. Steinberg, Conjugacy Classes in Algebraic Groups (Notes by V. V. Deodhar). VI, 159 pages. 1974. DM 18,-

Vol. 367: K. Langmann und W. Lütkebohmert, Cousinverteilungen und Fortsetzungssätze. VI, 151 Seiten. 1974. DM 16,-

Vol. 368: R. J. Milgram, Unstable Homotopy from the Stable Point of View. V, 109 pages. 1974. DM 16,-

Vol. 369: Victoria Symposium on Nonstandard Analysis. Edited by A. Hurd and P. Loeb. XVIII, 339 pages. 1974. DM 26,-

Vol. 370: B. Mazur and W. Messing, Universal Extensions and One Dimensional Crystalline Cohomology. VII, 134 pages. 1974. DM 16,-

Vol. 371: V. Poenaru, Analyse Différentielle. V, 228 pages. 1974. DM 20,-

Vol. 372: Proceedings of the Second International Conference on the Theory of Groups 1973. Edited by M. F. Newman. VII, 740 pages. 1974. DM 48,-

Vol. 373: A. E. R. Woodcock and T. Poston, A Geometrical Study of the Elementary Catastrophes. V, 257 pages. 1974. DM 22,-

Vol. 374: S. Yamamuro, Differential Calculus in Topological Linear Spaces. IV, 179 pages. 1974. DM 18,-

Vol. 375: Topology Conference 1973. Edited by R. F. Dickman Jr. and P. Fletcher. X, 283 pages. 1974. DM 24,-

Vol. 376: D. B. Osteyee and I. J. Good, Information, Weight of Evidence, the Singularity between Probability Measures and Signal Detection. XI, 156 pages. 1974. DM 16,-

Vol. 377: A. M. Fink, Almost Periodic Differential Equations. VIII, 336 pages. 1974. DM 26,-

Vol. 378: TOPO 72 - General Topology and its Applications. Proceedings 1972. Edited by R. Alò, R. W. Heath and J. Nagata. XIV, 651 pages. 1974. DM 50,-

Vol. 379: A. Badrikian et S. Chevet, Mesures Cylindriques, Espaces de Wiener et Fonctions Aléatoires Gaussiennes. X, 383 pages. 1974. DM 32,-

Vol. 380: M. Petrich, Rings and Semigroups. VIII, 182 pages. 1974. DM 18,-

Vol. 381: Séminaire de Probabilités VIII. Edité par P. A. Meyer. IX, 354 pages. 1974. DM 32,-

Vol. 382: J. H. van Lint, Combinatorial Theory Seminar Eindhoven University of Technology. VI, 131 pages. 1974. DM 18,-

Vol. 383: Séminaire Bourbaki - vol. 1972/73. Exposés 418-435 IV, 334 pages. 1974. DM 30,-

Vol. 384: Functional Analysis and Applications, Proceedings 1972. Edited by L. Nachbin. V, 270 pages. 1974. DM 22,-

Vol. 385: J. Douglas Jr. and T. Dupont, Collocation Methods for Parabolic Equations in a Single Space Variable (Based on C¹-Piecewise-Polynomial Spaces). V, 147 pages. 1974. DM 16,-

Vol. 386: J. Tits, Buildings of Spherical Type and Finite BN-Pairs. IX, 299 pages. 1974. DM 24,-

Vol. 387: C. P. Bruter, Eléments de la Théorie des Matroïdes. V, 138 pages. 1974. DM 18,-

Vol. 388: R. L. Lipsman, Group Representations. X, 166 pages. 1974. DM 20,-

Vol. 389: M.-A. Knus et M. Ojanguren, Théorie de la Descente et Algèbres d' Azumaya. IV, 163 pages. 1974. DM 20,-

Vol. 390: P. A. Meyer, P. Priouret et F. Spitzer, Ecole d'Eté de Probabilités de Saint-Flour III - 1973. Edité par A. Badrikian et P.-L. Hennequin. VIII, 189 pages. 1974. DM 20,-

Vol. 391: J. Gray, Formal Category Theory: Adjointness for 2-Categories. XII, 282 pages. 1974. DM 24,-

Vol. 392: Géométrie Différentielle, Colloque, Santiago de Compostela, Espagne 1972. Edité par E. Vidal. VI, 225 pages. 1974. DM 20,-

Vol. 393: G. Wassermann, Stability of Unfoldings. IX, 164 pages. 1974. DM 20,-

Vol. 394: W. M. Patterson 3rd, Iterative Methods for the Solution of a Linear Operator Equation in Hilbert Space - A Survey. III, 183 pages. 1974. DM 20,-

Vol. 395: Numerische Behandlung nichtlinearer Integrodifferential- und Differentialgleichungen. Tagung 1973. Herausgegeben von R. Ansorge und W. Törnig. VII, 313 Seiten. 1974. DM 28,-

Vol. 396: K. H. Hofmann, M. Mislove and A. Stralka, The Pontryagin Duality of Compact O-Dimensional Semilattices and its Applications. XVI, 122 pages. 1974. DM 18,-

Vol. 397: T. Yamada, The Schur Subgroup of the Brauer Group. V, 159 pages. 1974. DM 18,-

Vol. 398: Théories de l'Information, Actes des Rencontres de Marseille-Luminy, 1973. Edité par J. Kampé de Fériet et C. Picard. XII, 201 pages. 1974. DM 23,-

Vol. 399: Functional Analysis and its Applications, Proceedings 1973. Edited by H. G. Garnir, K. R. Unni and J. H. Williamson. XVII, 569 pages. 1974. DM 44,-

Vol. 400: A Crash Course on Kleinian Groups - San Francisco 1974. Edited by L. Bers and I. Kra. VII, 130 pages. 1974. DM 18,-

Vol. 401: F. Atiyah, Elliptic Operators and Compact Groups. V, 93 pages. 1974. DM 18,-

Vol. 402: M. Waldschmidt, Nombres Transcendants. VIII, 277 pages. 1974. DM 25,-

Vol. 403: Combinatorial Mathematics - Proceedings 1972. Edited by D. A. Holton. VIII, 148 pages. 1974. DM 18,-

Vol. 404: Théorie du Potentiel et Analyse Harmonique. Edité par J. Faraut. V, 245 pages. 1974. DM 25,-

Vol. 405: K. Devlin and H. Johnsbråten, The Souslin Problem. VIII, 132 pages. 1974. DM 18,-

Vol. 406: Graphs and Combinatorics - Proceedings 1973. Edited by R. A. Bari and F. Harary. VIII, 355 pages. 1974. DM 30,-

Vol. 407: P. Berthelot, Cohomologie Cristalline des Schémas de Caracteristique $p > o$. VIII, 598 pages. 1974. DM 44,-

Vol. 408: J. Wermer, Potential Theory. VIII, 146 pages. 1974. DM 18,-

Vol. 409: Fonctions de Plusieurs Variables Complexes, Séminaire François Norguet 1970-1973. XIII, 612 pages. 1974. DM 47,-

Vol. 410: Séminaire Pierre Lelong (Analyse) Année 1972-1973. VI, 181 pages. 1974. DM 18,-

Vol. 411: Hypergraph Seminar. Ohio State University, 1972. Edited by C. Berge and D. Ray-Chaudhuri. IX, 287 pages. 1974. DM 28,-

Vol. 412: Classification of Algebraic Varieties and Compact Complex Manifolds. Proceedings 1974. Edited by H. Popp. V, 333 pages. 1974. DM 30,-

Vol. 413: M. Bruneau, Variation Totale d'une Fonction. XIV, 332 pages. 1974. DM 30,-

Vol. 414: T. Kambayashi, M. Miyanishi and M. Takeuchi, Unipotent Algebraic Groups. VI, 165 pages. 1974. DM 20,-

Vol. 415: Ordinary and Partial Differential Equations, Proceedings of the Conference held at Dundee, 1974. XVII, 447 pages. 1974. DM 37,-

Vol. 416: M. E. Taylor, Pseudo Differential Operators. IV, 155 pages. 1974. DM 18,-

Vol. 417: H. H. Keller, Differential Calculus in Locally Convex Spaces. XVI, 131 pages. 1974. DM 18,-

Vol. 418: Localization in Group Theory and Homotopy Theory and Related Topics Battelle Seattle 1974 Seminar. Edited by P. J. Hilton. VI, 171 pages. 1974. DM 20,-

Vol. 419: Topics in Analysis - Proceedings 1970. Edited by O. E. Lehto, I. S. Louhivaara, and R. H. Nevanlinna. XIII, 391 pages. 1974. DM 35,-

Vol. 420: Category Seminar. Proceedings, Sydney Category Theory Seminar 1972/73. Edited by G. M. Kelly. VI, 375 pages. 1974. DM 32,-

Vol. 421: V. Poénaru, Groupes Discrets. VI, 216 pages. 1974. DM 23,-

Vol. 422: J.-M. Lemaire, Algèbres Connexes et Homologie des Espaces de Lacets. XIV, 133 pages. 1974. DM 23,-

Vol. 423: S. S. Abhyankar and A. M. Sathaye, Geometric Theory of Algebraic Space Curves. XIV, 302 pages. 1974. DM 28,-

Vol. 424: L. Weiss and J. Wolfowitz, Maximum Probability Estimators and Related Topics. V, 106 pages. 1974. DM 18,-

Vol. 425: P. R. Chernoff and J. E. Marsden, Properties of Infinite Dimensional Hamiltonian Systems. IV, 160 pages. 1974. DM 20,-

Vol. 426: M. L. Silverstein, Symmetric Markov Processes. IX, 287 pages. 1974. DM 28,-

Vol. 427: H. Omori, Infinite Dimensional Lie Transformation Groups. XII, 149 pages. 1974. DM 18,-

Vol. 428: Algebraic and Geometrical Methods in Topology, Proceedings 1973. Edited by L. F. McAuley. XI, 280 pages. 1974. DM 28,-

Vol. 429: L. Cohn, Analytic Theory of the Harish-Chandra C-Function. IV, 154 pages. 1974. DM 18,-

Vol. 430: Constructive and Computational Methods for Differential and Integral Equations. Symposium 1974. Edited by D. L. Colton and R. P. Gilbert. VII, 476 pages. 1974. DM 37,-

Vol. 431: Séminaire Bourbaki - vol. 1973/74. Exposés 436-452. IV, 347 pages. 1975. DM 30,-

Vol. 432: R. P. Pflug, Holomorphiegebiete, pseudokonvexe Gebiete und das Levi-Problem. VI, 210 Seiten. 1975. DM 23,-

Vol. 433: W. G. Faris, Self-Adjoint Operators. VII, 115 pages. 1975. DM 18,-

Vol. 434: P. Brenner, V. Thomée, and L. B. Wahlbin, Besov Spaces and Applications to Difference Methods for Initial Value Problems. II, 154 pages. 1975. DM 18,-

Vol. 435: C. F. Dunkl and D. E. Ramirez, Representations of Commutative Semitopological Semigroups. VI, 181 pages. 1975. DM 20,-

Vol. 436: L. Auslander and R. Tolimieri, Abelian Harmonic Analysis Theta Functions and Function Algebras on a Nilmanifold. V, 99 pages. 1975. DM 18,-

Vol. 437: D. W. Masser, Elliptic Functions and Transcendence. XIV, 143 pages. 1975. DM 18,-

Vol. 438: Geometric Topology, Proceedings 1974. Edited by L. C. Glaser and T. B. Rushing. X, 459 pages. 1975. DM 37,-

Vol. 439: K. Ueno, Classification Theory of Algebraic Varieties and Compact Complex Spaces. XIX, 278 pages. 1975 DM 28,-

Vol. 440: R. K. Getoor, Markov Processes: Ray Processes and Right Processes. V, 118 pages. 1975. DM 18,-

Vol. 441: N. Jacobson, PI-Algebras. An Introduction. V, 115 pages 1975. DM 18,-

Vol. 442: C. H. Wilcox, Scattering Theory for the d'Alember Equation in Exterior Domains. III, 184 pages. 1975. DM 20,-

Vol. 443: M. Lazard, Commutative Formal Groups. II, 236 pages 1975. DM 23,-

Vol. 444: F. van Oystaeyen, Prime Spectra in Non-Commutative Algebra. V, 128 pages. 1975. DM 18,-

Vol. 445: Model Theory and Topoi. Edited by F. W. Lawvere C. Maurer and G. C. Wraith. III, 354 pages. 1975. DM 30,-

Vol. 446: Partial Differential Equations and Related Topics Proceedings 1974. Edited by J. A. Goldstein. IV, 389 pages 1975. DM 32,-

Vol. 447: S. Toledo, Tableau Systems for First Order Numbe Theory and Certain Higher Order Theories. III, 339 pages. 197 DM 30,-

Vol. 448: Spectral Theory and Differential Equations. Proceeding 1974. Edited by W. N. Everitt. XII, 321 pages. 1975. DM 30,-

Vol. 449: Hyperfunctions and Theoretical Physics. Proceeding 1973. Edited by F. Pham. IV, 218 pages. 1975. DM 23,-

Vol. 450: Algebra and Logic. Proceedings 1974. Edited by J. N Crossley. VIII, 307 pages. 1975. DM 28,-

Vol. 451: Probabilistic Methods in Differential Equations. Proceedings 1974. Edited by M. A. Pinsky. VII, 162 pages. 1975. DM 20,-

Vol. 452: Combinatorial Mathematics III. Proceedings 197 Edited by Anne Penfold Street and W. D. Wallis. IX, 233 page 1975. DM 23,-

Vol. 453: Logic Colloquium. Symposium on Logic Held at Bosto 1972-73. Edited by R. Parikh. IV, 251 pages. 1975. DM 25,-

Vol. 454: J. Hirschfeld and W. H. Wheeler, Forcing, Arithmeti and Division Rings. VII, 266 pages 1975. DM 25,-

Vol. 455: H. Kraft, Kommutative algebraische Gruppen und Ring III, 163 Seiten. 1975. DM 20,-

Vol. 456: R. M. Fossum, P. A. Griffith, and I. Reiten, Trivial E tensions of Abelian Categories. Homological Algebra of Trivi Extensions of Abelian Categories with Applications to Rir Theory. XI, 122 pages. 1975. DM 18,-